Frege: Importance and Legacy

Edited by
Matthias Schirn

Walter de Gruyter · Berlin · New York

1996

Printed on acid-free paper which falls within the guidelines
of the ANSI to ensure permanence and durability.

Library of Congress Cataloging-in-Publication Data

Frege : importance and legacy / edited by Matthias Schirn,
p. cm. – (Perspektiven der Analytischen Philosophie =
Perspectives in analytical philosophy ; Bd. 13)
"Papers, most of which were presented at the international
conference Foundational Problems in Frege and in Modern Logic,
held in Munich from 8 to 13 July, 1991" – Pref.
Includes bibliographical references and indexes.
ISBN 3-11-015054-9
1. Frege, Gottlob, 1848–1925 – Congresses. 2. Logic, symbolic
and mathematical – Congresses. I. Schirn, Matthias. II. Series:
Perspectives in analytical philosophy ; Bd. 13.
B3245.F24F68 1996
193–dc21 96-39046
CIP

Die Deutsche Bibliothek – CIP-Einheitsaufnahme

Frege : importance and legacy / ed. by Matthias Schirn. – Berlin
; New York : de Gruyter, 1996
(Perspektives in analytical philosophy ; Bd. 13)
ISBN 3-11-015054-9
NE: Schirn, Matthias [Hrsg.]; Perspektiven der analytischen Philosophie

Printed in Germany
Typesetting and Printing: Saladruck, Berlin
Binding: Lüderitz & Bauer, Berlin
Cover design: Rudolf Hübler, Berlin

Contents

Preface VII
List of Contributors IX

Matthias Schirn: Introduction: Frege on the Foundations of Arithmetic and Geometry 1

Part I: Logic and Philosophy of Mathematics

1. Michael D. Resnik: On Positing Mathematical Objects 45
2. W. W. Tait: Frege versus Cantor and Dedekind: On the Concept of Number 70
3. Matthias Schirn: On Frege's Introduction of Cardinal Numbers as Logical Objects 114
4. Bob Hale and Crispin Wright: Nominalism and the Contingency of Abstract Objects 174
5. Richard G. Heck: Definition by Induction in Frege's *Grundgesetze der Arithmetik* 200
6. George Boolos: Whence the Contradiction? 234
7. Michael Dummett: Reply to Boolos 253
8. Christian Thiel: On the Structure of Frege's System of Logic . 261
9. Peter Simons: The Horizontal 280
10. Franz von Kutschera: Frege and Natural Deduction 301

Part II: Epistemology

11. Eva Picardi: Frege's Anti-Psychologism 307
12. Gottfried Gabriel: Frege's 'Epistemology in Disguise' 330
13. Tyler Burge: Frege on Knowing the Third Realm 347

Part III: Philosophy of Language

14. Terence Parsons: Fregean Theories of Truth and Meaning . . . 371
15. Richard L. Mendelsohn: Frege's Treatment of Indirect Reference 410
16. Bob Hale: Singular Terms (1) 438

Notes on the Contributors 459
Index of Subjects 461
Index of Names 464

Preface

This book is a mixed selection of papers, most of which were presented at the international conference "Foundational Problems in Frege and in Modern Logic", held in Munich from 8 to 13 July, 1991. The papers deriving directly from the conference are those of Michael Resnik, Bob Hale and Crispin Wright, Christian Thiel, Peter Simons, Franz von Kutschera, Eva Picardi, Gottfried Gabriel, Terence Parsons, and Bob Hale.

The conference was organized by the editor of this volume and sponsored by the Deutsche Forschungsgemeinschaft, to whom I express my gratitude. I am grateful to Bosch GmbH (especially to Dr. Marcus Bierich) and IBM Deutschland GmbH for their financial support. Special thanks are due to Uwe Lück for his valuable help during the entire conference. I would also like to thank the participants for much stimulating discussion throughout the conference. Last but not least, I wish to thank an anonymous referee for his detailed and useful comments on earlier versions of most of the essays collected in this volume.

The principal purpose of this collection is to display both the breadth and the significance of current Frege research. Frege's importance for the development of modern logic as well as for the formation of analytic philosophy can hardly be overestimated. And, what is more, his work continues to be much debated. There is, in particular, a renewed and profound interest in his philosophy of mathematics. Michael Dummett's superb study *Frege: Philosophy of Mathematics* (Duckworth, London 1991) is one prominent example. The second claim made by the title of this volume is that Frege's work has left a legacy, that is, a set of questions yet to be answered. I, for my part, believe that this claim is supported by most of the essays here collected.

I am very sad to have to report that George Boolos died of cancer on 27 May, 1996 at the age of 55. I knew George personally since July 1991 when he came to Munich to read a paper in our Institute and at the Frege conference. Afterwards we met in Boston and Cambridge, Massachusetts and again in Munich in the summer of 1993 during another conference I had organized. I very much admire George's work, especially his subtle and profound studies on Frege's logic and philosophy of mathematics. He will be greatly missed both as a person and as a scholar. I dedicate this volume to the memory of George Boolos.

Munich, 30 May, 1996 — Matthias Schirn

List of Contributors

George Boolos †, Department of Linguistics and Philosophy, M.I.T., Cambridge, MA 02139

Tyler Burge, Department of Philosophy, UCLA, Los Angeles, CA 90024

Michael Dummett, 54 Park Town, Oxford OX2 6SJ

Gottfried Gabriel, Institut für Philosophie, Friedrich-Schiller-Universität Jena, Zwätzengasse 9, 07743 Jena

Bob Hale, Department of Philosophy, University of Glasgow, Glasgow G12 8QQ

Richard G. Heck, Department of Philosophy, Harvard University, Cambridge, MA 02138

Franz von Kutschera, Institut für Philosophie, Universität Regensburg, 93053 Regensburg

Richard L. Mendelsohn, CUNY Graduate Center, 33 West, 42nd Street, New York, NY 10036

Terence Parsons, Department of Philosophy, UCI, Irvine, CA 927117

Eva Picardi, Dipartimento di Filosofia, Università di Bologna, Via Zamboni 38, 40126 Bologna

Michael D. Resnik, Department of Philosophy, University of North Carolina, Chapel Hill, NC 27599-3125

Matthias Schirn, Institut für Philosophie, Logik und Wissenschaftstheorie, Universität München, Ludwigstr. 31, 80539 München

Peter Simons, Department of Philosophy, University of Leeds, Leeds LS2 9JT

W. W. Tait, Department of Philosophy, University of Chicago, Chicago, IL 60637

Christian Thiel, Institut für Philosophie, Universität Erlangen-Nürnberg, Bismarckstr. 1, 91054 Erlangen

Crispin Wright, Department of Logic and Metaphysics, University of St. Andrews, Five KY 169 AL

Introduction: Frege on the Foundations of Arithmetic and Geometry

MATTHIAS SCHIRN

Logicians and philosophers nowadays largely agree that Frege did most important work in logic, the philosophy of mathematics and the philosophy of language. This is true even though some of his doctrines have been the target of sharp criticism. In my view, to reveal errors or shortcomings in Frege's work may well go hand in hand with admiration for his major achievements, the power and depth of his argument and the lucidity and precision of both his exposition and his style.

This collection is divided into three parts. The first is entitled "Logic and Philosophy of Mathematics", the second "Epistemology" and the third "Philosophy of Language". In what follows, I shall mainly give a kind of overview of some selected topics in Frege's philosophy of mathematics. In doing so, I shall place emphasis not only on his project of laying the logical foundations of number theory and analysis, but also on his remarks on the foundations of geometry which are scattered throughout a number of his writings. My motive for considering geometry is twofold. Firstly, Frege's "philosophy of geometry", although it is probably of minor importance compared with his philosophy of arithmetic, nonetheless deserves close attention. Moreover, investigating the former may shed considerable light on some aspects of the latter. Secondly, none of the papers collected in this volume actually deals with Frege's ideas about geometry. Of course, I do not mean to fill the gap here, but would be pleased if my account makes it feel a trifle narrower, at least for those who agree with me about the need to account for geometry in Frege's philosophy of mathematics. I follow my discussion of arithmetic and geometry with a brief assessment of current Frege research, and conclude with remarks about the essays in this collection.[1]

---

[1] I use the following abbreviations for references to Frege's works: BS: *Begriffsschrift. Eine der arithmetischen nachgebildete Formelsprache des reinen Denkens,* Louis Nebert, Halle a. S. 1879; GLA: *Die Grundlagen der Arithmetik. Eine logisch mathematische*

Number theory and analysis

To Frege we owe the first systematic construction of an axiomatized, complete and consistent calculus of first-order logic with identity which encompasses the classical propositional calculus. The calculus of first-order logic was developed in his *Begriffsschrift* of 1879. In this small book, Frege also in effect used second-order logic serving to introduce concepts such as *following in a sequence* and *heredity of a property in a sequence,* required for laying the logical foundations of arithmetic. By way of integrating the propositional and predicate calculus, and, in particular, by solving the problem of multiple quantification, Frege went far beyond Boole's logic.

The *Begriffsschrift* presented ground-breaking achievements in logic. Frege's second book, *Die Grundlagen der Arithmetik* of 1884, was intended to make plausible the claim that the truths of arithmetic are analytic, that is, that they can be derived exclusively from primitive truths of logic and definitions. The investigation of the concept of number is carried out within the framework of natural language and employs only a few technical devices. Frege's main concern in the critical part of *Grundlagen* is a vigorous attack on two rival theories of arithmetic, the physicalistic and the psychologistic. In the final chapter, he takes a third rival theory to task, Hankel's (and Kossak's) so-called formal theory of negative, fractional, irrational, and complex numbers.

In the constructive part of *Grundlagen,* Frege immediately turns to the central task he has set himself, namely to frame a definition of number in purely logical terms. After exploring two unsuccessful definitions, he proceeds to define the number which belongs to the concept F (symbolically: $N_xF(x)$) as an equivalence class of the second-level relation of equinumerosity. (Henceforth, "E" is to abbreviate "equinumerous".) However, this explicit definition (call it (D)) rests on the assumption that we intuitively know what the extension of a concept in general is. To be sure, at the time when Frege wrote *Grundlagen* he could not rely on a commonly accepted view of the nature of extensions of concepts, let alone of the nature of numbers. Thus, his assump-

---

Untersuchung über den Begriff der Zahl, W. Koebner, Breslau 1884; GGA: *Grundgesetze der Arithmetik. Begriffsschriftlich abgeleitet,* Vol. I, H. Pohle, Jena 1893, Vol. II, H. Pohle, Jena 1903; KS: *Kleine Schriften,* ed. I. Angelelli, G. Olms, Hildesheim 1967; NS: *Nachgelassene Schriften,* eds. H. Hermes, F. Kambartel, and F. Kaulbach, Felix Meiner, Hamburg 1969; WB: *Wissenschaftlicher Briefwechsel,* eds. G. Gabriel, H. Hermes, F. Kambartel, C. Thiel, and A. Veraart, Felix Meiner, Hamburg 1976.

tion seems to jeopardize the viability of his foundational project as outlined in *Grundlagen.* At any rate, (D) falls short of solving what has come to be known as "the Julius Caesar problem". The problem is this: the criterion of identity embodied in the tentative contextual definition (C) "$N_x(F(x) = N_xG(x) := E_x(F(x),G(x))$" is powerless to decide whether Julius Caesar is identical with, say, the number of churches in Rome. Unfortunately, (D) does not lead us out of the impasse, because we do not know whether Julius Caesar is a number unless we know whether or not he is an extension. In *Grundlagen,* Frege lays down identity-conditions not for extensions of concepts in general, but only for various kinds of equivalence classes. And, as I have just said, his assumption concerning extensions must be regarded as disputable. We are thus left with the problem that (D), like (C), fails to fix uniquely the reference of the cardinality operator "$N_x\varphi(x)$". I, for my part, believe that the prospects for removing the pervasive referential indeterminacy of "$N_x\varphi(x)$" within the setting of *Grundlagen* are poor, unless Frege were to have contrived appropriate additional stipulations.

Returning to the line of thought in *Grundlagen,* we see that Frege, once he has established definition (D), proceeds to define equinumerosity in terms of one-one correspondence. Note that the latter definition is conceptually prior to (D). Frege is then able to give his definition of *n is a number.* In what follows, he has to establish the systematic fruitfulness of his logicist definition of number by deriving from it (augmented by further definitions, for example, of the successor relation and of the ancestral) familiar laws of number theory. Somewhat surprisingly, Frege actually uses (D) only in his sketchy proof of the equivalence "$N_xF(x) = N_xG(x) \equiv E_x(F(x),G(x))$"; let us call this equivalence "(T)". The theory obtained by adjoining (T) to second-order logic enjoys consistency.[2] Having derived (T), Frege never appeals to (D) again. Instead, he derives fundamental theorems of number theory from (T) together with his other definitions. In the literature on Frege, the derivability of number theory from (T) within second-order logic was first noted by Charles Parsons in his essay 'Frege's Theory of

---

[2] Cf. J. P. Burgess, Review of C. Wright, *Frege's Conception of Numbers as Objects,* Aberdeen University Press, Aberdeen 1983, *The Philosophical Review* **93** (1984), 638–640; G. Boolos, 'The Consistency of Frege's Foundations of Arithmetic', in J. J. Thomson (ed.), *On Being and Saying. Essays for Richard Cartwright,* The MIT Press, Cambridge, MA 1987, 3–20.

Number'[3]. It has received due attention only in recent years, though, notably in Crispin Wright's study *Frege's Conception of Numbers as Objects* (1983) and in several articles by George Boolos[4].

In *Grundlagen,* Frege confined himself to indicating briefly how the proofs of certain arithmetical laws proceed. In *Grundgesetze der Arithmetik* (Vol. I 1893, Vol. II 1903), he set out to demonstrate the validity of logicism beyond doubt by producing gapless chains of inference. He firmly believed that only in this way do we gain a secure basis for assessing the epistemological status of the law that is proved. Frege emphasizes that his (so-called) Begriffsschrift has undergone a number of internal changes due to a far-reaching development of his logical views. The signs familiar from *Begriffsschrift* that appear outwardly unchanged in *Grundgesetze*[5], and whose "algorithm" has also scarcely changed, are said to be provided with different explanations. Here, then, are what appear to be more or less thoroughgoing changes: the decomposition of what was earlier called "judgeable content" into *thought* and *truth-value,* as a consequence of the distinction between the *sense* and the *reference* of a sign[6]; the introduction of the two truth-values as the references of assertoric sentences (or more exactly: as the references of object-names which have the syntactic structure of sentences) and the related conception of concepts and relations as (monadic or dyadic) functions from suitable arguments to truth-values; the clear-cut distinction between the functional expression and the function as its reference, which in turn goes hand in hand with a sharper characterization of the nature of functions as opposed to objects; the explicit distinction between functions of first, second, and third level as well as between equal-levelled and unequal-levelled functions (relations); the treatment of identity as an object language predicate; the introduction of the content-stroke "–" as a (primitive) function-sign (now called the "horizontal"), which undergoes a drastic reinterpretation; the introduction of a sign designed to serve as a substitute for the definite article of ordinary

---

3 In M. Black (ed.), *Philosophy in America,* Cornell University Press, Ithaca, NY 1965, 180–203.

4 'The Consistency of Frege's Foundations of Arithmetic', *op. cit.;* 'Saving Frege from Contradiction', *Proceedings of the Aristotelian Society,* 1986–87, 137–151; 'The Standard of Equality of Numbers', in Boolos (ed.), *Meaning and Method: Essays in Honor of Hilary Putnam,* Cambridge University Press, Cambridge 1990, 261–277.

5 There is one exception: "≡" is replaced by "=".

6 In retrospect, Frege is inclined to regard the judgeable content primarily as what he then calls the thought; cf. WB, 120.

language; and finally the introduction of the courses-of-values of functions, which, apart from achieving greater simplicity and flexibility, Frege considered to be of vital importance for carrying out his logicist programme: all numbers were to be defined as extensions of concepts.[7] Let us now take a closer look at some issues involved in his introduction of courses-of-values.

In *Grundgesetze*, Frege seems to be well aware that a methodologically sound introduction of courses-of-values as logical objects cannot proceed via an elucidation of the primitive, second-level function $\grave{\varepsilon}\varphi(\varepsilon)$, the so-called course-of-values function; more specifically, it cannot proceed via an elucidation modelled upon the pattern of the semantic explanations provided for the other primitive functions of his system. The simple reason is that an elucidation along these lines, let us say: "The value of $\grave{\varepsilon}\varphi(\varepsilon)$ for every monadic first-level function $\Phi(\xi)$ as argument shall be the course-of-values of $\Phi(\xi)$", presupposes that we already know what courses-of-values are. Remember that it is precisely this unwarranted assumption, relating to the more special case of extensions of concepts, that seems to overshadow the programme of *Grundlagen*. While Frege feels entitled to proceed from the assumption that in our ordinary practice of judging and asserting we are already familiar with the objects the True and the False, he refrains from assuming that the reader of *Grundgesetze* is sufficiently acquainted with courses-of-values. That the two truth-values are, indeed, distinguished logical objects for Frege emerges very clearly both from the way he introduces the eight primitive functions of his system and from his attempted proof of referentiality for all well-formed names of his formal language (cf. GGA I, § 31). If you allow an analogy: the True and the False are in the domain of objects of Frege's logical theory what the primitive functions are in the domain of functions. The truth-values may thus properly be called the *primitive objects of logic.*

In § 3 of *Grundgesetze,* Frege introduces courses-of-values of functions, which comprise extensions of concepts and of relations as special cases, by means of an informal stipulation corresponding to the ill-fated Axiom V of his logical system. Cardinal numbers are now defined as

[7] As we shall see later, Frege intended to define the real numbers as "*Relationen von Relationen*". He uses the term "Relation" to refer to special double courses-of-values, namely to courses-of-values of dyadic (first-level) functions whose values are, for every admissible pair of arguments, either the True or the False, in short: to extensions of (first-level) relations [*Beziehungen*]. Cf. GGA II, §§ 162, 245.

extensions of *first*-level concepts. Since, in contrast to Frege's view in *Grundlagen,* it is the extension of a concept that is construed as the "bearer" of number, the cardinality operator appears as a *first*-level function-name. While in *Grundlagen* Frege does not appeal to extensions of concepts, once he has derived (T) from (D)[8], in *Grundgesetze* matters stand differently. Throughout this book, he makes extensive use of courses-of-values, and he does so for reasons of economy and convenience. In § 34 of *Grundgesetze,* he defines membership in such a way that with its help one may employ, in place of second-level functions, first-level functions in his logical calculus. This "level-reduction" is rendered possible through the following device: first-level functions which appear as arguments of second-level functions are represented by their courses-of-values – "though of course not in such a way that they give up their places to them, for that is impossible" (GGA I, 52). Yet such "representational" uses of courses-of-values, however attractive they may have appeared to Frege, can easily be dispensed with. By contrast, the use of courses-of-values in the proof of (T) is indispensable.

In *Grundgesetze,* when it comes to the construction of arithmetic, Frege proceeds in much the same way as in *Grundlagen.* Here, too, he first derives (T) from his explicit definition of number and subsequently deduces the basic laws of cardinal arithmetic from (T) within axiomatic second-order logic. In *Grundgesetze,* (T) is presented in a somewhat different fashion, but the new version is otherwise equivalent to the old one in *Grundlagen.* Until recently it has not been noted, at least not in print, that Frege in fact provided an axiomatization for arithmetic. In his paper 'The Development of Arithmetic in Frege's *Grundgesetze der Arithmetik*'[9], Richard Heck draws attention to this. Although Frege proves each of the Dedekind-Peano axioms, he is primarily concerned to prove his own axioms for arithmetic. The latter are, of course, equivalent to the former. In his second paper on *Grundgesetze* (this volume), as well as in his third, entitled 'The Finite and the Infinite in Frege's *Grundgesetze der Arithmetik*'[10], Heck presents further important mathematical results of *Grundgesetze,* explains how

[8] It is not until Section 83 of *Grundlagen* that Frege employs the term "extension of concept" again. In this Section, he completes his sketch of the proof that every finite number has a successor.

[9] *The Journal of Symbolic Logic* **58** (1993), 579–601.

[10] Forthcoming in M. Schirn (ed.), *Philosophy of Mathematics Today,* Oxford University Press, Oxford 1997.

these are established by Frege and puts them into historical perspective. One point made in the last-mentioned paper is that Frege discovered, around 1892, at least the axiom of countable choice, if not the full axiom of choice.

When Frege introduces courses-of-values in the way hinted at earlier, he encounters a variation of his old indeterminacy problem from *Grundlagen*, now clad in formal garb. Construed as a means of fixing the references of certain abstract singular terms, Basic Law V, in view of its close affinity to (T) (actually, the exact structural analogue of (T)!), was likely to arouse suspicion anyway. The criterion of identity for courses-of-values incorporated in Basic Law V takes care of the truth-conditions of only those equations in which both related terms are of the form "$\grave{\varepsilon}\Phi(\varepsilon)$". Yet the criterion fails to determine the truth-value of "$\grave{\varepsilon}\Phi(\varepsilon) = q$" if "q" is not of the form "$\grave{\varepsilon}\Psi(\varepsilon)$". Frege proposes to remove the referential indeterminacy of course-of-values terms "by determining for every function, as it is introduced, what values it obtains for courses-of-values as arguments, just as for all other arguments" (GGA I, § 10). At the stage of § 10, the proposed procedure eventually boils down to the determination of the values of the identity relation. Somewhat surprisingly, Frege confines himself to determining its values only for courses-of-values and the two truth-values as arguments, contrary to what the phrase "just as for all other arguments" seems to suggest. He explains (or attempts to justify) this restriction by mentioning the fact that up to § 10 no other objects have been introduced. At any rate, Axiom V is powerless to decide whether or not either truth-value is a course-of-values.

Having set out what has come to be known as his *permutation argument*, Frege feels free to make a stipulation which he regards as the key for resolving the referential indeterminacy of a term "$\grave{\varepsilon}\Phi(\varepsilon)$": he identifies the True and the False with their own unit classes. The formal legitimacy of this "transsortal" identification, namely its consistency with Axiom V, is established by the permutation argument.[11] However, the technical details of this argument need not concern us here. Rather, I wish to consider the famous second footnote to § 10 of *Grundgesetze* as well as to provide a response to the question whether Frege succeeds

[11] T. Parsons ('On the Consistency of the First-Order Portion of Frege's Logical System', *Notre Dame Journal of Formal Logic* **28** (1987), 161–168, here 165) shows that, contrary to what Frege claims, it is not always possible to stipulate that an arbitrary course-of-values shall be the True and another the False.

in fixing completely the references of course-of-values terms, and thus, in justifying the use of these terms in his formal theory. In doing so, I shall prescind from the inconsistency of that theory.

The second footnote is both intricate and puzzling. Moreover, it is of considerable importance for assessing the overall strategy in § 10. Frege considers here the possibility of generalizing his stipulation concerning the two truth-values so that all objects whatsoever, including those already given to us as courses-of-values (i.e., referred to by course-of-values names), are identified with their own unit classes. The suggestion goes awry. Frege rejects it on the grounds that it may contradict the criterion of identity embodied in Axiom V, if the object to be identified with its own unit class is already given to us as a course-of-values. At the same time, he jettisons the intuitively appealing proposal of identifying with their unit classes all and only those objects which are given independently of courses-of-values.[12] He does so by using what is basically the same argument as that in § 67 of *Grundlagen,* saying that the mode of presentation of an object must not be regarded as an invariant property of it, since the same object can be given in different ways.[13]

There are several difficulties looming here. To begin with, Frege fails to spell out his motive for examining the possibility of generalizing the stipulation of § 10. We are only told that the identification of every object Δ with $\grave{\varepsilon}(\Delta = \varepsilon)$ would be a natural suggestion. But in what sense does it suggest itself? What would we achieve with it if it were formally sound? Now, it is quite true that even if Frege does regard the domain of the first-order variables as limited to such objects as are required to exist by the axioms of his logical system, and thus to truth-values and courses-of-values, he appears to have envisaged extensions of the system to include, say, geometry or physics. Such an extension having been made, versions of the Julius Caesar problem would then emerge at once.

Likewise, Frege appears to have been aware of what we might describe as follows. Suppose that his envisaged foundation of number theory and analysis required the introduction of a new, third type of

---

[12] The proposal is intuitively appealing, because it may seem that we can decide, by invoking Axiom V, whether a given course-of-values is identical with an object Δ, not referred to by a course-of-values name, once we have identified Δ with its own unit class.

[13] In fact, within the system of *Grundgesetze,* the extension of the concept $-\xi$, for instance – under it falls the True alone – could be designated not only by course-of-values terms (e.g., by "$\grave{\varepsilon}(\varepsilon = (\varepsilon = \varepsilon))$"), but also by truth-value names (e.g., by "$\forall x(x = x)$"), and definite descriptions (e.g., by "$\backslash\grave{\varepsilon}(-\varepsilon)$"). Outside the system, $\grave{\varepsilon}(-\varepsilon)$ could be referred to as "Frege's favourite object", for example.

objects and, correspondingly, a new category of singular terms to refer to those objects. In that case, too, the referential indeterminacy of course-of-values terms would arise anew. An extension of the domain could proceed in the following way. In addition to the second-level concept of generality $\forall x\varphi(x)$ and the course-of-values function $\grave{\varepsilon}\varphi(\varepsilon)$, a third primitive second-level function, say, $\hat{\alpha}\varphi(\alpha)$, is introduced through the following stipulation: its value for every monadic first-level function as argument shall be an object of the third kind. I am aware that Frege might be reluctant to endorse this way of introducing a function $\hat{\alpha}\varphi(\alpha)$, on the grounds that it illegitimately presupposes an acquaintance with the function-values qua objects of a third kind. However, let us assume, for the sake of argument, that he would accept the proposed introduction of objects of a hypothetical third kind.

Plainly, from what I have just speculated to be an extension of the domain of Frege's formal theory, it would follow that the intended complete specification of the references of course-of-values terms by adjoining the stipulation of § 10 to Axiom V could not be secured.[14] For, from Frege's point of view, we could not rule out that an object of the third kind, which $\hat{\alpha}\varphi(\alpha)$ assigns to a suitable argument, in fact coincides with a course-of-values. In order to overcome this iterated indeterminacy, Frege would be compelled to make a further stipulation, designed to guarantee that the truth-value of every equation in which the sign of identity connects a course-of-values term with a term of the new category is determined. And, of course, every additional extension of the domain would require further stipulation.

If, on the one hand, we focus on the way Frege actually proceeds in § 10, and more importantly in § 31 of *Grundgesetze*, we should assume that he does regard the domain of first-order quantification as restricted to truth-values and courses-of-values. Canvassing the possibility of generalizing the stipulation of § 10 could then be construed as an attempt to supply a general solution to versions of the Julius Caesar problem that would inevitably arise for every possible extension of the domain. On the other hand, there are remarks in *Grundgesetze* suggesting that Frege is after all operating with an all-encompassing domain. Yet if the domain comprises all the objects there are, then the second footnote to § 10 might rather be seen as reflecting Frege's resi-

---

[14] I assume here, for the sake of argument, that the domain of Frege's system is (initially) limited to truth-values and courses-of-values. The question whether it is plausible to assume that the domain is all-inclusive anyway will be discussed later.

dual uneasiness about the restriction he imposes on the range of the arguments when he comes to determine the values of $\xi = \zeta$. So, we face what appears to be a head-on conflict; I shall spell it out more fully in a moment. But let me first draw attention to two further difficulties to which Frege's line of argument in the footnote gives rise.

Suppose that Frege considers the domain of his formal theory to contain only truth-values and courses-of-values. Suppose further that he regards course-of-values terms as referentially indeterminate, even after having identified the True and the False with their own unit classes. One obvious reason for his taking their references to be indeterminate would be that at a later stage the domain has actually undergone an extension. Consider now, in the light of these assumptions, the tentative stipulation $\dot{\varepsilon}(\Delta = \varepsilon) = \Delta$, and recall that it is supposed to embrace not only objects given independently of courses-of-values, but also objects referred to by course-of-values terms. But why should objects already given to us as courses-of-values be taken into account at all? If I am right, then the question as to whether the stipulation $\dot{\varepsilon}(\Delta = \varepsilon) = \Delta$ can be consistently extended to objects already given to us as courses-of-values proves to be irrelevant for any attempt to eliminate the referential indeterminacy of course-of-values terms. For clearly, the question whether, e.g., $\dot{\varepsilon}(\dot{\varepsilon}(\varepsilon = (\varepsilon = \varepsilon)) = \varepsilon)$ coincides with Julius Caesar poses the same problem as the question whether $\dot{\varepsilon}(\varepsilon = (\varepsilon = \varepsilon))$ is identical with the Roman general who crossed the Rubicon.[15]

Another difficulty may be described as follows. At the outset of the footnote, Frege claims that the stipulation $\dot{\varepsilon}(\Delta = \varepsilon) = \Delta$ is possible for every object given to us independently of courses-of-values on the same basis as he has observed with the truth-values. His subsequent argument calls this claim into question, however. For the argument seems to convey that we may have to recognize any object Δ not given to us as a course-of-values as being a course-of-values. Yet if Δ is a course-of-values, then we cannot identify Δ with its own unit class without offending against Basic Law V, unless Δ is the extension of a concept under which Δ alone falls.

---

[15] We cannot rule out, prior to the stipulation made in § 10, that, e.g., $\dot{\varepsilon}(\varepsilon = (\varepsilon = \varepsilon))$ coincides with Julius Caesar. But are we better off after the True has been identified with its own unit class? Frege would presumably insist that it is by virtue of our intuitive familiarity with the two truth-values – something which we are lacking with courses-of-values – that we can distinguish safely the True and the False from Julius Caesar. However this may be, it seems awkward to say that once the True has been identified with $\dot{\varepsilon}(\varepsilon = (\varepsilon = \varepsilon))$ we know that $\dot{\varepsilon}(\varepsilon = (\varepsilon = \varepsilon))$ is distinct from Julius Caesar.

How does the stipulation in § 10 fare in the light of what appears to be Frege's argument in the footnote? Should we deny that the latter has any repercussions at all on the former? Or should we maintain the other extreme, and perhaps say that the identification of the True and the False with their own unit classes is after all nothing but a flash in the pan? On the face of it, it seems consistent for Frege (1) to dismiss as indefensible the general proposal of identifying with their unit classes all and only those objects which are not given to us as courses-of-values and yet (2) to allow certain particular identifications which the general proposal, if accepted, would also license. On closer reflection, however, this is less clear. As was said earlier, the stipulation in § 10 is indeed consistent with Axiom V; yet following Frege's line of thought in the footnote, it seems that, before we make the stipulation, we are bound to rule out that the True and the False are courses-of-values "containing more than one object". For according to the argument given there, the fact that an object Δ is not given to us as a course-of-values does not imply that it is not one; in particular, we have no guarantee that Δ is not a course-of-values distinct from its unit class. But why should this argument not apply to the object referred to by "$\forall x(x = x)$", for instance? And if it does, how can we then legitimately identify the True with $\grave{\varepsilon}(-\varepsilon)$, i.e., with its own unit class?

It is generally agreed that it is essential for Frege's foundational project, resting on a classical logic with a classical semantics as it does, to secure a reference for every well-formed expression, especially for every well-formed formula of his Begriffsschrift. To establish beyond doubt that every well-formed expression is referential is precisely what his proof in § 31 of *Grundgesetze* is supposed to accomplish. As a matter of fact, Frege confines himself to demonstrating that every primitive function-name of his system has a reference, apparently relying on the assumption that if the primitive function-names are referential, his formation rules will bequeath the property of being referential to every name formed in accordance with them. Now, we may certainly grant that in attempting to secure a reference for every well-formed Begriffsschrift expression, Frege is at liberty to make certain additional stipulations. Yet, whenever he thinks it is convenient to make one, he ought to take care that it tallies with any thesis he propounds or advocates in the relevant context.

To conclude my assessment of § 10 and its attendant footnote, let me return to what I called a head-on conflict. Frege's remark in § 9 of *Grundgesetze,* that by introducing his notation for courses-of-values we also extend the domain of what can appear as the argument of a (first-level) function, might be taken to suggest that it is the expressive power or the referential repertoire of his Begriffsschrift that determines the range of what can appear as the argument of a first-level function. Seen from this angle, Frege probably thinks it is sufficient to determine the values of the relation of identity for courses-of-values and truth-values as arguments, because the language of his formal theory does not and need not contain any means to refer to other objects. Admittedly, it was said above that the existence of any objects other than truth-values and courses-of-values is not required by the axioms of Frege's theory. Yet limiting the determination of the values of $\xi = \zeta$ to just those objects to which we can refer by means of well-formed function-value names of his Begriffsschrift seems to be incompatible with several remarks by Frege, which I shall now present.

In § 2 of *Grundgesetze,* Frege emphasizes that the domain of what is admitted as arguments of type 1, i.e., objects, must be extended to objects in general. I term this demand "Frege's non-exclusiveness doctrine". Correspondingly, he elucidates the primitive first-level functions of his system for all suitable arguments whatsoever, i.e., for an all-embracing domain of objects, and defines certain logically complex functions of first level, in accordance with his principle of completeness, "for all possible objects as arguments" (GGA I, 52 f.). Finally, the free individual variables a, b, ..., which Frege uses in his formal theory, have the task of indicating objects in general, not only those of a domain with fixed boundaries (cf. GGA II, 78). If we take these remarks at face value – and I think we should – then we may plausibly assume that in *Grundgesetze* Frege takes the individual variables to range over all the objects there are. To respond to this by saying that the formal language of *Grundgesetze* (presumably) does not contain names for Julius Caesar or the Eiffel Tower, or that such spatio-temporal objects need not fall within the domain of a model of the axioms of the system would lack any force. For if the domain of first-order quantification encompasses all objects whatsoever, then Frege faces the question (and, indeed, has to answer it) whether a course-of-values included in the domain coincides with, say, Julius Caesar or the Eiffel Tower. The reason is that formulae of the form "$\forall x(x = \grave{\varepsilon}\Phi(\varepsilon) \rightarrow p)$" will not have been provided with a determinate reference, i.e., truth-value, unless "$x = \grave{\varepsilon}\Phi(\varepsilon)$", for Julius Caesar (or the Eiffel Tower) taken as value of "x", has been given a reference.[16] Thus, if the domain embraces all the objects there are, Axiom V together with the additional stipulation in § 10 fails to fix completely the references of course-of-values terms.

The conflict under discussion is even exacerbated, when we take a look at Frege's (abortive) attempt to prove that every well-formed name of his formal language has a reference. The proof, in which he invokes both Axiom V and the stipulation of § 10 when he attempts to show that the primitive function-name "$\grave{\varepsilon}\varphi(\varepsilon)$" has a reference, relies crucially on the assumption that only truth-values and courses-of-values are in the domain. For, if we do not make this assumption, then sentences of the form "$x = \grave{\varepsilon}\Phi(\varepsilon)$" will not have been given a reference for every assignment of members of the domain to "x", and the proof would not even get off the ground.[17] I, for my part, fail to see how the two positions regarding the size of the domain, which Frege appears to endorse in *Grundgesetze*, could be reconciled. To claim that he cannot have been blind to the above-mentioned aspect of his attempted proof of referentiality, and, therefore, did not consider the domain of his system to include all the objects there are, hardly resolves the discrepancy.

So much to Frege's introduction of courses-of-values and his indeterminacy problem.[18] It remains to cast a glance at his work on the foundations of analysis in the second volume of *Grundgesetze*. Let us then do this.

Frege's way of discussing analysis in *Grundgesetze* is, in a sense, akin to his treatment of the natural numbers in *Grundlagen*. In both cases, he does not begin by presenting his own theory, but rather by launching an attack on rival theories. In the second volume of *Grundgesetze*, the main targets are Heine's and Thomae's radical version of formalism, Cantor's theory of real numbers as well as Weierstraß's conception of numbers. The reader of this volume who is expecting a thorough examination of the theory of irrational numbers "of such a distinguished mathematician as Weierstraß" (GGA II, 149) will be disappointed. Frege takes the easy route. He basically confines himself to making critical remarks, spiced with plenty of irony, about Weierstraß's treatment of the natural numbers[19] and eventually tries to convince us

---

16 Thanks to Richard Heck at this point.

17 I am grateful to Richard Heck for pointing this out to me.

18 For a more detailed investigation see my paper 'Referential Indeterminacy in Frege's Philosophy of Arithmetic' (forthcoming).

19 One might speculate what Frege himself would have answered if he had been roused in the night with the question: "What is a number?"

that, due to its shaky foundations, Weierstraß's theory of irrational numbers need not be discussed in greater detail. Likewise, Frege pays comparatively little attention to Dedekind's theory, though he praises his sharp distinction between sign and reference [*Bedeutung*] and his taking numbers to be what numerical signs refer to. However, himself endorsing arithmetical platonism, Frege naturally finds fault with Dedekind's creation of new mathematical objects.

Frege begins the constructive part of *Grundgesetze II* by contrasting the reals with the cardinal numbers. He claims that the domain of the latter cannot be extended to that of the former. Cardinal numbers, by their very nature, answer to questions of the kind "How many objects of a certain kind are there?". By contrast, the reals are to be construed as ratios of quantities; they measure how large a given quantity is compared with a unit quantity. Thus, in Frege's view, the mode of application of the reals differs fundamentally from that of the cardinal numbers. And just as in *Grundlagen* he attempted to account for the application of the natural numbers in counting in their definition, so in *Grundgesetze* he takes pain to secure that the application of the real numbers in measurement is appropriately built into their definition.

The method of introducing the real numbers proposed by Frege lies between the traditional geometrical approach and the theories developed by Cantor, Weierstraß, and Dedekind. From the geometrical approach Frege adopts the characterization of the reals as ratios of quantities or, as he also says, as "measurement-numbers". Following Cantor, Weierstraß, and Dedekind, he detaches the reals from all special types of quantity. The rationale for doing this, so we are told, is that the application of the real numbers is not restricted to any special types of quantity, but rather relates to the domain of the measurable which encompasses all types of quantity whatsoever.

Frege observes that all previous attempts to define the notion of quantity have miscarried, because they posed the question wrongly. Instead of asking "What properties must an object have in order to be a quantity?" we ought to ask: What properties must a class (extension of a concept) have in order to be a quantitative domain? Frege adds that something is a quantity not in itself, but only in so far as it belongs with other objects to a class which is a quantitative domain.

The quantities to be considered are extensions of relations [*Beziehungen*] which Frege calls for short "*Relationen*". (Henceforth, I use the term "Relation" for "extension of a relation".) Frege thus takes the reals or ratios of quantities to be Relations of Relations. Quantitative domains are classes of Relations, namely extensions of concepts subordinate to the concept *Relation.* Frege then turns to the question of where to obtain the quantities whose ratios are irrational numbers. Plainly, they have to be non-empty Relations, since with the aid of the empty Relation one cannot define a real number at all. If q is the empty Relation, then both the inverse of q and the composition of q with its inverse coincide with q. The upshot so far is this: "We thus need a class of objects which stand to one another in the Relations of our quantitative domain, that is, this class must contain infinitely many objects" (GGA II, § 164). Frege observes that the required class must have a cardinality greater than the class of natural numbers, and further that the number of the concept *class of natural numbers* is in effect greater than the number of the concept *natural number.* Surprisingly, Cantor's proof that for any set M, 𝔓M is always of a power greater than M itself is passed over in silence.

Having arrived at this point, Frege sketches his plan for the envisaged introduction of the real numbers. In order to make his exposition more readily understood, he temporarily assumes the irrational numbers known. Every positive real number a can be represented in the form

$$r + \sum_{k=1}^{k=\infty} \left\{ \frac{1}{2^{n_k}} \right\}$$

where r is a positive integer or 0, and n_1, n_2, ... form an infinite monotone increasing sequence of positive integers. To every positive rational or irrational number a there belongs an ordered pair <r,R>, where r is a positive integer or 0, and R an infinite class of positive integers (class of the n_k). If instead of the integers we take cardinal numbers, then to every positive real number there belongs an ordered pair whose first member is a cardinal number and whose second member is a class of cardinal numbers which does not contain the cardinal number 0.[20] Suppose that a, b and c are positive real numbers and that a + b = c holds. Then for every b there is a relation holding between the pairs belonging to a and c. This relation is said to be definable without presupposing the real numbers. Thus, we have relations, each of which in turn is characterized by a pair (belonging to b), to which we add their inverses. As

[20] For simplicity's sake, I do not use Frege's special notation for the cardinal numbers (cf. GGA II, § 157).

Frege points out, the extensions of these relations (i.e., these Relations) correspond one-one to the positive and negative real numbers; and to the addition of the numbers b and b′ corresponds the composition of the corresponding Relations. He further observes that the class of these Relations is a domain which suffices for his plan, but hastens to add that "it is not thereby said that we shall hold precisely to this route".

Before embarking upon the formal development – that part is entitled "The theory of quantity" [*Die Größenlehre*] – Frege draws attention to two points he considers to be important. First, neither the classes of natural numbers nor the ordered pairs mentioned above, nor the Relations between these pairs are irrational numbers. Second, it is possible to define the aforementioned relations between the pairs without invoking any acquaintance with the irrational numbers. "In this way, we shall succeed in defining the real numbers purely arithmetically or logically as ratios of quantities which can be shown to be available, so that no doubt can remain that there are irrational numbers" (GGA II, 162). Frege's formal account, as far as it goes, can only be presented in crude outline here.

Frege begins by asking: What properties must a class of Relations possess in order that in it the commutative and associative laws for the composition of Relations hold? The proof of the associative law (Theorem 489) requires that several sentences about the identity of Relations be derived (Theorems 485–487). Unlike the associative law, the commutative law does not hold in general. Frege first proves it for members of a sequence like K, K|K, K|(K|K), ... (Theorem 501). (Here I use the symbol | for composition.) On the strength of Theorem 501, the class of the members of such a sequence can be taken to be a quantitative domain, and every positive rational number can be defined as a ratio of two quantities belonging to that domain. The negative rational numbers could be introduced by adding the inverses of Relations. When it comes to the irrational numbers, Frege observes that "they can be obtained only as limit", which in turn can be defined only in terms of the *greater than* relation. For reasons of convenience, he wants to reduce this relation to the notion of the positive: a is greater than b if and only if the Relation composed of a and the inverse of b is positive. Now, if a positive class P is at hand, the quantitative domain associated with it can be determined as follows: to it belongs every Relation which either is a member of P, or is the inverse of a Relation belonging to P, or is composed of a Relation belonging to P and its inverse (cf. GGA II, § 173).

Having defined the notion of a quantitative domain, Frege outlines the following strategy. The central concept of a positive class cannot be defined directly. One must rather take a roundabout route, that is, one must first introduce the wider concept of what Frege calls a *positival class.* Equipped with the latter we can define the notion of least upper bound, and with the least upper bound we arrive at the notion of a positive class. A positival class S is a class of Relations satisfying the following five conditions: (1) Each Relation belonging to S is one-one; (2) the composition of such a Relation with its inverse does not belong to S; (3) if the Relations p and q are members of S, then the composition of p with q is in S; (4) if p and q are in S [and p ≠ q], then the Relation composed of p and the inverse of q belongs to the quantitative domain of S; (5) if p and q are in S, then the composition of the inverse of p with q is in the domain of S (GGA II, § 175, Definition Ψ).

Frege emphasizes that in his definition of the notion of a positival class he has taken pains to include only those clauses that are independent of one another. He claims, however, that their mutual independence cannot be proved, and expresses the belief that especially clause (5) cannot be dispensed with. Naturally, the question arises here as to whether the mutual independence of the clauses, in particular the independence of clause (5) of the other four (if it does exist), can in fact be proved, contrary to what Frege claims. It seems unsatisfactory to say: I have tried repeatedly, but in vain, to reduce, say, clause (5) to any of the other clauses. Hence, clause (5) is likely to be independent of the other four (cf. GGA II, 172).[21]

Frege must have felt uneasy about the way he presents his "independence problem", as is evident from a footnote at the very end of his formal account (cf. GGA II, 243). There he suggests that his earlier claim that the mutual independence of the clauses of definition Ψ is unprovable ought not to be construed in an absolute sense. He doubts, however, that at the stage he has reached in § 175 it should be possible to give examples of classes of Relations to which all clauses of definition Ψ except one apply, without presupposing geometry, the rational and irrational numbers, or even empirical facts.

Frege proves a number of theorems concerning the notion of a positival class and then proceeds to define the notion of least upper

[21] On Frege's "independence problem" see S. A. Adeleke, M. Dummett, and P. Neumann, 'On a Question of Frege's about Right-Ordered Groups', in Dummett, *Frege and Other Philosophers,* Oxford University Press, Oxford 1991, 53–64.

bound of a class of relations in a positival class (he says: d is an S-limit of Q). The Relation d is a least upper bound of a class Q in a positival class S if and only if the following conditions are satisfied (GGA II, § 193, Definition *AA*): (1) S is a positival class; (2) d belongs to S, (3) every Relation in S which is smaller than d belongs to Q; (4) every Relation in S which is greater than d is greater than at least one Relation in S which does not belong to Q.

The requirements for defining the notion of a positive class are now met. A class S must have the following properties to be a positive class (GGA II, § 197, Definition *AB*): (1) S must be a positival class; (2) for every Relation in S there must be a smaller Relation in S; (3) if there is a Relation in S which is such that in S every smaller Relation belongs to a class Q, while there is a Relation in S which is not a member of Q, then Q must have a least upper bound in S.

Frege's next objective is to prove the Archimedian Law: For any two Relations in a positive class there is a multiple of one which is not smaller than the other (Theorem 635). (The proof is carried out in GGA II, §§ 199–214). In what follows, Frege turns to the task of proving the commutative law. He first proves it in a positive class (Theorem 674) and then for the entire quantitative domain of a positive class (Theorem 689) (cf. GGA II, §§ 215–243). In the concluding § 245, Frege describes in a few sentences what he thinks has to be done next. First, he plans to prove the existence of a positive class, as indicated in § 164, and then he will define the real numbers as ratios of quantities of a domain belonging to a positive class. He adds that this will enable him to prove that the real numbers themselves belong as quantities to the domain of a positive class. With these remarks, Frege's formal account breaks off, overshadowed by Russell's paradox.[22]

In their paper 'On a Question of Frege's about Right-ordered Groups' (*op. cit.*, 64), S. A. Adeleke, M. Dummett, and P. M. Neumann have observed that Frege "treated the applications of the real numbers as far more decisive for the way they should be defined than they are in other theories of the foundations of analysis. Mathematically, his construction of the real numbers, uncompleted because of the disaster

---

[22] For more information about Frege's theory of real numbers and, in particular, for possible ways of completing that theory see F. von Kutschera, 'Frege's Begründung der Analysis', *Archiv für mathematische Logik und Grundlagenforschung* 9 (1966), 102–111; P. Simons, 'Frege's Theory of Real Numbers', *History and Philosophy of Logic* 8 (1987), 25–44; M. Dummett, *Frege: Philosophy of Mathematics, op. cit.*, chapter 22.

wrought by Russell's contradiction, was a pioneering investigation of groups with orderings. [...] It is an unjustice that, in the literature on group theory, Frege is left unmentioned and denied credit for his discoveries."

Besides Frege's attempt to demonstrate that arithmetic (number theory and analysis) is a branch of logic, the debate with his antagonist Hilbert on the axiomatic method has gained a fair amount of attention in the history and philosophy of mathematics. The key issues of this controversy are the methodological status of the axioms and definitions of a mathematical theory, the (alleged) necessity to distinguish sharply between these two types of sentence, the consistency and independence of an axiom system, and the problem of how the primitive terms of a mathematical theory are to be given a meaning. Some interpreters have maintained that Frege failed to appreciate the innovative character of the axiomatic method, while others have claimed that in his correspondence with Hilbert and his series of papers entitled 'Über die Grundlagen der Geometrie' (1903 and 1906) he demonstrated a subtle understanding of some important features of this method. The debate between Frege and Hilbert is only one of several issues controversially discussed by Frege scholars.

Geometry

As promised at the outset, I now turn to geometry. I begin by listing several theses which Frege propounded in his writings and philosophical correspondence between 1873 and about 1914; the theses partly overlap. Let me add that the following list does not aim at completeness. As will become obvious, the most important source for assessing Frege's philosophy of geometry is *Grundlagen.* In this book, his observations about geometry are almost exclusively motivated by the desire to contrast it with arithmetic. This may partly explain why he does not go to great lengths to explain the nature of geometrical knowledge *per se.*

(1) The whole of geometry rests ultimately on axioms which derive their validity from the nature of our intuitive faculty (KS, 1).

(2) There is a remarkable difference between geometry and arithmetic in the way in which their fundamental principles are grounded (KS, 50).

(3) In geometry general sentences are derived from intuition (GLA, § 13).

(4) Geometrical truths govern the domain of the spatially intuitable (GLA, § 14; cf. KS, 104; WB, 164).

(5) Euclidean space is the only space of whose structure we have any intuition. In non-Euclidean geometry we completely abandon the base of intuition (GLA, § 14).

(6) Everything geometrical must be originally intuitable (GLA, § 64).

(7) The axioms and theorems of Euclidean geometry are synthetic a priori (GLA, §§ 14, 89; cf. WB, 163).

(8) We cannot know whether space appears the same to one man as to another. Yet there is something objective in it all the same; everyone recognizes the same geometrical axioms, and must do so if he is to find his way about the world. What is objective in space is what is subject to laws, what can be conceived, judged, expressed in words. What is purely intuitable is not communicable (GLA, § 26).

(9) Our knowledge of the axioms of geometry flows from a source very different from the logical source, a source which might be called spatial intuition (WB, 63, cf. 70; KS, 262).

(10) The sense of the geometrical terms "straight line", "parallel" and "intersect" is inseparably connected with Euclid's parallels axiom (NS, 266).

(11) No man can serve two masters. Whoever acknowledges Euclidean geometry to be true must reject non-Euclidean geometry as false, and whoever acknowledges non-Euclidean geometry to be true must reject Euclidean geometry (NS, 183 f.).

Near the end of his life, Frege abandoned the idea of logicism, having convinced himself of its irremediable failure. He then turned, albeit in a rather fragmentary fashion, to a geometrical foundation of arithmetic, thus giving up another conviction he had defended from the beginning of his career, namely that the principles of arithmetic and geometry are to be justified in fundamentally different ways. Apart from (2), he did not expressly relinquish any of the remaining theses listed above. In particular, it is plausible to assume that Frege always held that geometrical truths are synthetic a priori and that our knowledge of them is based upon spatial intuition.

Frege opens his doctoral thesis *Über eine geometrische Darstellung der imaginären Gebilde der Ebene* (1873) by propounding thesis (1). This raises the question as to the sense we may attach to imaginary forms, since we attribute to them properties which clash with our intuitions. To make this plain, Frege appeals to points at infinity which likewise are non-intuitable. He not only seeks to treat these "improper elements" in the same way as the proper ones, (i.e., to calculate with them in the same way), but he also wishes to make them amenable to intuition, to have them before his eyes. For points at infinity in the plane this is easily achieved by projecting the plane on a sphere from a point of the sphere which is neither the nearest nor the furthest.

Reading through Frege's dissertation, it springs to mind that by the act of making visible ("sichtbar machen") or of illustrating ("veranschaulichen") improper elements he understands a geometrical construction with a pair of compasses and ruler.[23] I venture to surmise that the word "intuitive faculty", as it is used in thesis (1), is meant to refer to a faculty of visualizing geometrical configurations in a way which is essentially the same for all or most human beings. The particular intuitions which we have of geometrical figures or our making these figures visible are to be construed as realizations of this faculty. If this is correct, one might suggest that for the early Frege our intuitive ability as regards geometry consists in the ability to visualize with closed eyes, as it were, spatial configurations (call this faculty visual imagination or imaginative representation) as well as to carry out visualizations according to the rules of geometrical constructions. Undoubtedly, thesis (1) has a Kantian ring to it. It would be unjustified, however, to infer from this that in his doctoral dissertation Frege had adopted the point of view of transcendental idealism.

At the outset of his post-doctoral dissertation *Rechnungsmethoden, die sich auf eine Erweiterung des Größenbegriffs gründen* (1874), Frege aims at illustrating what I referred to as thesis (2) by investigating the notion of quantity. This notion is said to have gradually freed itself from intuition and made itself independent. Its range of application is indeed so comprehensive that Frege is certainly right in denying that it derives from intuition. "The elements of all geometrical constructions are intuitions, and geometry refers to intuition as the source of its axioms" (KS, 50).

In his first dissertation, Frege mentions these elements by appeal to the foundations of analytic geometry. "The equation of a straight line is derived with the aid of sentences about the similarity of triangles and about the angles formed by parallel lines. From the same sentences we can infer Pythagoras's theorem which in turn gives us the expression for the distance between two points. These are the elements from which

[23] Frege uses "sichtbar machen" and "veranschaulichen" in the same sense; cf. KS, 26, 31, 37. For lack of a better word, I have rendered "veranschaulichen" as "illustrate".

all geometrical constructions are composed" (KS, 2). If we disregard the real numbers, it is only the similarity of triangles and the parallel lines that remain as purely geometrical elements. The existence of similar triangles can be inferred from Euclid's parallels postulate, and vice versa. Yet Frege holds that everything geometrical must originally be given in intuition. In § 64 of *Grundlagen,* he disavows that anyone has an intuition of the direction of a straight line, but claims that we do have an intuition of parallel straight lines. This claim may be doubted, on the grounds that to obtain the concept of parallelism concerning straight lines some mental activity connected with intuition is required, as is the case when the concept of direction is to be obtained. Especially when it comes to Euclid's parallels postulate, the question arises as to whether our spatial intuition is exact enough to yield it.

So much to Frege's early views on geometry. How about his early views on the foundations of arithmetic? His answer in *Rechnungsmethoden* is this. Since we have no intuition of the object of arithmetic, its principles cannot rest on intuition either. We are, however, not told from which source of knowledge they are supposed to originate. Although logic is not even mentioned in this work, stressing the comprehensive range of application of the concept of quantity, as Frege does, seems to foreshadow his later argument from the universal applicability of arithmetic to its purely logical nature. In part III of *Begriffsschrift,* entitled "Einiges aus der allgemeinen Reihenlehre", Frege derives a number of sentences about sequences to provide a general idea of how to handle his Begriffsschrift and underscores the extensive applicability of the sentences obtained. He makes it clear that the range of validity or application of truths is as wide as the scope of the source of knowledge from which they derive. Finally, in *Grundlagen* and 'Über formale Theorien der Arithmetik' (1885) the truths of arithmetic are said to govern the domain of what is countable. According to Frege, this is the widest domain of all; in fact, it is all-embracing, because everything thinkable belongs to it. A source of knowledge more restricted in scope, like sense perception or spatial intuition, would not suffice to guarantee the universal applicability of arithmetical truths.

I shall now focus on the theses on geometry which Frege puts forward in *Grundlagen.* Let me begin with thesis (5). Does Frege's claim that in non-Euclidean geometry we leave the base of intuition entirely behind carry as much conviction as he wishes to make us believe? One might question his view by drawing attention to an argument of his contemporary Hermann von Helmholtz, for example. By way of describing several non-Euclidean situations, Helmholtz seeks to demonstrate that the objects in a space of negative curvature (he calls it "pseudospherical space") can well be intuited, visualized or represented by the mind's eye, or more specifically: that they satisfy his definition of what it means to visualize or imagine an object that we have never encountered in our visual experience.[24] Thus he assumes, for instance, that a convex mirror maps an open region of ordinary space into an imaginary space. The mapping is injective, and every straight line of the outer world is represented by a straight line in the image, and likewise every plane by a plane.[25] A similar example is Eugenio Beltrami's representation of pseudospherical (hyperbolic) space in a sphere of Euclidean space to which Helmholtz appeals repeatedly and with predilection.[26] Beltrami's model enables us indeed to describe in fairly precise terms how the objects of a pseudospherical world would appear to an observer who could enter in it, assuming that he has gained both his sense of proportion and his visual experiences in Euclidean space. It enables us to do this, because the metric in the centre of the "Beltrami-sphere" is approximatively Euclidean and straight lines actually appear as such. Helmholtz emphatically gainsays that we should be able to visualize a four-dimensional space, however.[27] He points out, moreover, that if the thesis that the Euclidean axioms provide the only proper foundation of geometry is to be sustained, our inner intuition of the straightness of the lines, of the equality of distances or of angles ought to be absolutely exact.[28] It is undeniable, however, that our visualiz-

[24] Cf. H. von Helmholtz, 'Über den Ursprung und die Bedeutung der geometrischen Axiome', 'Über die Tatsachen, die der Geometrie zum Grunde liegen' and 'Über den Ursprung und den Sinn der geometrischen Axiome: Antwort gegen Professor Land', all reprinted in von Helmholtz, *Über Geometrie,* Wissenschaftliche Buchgesellschaft, Darmstadt 1968; see especially 25 f., 28, 64, 73.

[25] Cf. von Helmholtz, 'Über den Ursprung und die Bedeutung der geometrischen Axiome', 28.

[26] Cf. E. Beltrami, 'Saggio di interpretazione della geometria non-euclidea', *Giornale di matematiche* **6** (1868), 284–312; reprinted in Beltrami, *Opere matematiche,* Ulrico Hoepli, Vol. I, Milan 1902, 374–405.

[27] See in this connection Hans Reichenbach's arguments regarding the visualization of non-Euclidean geometries in his book *Philosophie der Raum-Zeit-Lehre,* Berlin, Leizpig 1928, § 11. He also deals with the question as to whether we could, in principle, visualize a space of say, four dimensions (cf. 329).

[28] In my view, there are good reasons for assuming that Kant, if he had been confronted with non-Euclidean geometry, would have rejected it; see my 'Kants Theorie der geometrischen Erkenntnis und die nichteuklidische Geometrie', *Kant-Studien* **82** (1991), 1–28.

ation of geometrical objects lacks the absolute precision required by Helmholtz, especially concerning their metrical properties.[29]

It is time to say a little about Frege's relation to Kant as far as geometrical knowledge is concerned. In § 13 of *Grundlagen,* Frege maintains that it is only when several points, lines or planes are simultaneously grasped in a single intuition that we distinguish them. "When in geometry general sentences are derived from intuition, it is evident from this that the points, lines, or planes that are intuited are not really particular ones and hence can serve as representatives for the whole of their kind." At first glance, this may be reminiscent of Kant's dictum that mathematical knowledge, construed as knowledge gained by reason from the construction of concepts, considers the universal in the particular. One might thus be tempted to establish a parallel between the view Frege expresses in the passage quoted above and Kant's construction of geometrical concepts in spatial intuition, conceived of as an exhibition a priori of the intuition (or object) which corresponds to the concept. Let us see whether Frege follows indeed in Kant's footsteps.

To attain synthetic a priori knowledge it is not mandatory, according to Kant, that the construction qua exhibition a priori be carried out in pure intuition; under certain conditions, empirical intuition may serve the purpose as well. The particular geometrical figure, say an obtuse-angled triangle, which we draw is empirical; nonetheless, it expresses "universal validity" for all possible intuitions which fall under the concept *triangle* (cf. *Kritik der reinen Vernunft,* B 741 f.), or in Frege's wording: it serves as a representative for the whole of its kind. Kant tells us that this is possible, because the geometer abstracts from the accidental properties of the particular triangle (magnitudes of the sides and of the angles) and focuses entirely on his act of construction as determined by certain general conditions. In this way, he is supposed to arrive at *general* synthetic sentences. Now, despite first appearances it seems rather unlikely that Frege tacitly adopted Kant's method to form his

[29] Felix Klein regards spatial intuition as something that is essentially imprecise. By a geometrical axiom he understands the demand by virtue of which he makes exact statements out of inexact intuition; cf. Klein, *Gesammelte mathematische Abhandlungen,* Vol. I, eds. R. Fricke and A. Ostrowski, Berlin 1921, 381 f. Klein jettisons here Moritz Pasch's idea developed in *Vorlesungen über neuere Geometrie* (Leipzig 1882) that the geometrical axioms express the "facts" of spatial intuition in a way so complete that in our geometrical considerations we need not rely on intuition. Klein, for his part, considers geometrical considerations to be impossible unless he has constantly before his eyes the figure in question. However, his view about the relation between geometrical axioms and (inexact) spatial intuition lacks, I think, persuasive power.

own conception of how we attain geometrical knowledge. First, neither in *Grundlagen* nor in any other of Frege's writings is the construction of geometrical concepts à la Kant at issue. Second, and more important, Kant's method rests crucially on the results achieved in his Transcendental Aesthetics. Yet Frege's remarks on space and spatial intuition in § 26 of *Grundlagen* are clearly at variance with Kant's metaphysical and transcendental exposition of the "concept" of space.

Although in *Grundlagen* Frege expressly endorses Kant's view that the truths of Euclidean geometry are synthetic a priori and our knowledge of its axioms rests on pure spatial intuition, he in no way subscribes to Kant's transcendental idealism.[30] According to Kant, space is a pure intuition; it is facts of the world of appearances that make geometrical sentences true, facts that do not exist independently of human beings. For Frege, by contrast, space is objective in so far as it is independent of our sensation, intuition, and imagination; it is objective, because we can express its properties in words possessing a meaning which is the same for all who are able to grasp it.[31] Frege maintains that the axioms of Euclidean geometry do not state facts about our intuition, but express states of affairs about space obtaining quite independently of our spatial intuition. Otherwise, it could well happen that one and the same geometrical axiom is acknowledged to be true by one person and rejected to be false by another.

The question forcing itself upon us is whether Frege regarded intuition as justifying geometrical knowledge. As a matter of fact, he makes some remarks that seem to suggest a positive answer (cf. WB, 63, 70 and theses (1) and (3)). Moreover, in two of his late fragments Frege understands by (spatial) intuition the *geometrical source of knowledge*, that is, the source from which the axioms of geometry flow (cf. NS, 286, 292 ff., 297 ff.). Yet a source of knowledge is explained as justifying the recognition of truth, the judgment (cf. also WB, 63, 163 f.).

---

[30] In Frege's view, the fact that we can always consistently deny the axioms of Euclidean geometry suggests that they are independent of one another and the primitive laws of logic, and are therefore synthetic. He also regards them as a priori, and he probably does so on the grounds that they rest on pure intuition.

[31] Note in this connection that in *Grundlagen* and subsequent writings Frege by no means uses the word "objective" in the same sense as Kant. For Kant, space is a subjective condition of our outer intuition; it is ideal as regards objects when they are considered in themselves through reason, but at the same time it is objective, i.e., empirically real with respect to outer appearances. If that were not so, Kant would be at a loss to explain the universal and necessary validity of the truths of geometry. Here I naturally cannot analyze further his notion of objectivity.

In *Grundlagen* Frege underlines the subjectivity, privacy and incommunicability of our spatial intuitions, which they share with our ideas, sensations and imagination. We cannot, in order to compare them, lay one man's intuition of space beside another's. To regard spatial intuition thus characterized as justifying our geometrical knowledge appears to be at odds or at least in tension with Frege's firm belief that the axioms of geometry are objective. If one man did not intuit or visualize spatial configurations in essentially the same way as another, then it would be hard to understand how the axioms of three-dimensional Euclidean geometry could derive their validity from our intuitive ability. To be sure, in projective geometry, where the principle of duality holds, it is perfectly intelligible to suppose that two rational beings, who connected different intuitions with the word "plane", for instance, would nevertheless be in complete agreement over all geometrical theorems (cf. GLA, § 26). In Euclidean geometry, however, Frege wishes to rule out this possibility.

Let me add one more brushstroke to the picture we have gained so far of Frege's views of geometry. Unfortunately, both his writings and his philosophical correspondence provide scarcely a clue about to what extent he kept abreast with developments in geometry in the second half of the nineteenth century. In all likelihood, he was familiar with Georg von Staudt's investigations on projective geometry[32] as well as with Bernhard Riemann's famous essay (*Habilitationsvortrag*) 'Über die Hypothesen, welche der Geometrie zugrunde liegen'[33]. We may further quite safely assume that Frege knew some of the work on geometry by Felix Klein. To my mind, it is astonishing that the influential work on the foundations of geometry by Riemann and Helmholtz, in particular their arguments against the a priori character of geometry, are completely passed over in silence in Frege's writings.[34] This might be due partly to the disparaging attitude, if not disdain, that Frege appears

[32] G. von Staudt, *Die Geometrie der Lage,* F. Korn, Nürnberg 1847 and *Beiträge zur Geometrie der Lage,* Bauer and Raspe, Nürnberg 1856–1860 (3 fascicles).

[33] *Göttinger Abhandlungen* **13** (1854), 133–152.

[34] As far as I know, Frege refers only in one place to a work by Helmholtz. In the second volume of *Grundgesetze* (139 f., footnote 2), he finds fault with Helmholtz's intention to provide an empirical foundation of arithmetic in the essay 'Zählen und Messen erkenntnistheoretisch betrachtet'. Frege regards Helmholtz's approach as confused and concludes: "I have hardly ever seen anything less philosophical than this philosophical essay, and hardly ever has the sense of the epistemological problem been more misunderstood than here." So it seems that Frege had a low opinion of Helmholtz's philosophical talents.

to have entertained towards non-Euclidean geometry. Speculations aside, it is true that he fails to work out any solid argument for the claim that geometrical truths are known a priori. By contrast, Riemann and Helmholtz in effect adduce powerful arguments in favour of the empirical nature of geometry. The idea that the existence of consistent theories of non-Euclidean geometry had an impact on their view seems to make sense, and I do not think that it can be dismissed.

In his aforementioned essay, Riemann holds that an n-fold extended quantity admits different metric relations, so that space constitutes only a special case of a threefold extended quantity. He concludes from this that the sentences of geometry cannot be derived from general quantitative concepts. Instead, there is supposed to be ample evidence that those properties by virtue of which space differs from other threefold extended quantities can only be ascertained by experience. Riemann tries to make it plain that the simplest facts which serve to determine the metric relations of space are not necessary (or a priori), but possess only empirical certainty. This is why he calls them *hypotheses*. According to Riemann, all we can say about space without invoking experience is that it is one among many possible kinds of manifolds.

I have already mentioned two objections which Helmholtz directs against Kant's thesis that the geometrical axioms originate from an a priori source of knowledge. In addition, Helmholtz argues that the geometrical principles belong not only to the pure theory of space, but deal also with quantities. Yet the introduction of quantities is said to make sense only if we provide suitable procedures of measurement for them. Every measurement of space, and therefore every quantitative concept applied to space, presupposes the possibility of spatial figures moving without change of form or size. By way of adjoining to the geometrical axioms sentences relating to the mechanical properties of natural bodies we arrive, Helmholtz claims, at a set of sentences which can be confirmed or refuted by experience, but just for the same reason can also be gained by experience.

I leave it to the reader to judge whether Frege's characterization of our knowledge of Euclidean geometry as synthetic a priori must be regarded as a retrograde step, especially in the light of the work of Riemann, Helmholtz and other contemporaries. To repeat, unlike Kant, Frege spares himself the trouble of buttressing such characterization by detailed argument; instead, he seems to take it more or less for granted. Furthermore, to stigmatize non-Euclidean geometry as a pseudoscience, as Frege tends to do in the fragment 'Über Euklidische Geometrie' (cf. NS, 182–184), appears to have been an aberration on his part. It seems that Frege was unwilling to realize that the existence of consistent theories of non-Euclidean geometry by no means compels us to acknowledge one geometry – Euclidean or non-Euclidean geometry – as the only true one. His belief that one cannot consistently recognize both Euclidean and non-Euclidean geometry as true presumably derives from his assumption that the primitive geometrical terms allow only for one interpretation, namely the Euclidean one. Yet even if Frege had accepted the legitimacy of endowing geometrical terms with a non-Euclidean interpretation, he would probably have been inclined to assign a definite priority to Euclidean over non-Euclidean geometry. For it seems that he never abandoned his conviction that everything geometrical must be originally intuitable and that non-Euclidean geometry leaves the basis of intuition entirely behind.[35]

Frege research

Of course, I cannot claim complete knowledge of the developments in Frege research over, say, the last fifteen years. It seems clear, also from what I have said before, that the discussion of Frege's work during that period has various motives. I should like to mention the following six: (i) to locate his work more accurately in the history of logic, mathematics and philosophy; (ii) to bring into sharp focus and reassess both his logicism and his arithmetical platonism, also in the light of most recent work in the philosophy of mathematics; (iii) to examine thoroughly particular aspects of his logical theory, such as his so-called permutation argument, his attempted proof of referentiality for the formal language of *Grundgesetze* or the question as to what really caused the inconsistency of his system; (iv) to analyze his mathematical work in *Grundgesetze;* (v) to investigate the various facets of his epistemology; (vi) to provide a systematic account of his semantics and to develop further certain central ideas of it.

[35] On Frege's reflections on geometry see M. Dummett, 'Frege and Kant on Geometry', *Inquiry* **25** (1982), 233–254; M. Wilson, 'Frege: The Royal Road from Geometry', *Noûs* **26** (1992), 149–180; J. Tappenden, 'Geometry and Generality in Frege's Philosophy of Arithmetic', *Synthese* **102** (1995), 319–361.

To my mind, the most important work on Frege in recent years has been on his philosophy of mathematics.[36] Despite his pioneering work in the field of semantics, which still has a considerable bearing upon present discussions, Frege was, in his own express opinion, a logician and philosopher of mathematics, albeit a logician with a profound interest in dealing with topics we nowadays assign to the philosophy of language. Compared with the elaboration of the logicist programme, to which Frege devoted more than twenty years of his academic career, his work on the philosophy of language in a narrow sense plays a rather subordinate role. It would be short-sighted, however, to regard his theory of sense and reference merely as an appendix to his philosophy of mathematics. Undoubtedly, this theory plays a crucial role in the construction of the logical system of *Grundgesetze*.

Needless to say, to stress the importance of Frege's philosophy of mathematics is by no means to disparage the achievements brought about in other areas of Frege research. And to be sure, especially questions concerning Frege's epistemology have quite naturally been involved in one or the other investigation of his philosophy of mathematics. In any case, I do think that the distribution of the papers in the present volume can be seen as reflecting altogether the prevailing tendency in current Fregean scholarship. It is mainly for this reason that in my preceding account I have tried to throw some light on what I take to be interesting topics in Frege's philosophy of mathematics.

The essays in this collection

Michael Resnik's essay 'On Positing Mathematical Objects' opens the section on logic and philosophy of mathematics. It connects his own postulational version of structuralism (and the modal approaches of Hartry Field and Geoffrey Hellman) with the views of Dedekind, Cantor and Hilbert which Frege criticized with great effect. Resnik starts with a review of the criticisms levelled by Frege against the

---

36 I should like to draw attention to William Demopoulos's very useful collection *Frege's Philosophy of Mathematics*, Harvard University Press, Cambridge, MA 1995. Several of the aforementioned essays on Frege are reprinted in this volume. See also the section 'Frege and the Foundations of Arithmetic' in my forthcoming collection of articles by different hands *Philosophy of Mathematics Today* (*op. cit.*).

method of creating or postulating mathematical objects by means of definition or abstraction. It is argued that despite the force of some of Frege's objections he apparently failed to distinguish clearly between two views of introducing mathematical objects. The first endorses the erroneous device of creative definitions, while the second subscribes to the defensible method of mental constructions or acts of postulation. Resnik believes that Frege's inclination to lump together these views kept him from appreciating fully the mathematical achievements of Cantor, Dedekind, and Hilbert. As to Dedekind, it is held that he suggested creating number systems by creating new structures in which the numbers in question are nothing but positions in those structures. In Resnik's interpretation, Dedekind thought that the creation of such structures needs to be justified by showing that they can be abstracted from a suitable system of "thought objects". The serious drawback of this position, we are told, is that it merely transfers the problem of mathematical existence to that of sets. As regards Hilbert's axiomatic method, Resnik maintains that although it appears to be akin to Dedekind's structuralism, a more thorough investigation would reveal significant differences between their views. It is further suggested that (the early) Hilbert could be seen as attempting to replace Dedekind's ontological thesis that we create mathematical structures by means of abstraction with the epistemological thesis that we can recognize new structures by postulating them through axioms.

As far as Resnik's own position is concerned, he weds a realist mathematical structuralism to a postulational epistemology. From Dedekind he takes structuralism which in his hands takes on a form that parallels Hilbert's view on mathematical existence and truth. Since Resnik's postulational epistemology commits him to acknowledge the independent existence of mathematical objects, it seems that postulationism is compatible with realism. In the remainder of his paper, he shows in what sense the two doctrines can actually be combined.

In the secondary literature, Frege's polemical remarks in *Grundlagen* on various doctrines of his predecessors and contemporaries such as Baumann, Cantor, Jevons, Locke, Leibniz, Mill, Schröder, Thomae, and others on the concept of number and on unity are often said to be devastating. W. W. Tait, in his essay, does not share this view. On the contrary, he holds that Frege's scrutiny of the work of his fellow mathematicians is often characterized by lack of charity, and, what is worse, marred by serious defects and misinterpretation. Tait has accordingly set himself the task of reassessing some of Frege's criticisms and to compare Frege's views on the concept of number with those of Cantor and Dedekind. In addition, Tait juxtaposes a number of critical observations on Dummett's views in *Frege: Philosophy of Mathematics.* The central objection against Dummett seems to be that in comparing Frege's and Dedekind's treatment of the foundations of arithmetic he has failed to do justice to Dedekind. In what follows, I shall confine myself to summarizing some of the arguments that Tait advances against Frege's critique of a prominent conception of number, a version of which was also defended by Cantor: it is the conception that identifies cardinal numbers with sets of pure or featureless units (cf. Frege, GLA, §§ 29–45; KS, 163–166; NS, 76–80).

The vulnerable spot of this conception is, according to Frege, that it fails to reconcile identity of units with distinguishability and, furthermore, that every attempt to resolve this difficulty is doomed to failure: "If we try to make the number originate from the combination of distinct objects, we obtain an agglomeration comprising the objects with just those properties which serve to distinguish them from one another; and that is not the number. If, on the other hand, we try to form the number by combining identicals, this constantly coalesces into one, and we never arrive at a plurality" (GLA, § 39). In this connection, Tait criticizes Frege for conflating two questions which ought to be distinguished clearly: (1) What are the things to which numbers apply? (2) What are numbers? Tait claims that the first horn of the dilemma as laid out by Frege concerns (1). He argues further that the things to be numbered are not agglomerations, but sets, which indeed originate from the combination of distinct objects. The fact that these sets were called *numbers* by some of Frege's contemporaries, he considers to be one source of Frege's (alleged) confusion. The second horn of the dilemma concerns the conception of numbers as sets of pure units. Tait finds it ill-conceived, but, unlike Frege, in no way incoherent. Frege discusses several suggestions which might lead out of the quandary, but rejects them all. One proposal is to invoke instead of spatial or temporal order a more generalized concept of series (cf. GLA, § 42). Tait accuses Frege of conflating here the notion of a series with that of a linearly ordered set $\langle A,>\rangle$ (where, for x and y in A, $x < y$ implies that x and y are distinct).

Cantor construed the cardinal number $\overline{\overline{M}}$ of a given set M as a definite set, comprised of nothing but units, which exist in our mind as an intellectual copy or a projection of M. $\overline{\overline{M}}$ is obtained by carrying out the process of abstraction from both the nature of the elements of M and their order.[37] Not surprisingly, Frege considered this view to be a thorn in the flesh and commented on it with sarcasm. Tait, for his part, regards Cantor's view as unattractive, but not as indefensible, provided that it is interpreted as follows: the abstraction concerns not the individuating properties of the elements relative to one another, as Frege assumes, but rather the individuating property of the set M itself. On this interpretation, the "cardinal set" (i.e., the cardinal qua set of pure units) C "corresponding to a set M is to be constituted of unique elements, specified in no way other than that they are elements of C and that C is equipollent to M. Thus, the cardinal sets are not sets of points in Euclidean space or of numbers or of sets, or of apples or etc." So much to Tait's vindication of Cantor.

In my own contribution, I examine Frege's conception of numbers as objects in *Grundlagen* and argue that it suffers from a number of defects. One objection is that he fails in his attempt to analyze what he calls "ascriptions of number" [*Zahlangaben*] such as "The number 9 belongs to the concept planet" in such a way that cardinal numbers emerge as self-subsistent objects. Another point made is that Frege could have acknowledged ascriptions of number as numerical statements in their own right instead of construing them as equations in which the number words flanking "=" function as singular terms. Harold Hodes has claimed that Frege's method of bestowing a definite reference upon a numerical singular term, by fixing the sense of all relevant sentences in which it occurs, fails to explain the "microstructure" of reference, for instance, to cardinal numbers. I try to show that this argument rests partially on an outright misinterpretation of Frege's context principle and its relation to his thesis that a thought is built up out of parts which correspond to the parts out of which the sentence expressing the thought is built up. I conclude the first half of my paper by claiming that unless someone has succeeded in refuting Paul Benacerraf's ontological argument against number-theoretic platonism, the conception of numbers as objects remains a dogma bequeathed to us by Frege. In the second half, I canvass Frege's three attempts to define number in *Grundlagen* and argue that he falls short of resolving the pervasive indeterminacy of reference affecting the cardinality operator.

Although Bob Hale's and Crispin Wright's essay is not directly concerned with Frege, it can be seen as standing in the tradition of his phil-

---

[37] Cf. G. Cantor, *Gesammelte Abhandlungen mathematischen und philosophischen Inhalts,* ed. E. Zermelo, Berlin 1932, 283, 387, 411 f.

osophy of arithmetic. It is probably correct to say that in their attempt to undermine the viability of Hartry Field's nominalism, a version of Fregean mathematical platonism which they accept figures in the background, as it were. Field's rather unorthodox defence of nominalism as a philosophy of mathematics accepts a platonist account of the truth-conditions of purely mathematical statements, that is, an account which discerns in them purported reference to or quantification over mathematical entities of various kinds. In the same breath, Field maintains that such statements are false – or at least never non-vacuously true – one the grounds that there are simply no such objects as numbers, sets and the like. The key idea in his advocacy of this thesis is that we can avoid wholesale rejection of standard mathematical theories by holding that such theories have a property falling well short of truth, but akin to consistency, namely *conservativeness.* According to Field, a mathematical theory T is conservative if, for any nominalistic assertion A and any body of such assertions N, A is not a consequence of N + T, unless A is a consequence of N alone. A seemingly serious difficulty for this position, pointed out by Hale and Wright in earlier writings, is that it appears to commit Field to maintaining that the falsehood of standard mathematical theories is at worst a contingent matter. This appears open to the objection, crudely stated, that Field should be able to explain, in a nominalistically acceptable way, why the putative contingency is resolved, as, in his opinion, it is, but can provide no suitable such explanation. Field has sought to fend off this line of objection by claiming that it rests on an equivocation over the notion of contingency. Hale and Wright, for their part, try to show that this response is ineffective, though they admit that the objection, as originally presented, is unsatisfactory. Their principal aim is to arrive at a reformulation of the objection which captures its core, while relying on principles governing the notion of contingency which should command general assent.

The mathematical argumentation in Frege's *Grundgesetze* has been largely ignored by his interpreters. Richard Heck is convinced that this lack of esteem or interest is unjust, and in his paper 'Definition by Induction in Frege's *Grundgesetze der Arithmetik*' he tells us why. In it, he discusses at length Frege's proof of Theorem 263, which amounts to a proof that all structures satisfying certain conditions are isomorphic. These conditions Heck takes to be Frege's own axioms for arithmetic. It is argued that Theorem 263 is one of the central results of *Grundgesetze* and that Frege's proof of it can be reconstructed in

Fregean Arithmetic (FA), with or without the use of the ordered pair axiom. As a matter of fact, Frege proves Theorem 263 in FA, augmented by the ordered pair axiom, although, in Heck's opinion, he knew that he could have carried out the proof without it. (Heck reconstructs Frege's proof of Theorem 263 using the definition of the 2-ancestral, rather than ordered pairs.) Heck speculates that Frege's reluctance to dispense with ordered pairs when he comes to prove Theorem 263 may have had two reasons. Firstly, although the use of ordered pairs was fairly common among mathematicians of Frege's day, no suitable definit of them was at hand when he set about writing the first volume of *Grundgesetze.* Providing such a definition is in the spirit of Frege's claim to be able to formalize, in his Begriffsschrift, classical mathematics in its entirety. Secondly, using ordered pairs in this context spares him the trouble of having to do two things: first, to set up a new definition of the ancestral and, second, to prove several theorems which are analogues of ones he had already proven. The result that all "simple" and "endless" sequences, which are models of Frege's own axioms for arithmetic, are isomorphic is closely analogous to one of the theorems for which Dedekind's study *Was sind und was sollen die Zahlen*[38] is celebrated; it is Theorem 132 (§ 10) stating that all "simply infinite systems", that is, structures which satisfy the Dedekind-Peano axioms, are isomorphic. In the course of the proof of his Theorem 263, Frege proves a generalization of another of Dedekind's important results, namely the so-called recursion theorem for ω (referred to by Dedekind as Theorem 126; cf. § 9), justifying the definition, by induction, of a function defined on the natural numbers. Heck concludes that the fact that Frege proved such results may have a considerable impact on our understanding of his philosophy of mathematics.

The paper by George Boolos addresses the question as to what gave rise to the inconsistency of Frege's logical system. For many years, it has been taken more or less for granted, following Frege's own assessment, that it is Axiom V of *Grundgesetze* which is to be held responsible for the contradiction. As far as I know, this view was challenged for the first time by Christian Thiel in his largely neglected article 'Zur Inkonsistenz der Fregeschen Mengenlehre'[39]. In chapter 17 of his book *Frege: Philosophy of Mathematics,* Michael Dummett has argued that

[38] Vieweg, Braunschweig 1888.

[39] In C. Thiel (ed.), *Frege und die moderne Grundlagenforschung,* Anton Hain, Meisenheim am Glan 1975, 134–159.

the contradiction in Frege's system is primarily due to his careless treatment of the second-order quantifier in his attempted proof of referentiality for all well-formed names of his formal language (cf. GGA, § 31). Dummett regards this proof as an attempted consistency proof. In fact, Frege's first reaction to Russell's startling discovery suggests that he had more than an inkling of this interconnection (see WB, 213). At first glance, Dummett's diagnosis of what led to the inconsistency appears to be buttressed by the result that the first-order fragment of the system of *Grundgesetze* is consistent, as first established by Terence Parsons in his article 'On the Consistency of the First-Order Portion of Frege's Logical System' (*op. cit.*). Without second-order quantification, Frege's system would be "paralyzed", however, because membership would be indefinable for him. Boolos maintains that the greater the paralysis, the less plausible Dummett's view about the primary source of the inconsistency of Frege's system appears. Only if the first-order fragment had been strong enough to yield arithmetic or an interesting portion of it would it be tempting, Boolos thinks, to trace the inconsistency back to the presence of the second-order quantifier. Let me add that in Dummett's opinion Frege failed to carry out a valid consistency proof even for the first-order fragment of the system of *Grundgesetze*.

As I read him, Boolos is chiefly concerned to convince us that, contrary to what Dummett claims, everything stands as it was, though it must probably be seen in the light of a more profound and more subtle analysis: the culprit for the breakdown of the system of *Grundgesetze* is what Frege took it to be, namely Axiom V. Boolos argues, in particular, that we should not put the blame for Frege's error on the stipulations he made regarding the truth-conditions of sentences beginning with second-order quantifiers, but rather on those concerning course-of-values equations. A number of doubts in connection with Dummett's account of what caused the fatal flaw in Frege's logicist project are expressed. One point made is that Dummett has taken a "background condition" to be the cause of the contradiction. Another proviso relates to his contention that Frege appears to favour a substitutional interpretation of the second-order quantifier rather than an objectual one. Boolos thinks that to see why T. Parsons's construction of a model for the first-order portion of Frege's system cannot be extended to the full system provides further evidence for the claim that it is not any deficiency in Frege's stipulations concerning the second-order quantifier that caused the inconsistency. It is further pointed out that it is due to the "Cantor-Russell *aporia*" that any attempt to construct a model for Axiom V within Frege's full second-order language is bound to fail. Boolos concludes that in the light of possible revisions of the system of *Grundgesetze* which yield a consistent version of it, allow the construction of arithmetic and prove to be less thoroughgoing than the idea of dispensing altogether with second-order quantification, it is not the use of the latter that is to be blamed for the inconsistency of Frege's formal theory.

In spite of the arguments advanced by Boolos, Dummett, in his reply, reiterates his former contention that second-order quantification was essential for the inconsistency of Frege's logical system. Dummett is willing to concede, though, that he should not have firmly ascribed to Frege a substitutional interpretation of quantification. Nonetheless, he does not follow Boolos in attributing to Frege an objectual interpretation, because he holds that there is nothing in the system of *Grundgesetze* that compels us to do this. To demonstrate how Frege's consistency proof founders in the presence of the second-order quantifier, Dummett recalls the strategy pursued in *Frege: Philosophy of Mathematics.* It was this: to show, first, by example, without invoking Axiom V, that Frege's inductive line of argument in § 31 of *Grundgesetze* is faulty; and to show, second, that Russell's paradox can be obtained by a "modest appeal" to Axiom V. Dummett finds himself in considerable disagreement with Boolos about domains of quantification in general. Boolos repudiates Dummett's conception of "indefinitely extensible concepts", claiming that Frege "did not have the glimmering of a suspicion of the existence" of such concepts. According to Dummett, Boolos takes this denial to follow from his rejection of the view that the objects over which the individual variables of a mathematical theory range form a collection. Dummett, for his part, maintains that it is precisely the indefinite extensibility of the concept of set or class which suggests taking the objects over which the individual variables of a theory range as forming a domain or totality. Dummett concludes by emphasizing that completely unrestricted quantification is not illegitimate; what, in his view, is illegitimate is a truth-conditional interpretation of sentences involving it.

Christian Thiel, in his essay, deals with some problems deriving from the logical system of *Grundgesetze.* He begins by considering three *desiderata* of current Frege research. The first concerns Peter Aczel's claim that it is Frege's horizontal function that is to be held responsible for the derivability of Russell's paradox in the system of

Grundgesetze. The second *desideratum* is that the so-called permutation argument in § 10 of *Grundgesetze* ought to be reconsidered, even though several analyses of that argument have recently been put forward. The third *desideratum* is a more thorough examination of Frege's attempted proof in § 31 of *Grundgesetze* that every well-formed term of his formal language has a reference. Contrary to what Resnik claims in his article 'Frege's proof of referentiality'[40], Thiel maintains that a proof of referentiality, if successful, would imply the consistency of the formal theory of *Grundgesetze.* Thiel discusses the failure of Frege's attempted proof of referentiality by appeal to his essay 'Zur Inkonsistenz der Fregeschen Mengenlehre' (*op. cit.*). The pivotal point of his assessment is that the proof miscarries due to the "impredicative" nature of the rules governing the correct formation of function-names which Frege states in § 26 of *Grundgesetze,* the so-called "gap-formation principles". The latter are said to block the inductive process of transferring a reference to certain newly formed names. Thiel further analyzes, from the point of view of the inconsistency of the system of *Grundgesetze,* Frege's derivation of theorem (χ) of *Grundgesetze* which appears in the Appendix to the second volume.

One of the significant changes that Frege's Begriffsschrift had undergone between 1879 and 1893 concerned the interpretation of "–". In *Grundgesetze,* Frege introduces $-\xi$ as a primitive function (concept) mapping the True as argument on the True and every argument of type 1 (i.e., every object) distinct from the True on the False. Peter Simons, in his essay, examines both the nature and the role of "–" in *Begriffsschrift* (where "–" is called the content stroke) and especially in *Grundgesetze* (where "–" is referred to as the horizontal). In addition, he sheds light on a number of special features of the logical system of *Grundgesetze.* Simons argues that despite its obscure theoretical status in *Begriffsschrift,* "–" played an important heuristic role in Frege's logic of 1879. Matters are said to stand differently in *Grundgesetze.* Due to the thoroughgoing reinterpretation that Frege imposed in that work on the notation stemming from his *Begriffsschrift,* "–" is now given substantial work to do in the logical calculus. Simons points out that Frege was quite aware that he could have eliminated the horizontal function (and also negation). In his doctoral dissertation *On the Primitive Term*

[40] In L. Haaparanta and J. Hintikka (eds.), *Frege Synthesized. Essays on the Philosophical and Foundational Work of Gottlob Frege,* D. Reidel, Dordrecht, Boston 1986.

of Logistic of 1923[41], Tarski showed how to define conjunction in Lesniewski's protothetic in terms of material equivalence and universal quantification. Simons uses an analogue of Tarski's simplest definition and shows that Frege in effect could have reduced the number of his primitives to four, dispensing entirely with the horizontal, negation and the conditional. It is claimed that nothing illustrates more conspicuously how much Frege's logic had lost the hierarchical structure of *Begriffsschrift* and had become much more of a piece in *Grundgesetze*. In the remainder of his paper, Simons presents several further results about Frege's logic. One is that Frege's treatment of the two truth-values as objects imparts features to his formal system which brings it into the vicinity of many-valued propositional logics.

Franz von Kutschera's contribution is an historical footnote on the development of systems of natural deduction. He shows that in the first volume of *Grundgesetze* Frege formulates a calculus which, in a sense, is intimately related to Gentzen's classical calculus of sequents in his 'Untersuchungen über das logische Schließen'[42]. The crucial difference between the two approaches resides in the fact that while Frege states elimination rules for the succedent, Gentzen employs introduction rules for the antecedent. Clearly, Frege and Gentzen pursued different goals. Frege only aimed at establishing simple inference rules for the manipulation of antecedents and the succedent in implicational formulae. Gentzen, by contrast, intended to do justice to the "real" deductive practice in mathematical proofs. Von Kutschera emphasizes that the two methods rest, after all, on the same fundamental idea, namely of stating sufficient and necessary conditions for the introduction or elimination of logical operators. By confining himself to introduction rules, Gentzen paved the way for arriving at his *Hauptsatz*, which played an important role in proof theory.

Eva Picardi's essay 'Frege's Anti-Psychologism' is one of three in this volume dealing with issues belonging directly or indirectly to epistemology. Picardi explores what she believes Frege considered to be the main defect of psychologism: to rely on a picture of language which turns both the objectivity of sense and the communication of thoughts into a mystery. Her central thesis I take to be twofold. Firstly, there is a close link between the attack Frege mounts on psychologistic concep-

---

[41] In A. Tarski, *Logic, Semantics, Metamathematics*, second edition (ed. J. Corcoran), Hackett, Indianapolis 1983, 1–23.

[42] *Mathematische Zeitschrift* **39** (1934), 176–210, 405–431.

tions of logic on the one hand and the sharp criticisms he levels against psychologistic accounts of meaning on the other. Secondly, this link is to be found in Frege's realistic conception of truth. Picardi claims that due to the fact that Frege's anti-psychologism is essentially semantic in nature, little is to be gained by discussing it in the context of his conception of epistemology. In particular, she argues that a certain thesis propounded by Philip Kitcher and Hans Sluga should be rejected. The thesis is that Frege tacitly adopted a form of Kantian transcendentalism as a safeguard against psychologism.

In his paper 'Frege's 'Epistemology in Disguise'', Gottfried Gabriel attempts to determine the role that epistemology plays in Frege's philosophy. Gabriel holds that Frege had a direct interest both in logic and epistemology, but only an indirect interest in the philosophy of language. The main focus of the paper is the relationship between logic and epistemology. Gabriel pays close attention to the notion of apodictic statement in *Begriffsschrift* and the way it is related to the notion of a priori truth as defined in *Grundlagen.* It is argued that the former notion involves only necessity relative to general premises whose status may range from logically true to a posteriori true. Frege applies the latter notion or that of a priori knowledge, so we are told, only when the provability of the premises can be traced back to truths which neither need nor admit of proof. Gabriel puts it succinctly: the "proof-theoretical" difference between the metapredicates "apodictic" and "a priori" resides in the fact that the first embodies only relative provability, while the second expresses absolute provability. He points out that the difference between logic and epistemology comes out clearly in Frege's distinction between "reasons for something's being true" and "reasons for our taking something to be true".

Tyler Burge begins his essay by stating what he considers to be a puzzle: On the one hand, the principal aim of Frege's project is to explain the foundations of arithmetic in such a way as to enable us to understand the nature of our knowledge of arithmetic. On the other hand, Frege says strikingly little about our knowledge of the foundations of arithmetic. For Burge, the short solution of the puzzle, though leaving out a great deal, is that Frege thought he had little to add to the traditional view according to which the primitive truths of geometry and logic are taken to be self-evident. Burge argues in considerable detail for the claim that Frege was a platonist as regards abstract (i.e., non-spatial, atemporal, causally inert) entities such as logical objects, functions and thought contents. Frege's platonism is said to show itself in two ways: First, unlike an idealist, he takes the objectivity of abstract entities to be fundamental. Second, Frege believes that the assumption of the relevant abstract entities serves to explain both the objectivity and the success of science and communication. Burge stresses that there is nothing in Frege's work which might remotely indicate that he regarded either the physical world or the realm of abstract entities as dependent upon any activities of judgment, inference or linguistic practice. Turning to Frege's view about how we know the so-called "third realm" of entities that are neither physical nor mental, Burge raises the question: How could Frege believe that reason alone could supply knowledge of it? In the concluding part of his paper, Burge attempts to answer this question by explaining the role that the primitive laws of truth (or of logic) as well as our acknowledgement of them play in Frege's philosophy. It is suggested that in Frege's view, first, the justification for holding logical laws to be true rests on primitive laws of logic and, second, this dependence is manifest in two ways. (a) Any judgment by a particular person is necessarily subject to the primitive laws of logic conceived of as laws that prescribe how one ought to think (judge, infer) if one would attain truth. (b) Acknowledging certain basic laws of truth is a prerequisite for having reason and for engaging in rational thinking. Burge puts it in a nutshell: for Frege, reason and judgment are partly defined in terms of acknowledging the basic laws of truth. Questions of "access" to the third realm are said to be misconceived.

There are likewise three essays on Frege's philosophy of language. In his paper 'Fregean Theories of Truth and Meaning', Terence Parsons deals to a certain extent with a topic he had already explored in great detail in his almost classic paper 'Frege's Hierarchies of Indirect Senses and the Paradox of Analysis'[43]: the semantic analysis of indirect or oblique contexts along the lines of Frege's approach. The much broader objective is now to devise a semantic theory of natural language in terms of Frege's notions of referring and expressing, and to study how he thought natural language actually works as opposed to studying an ideal artificial language that works better. The theory designed by Parsons thus embodies Frege's view that in certain (nonextensional) contexts words refer to the senses which they express when they occur in

[43] *Midwest Studies in Philosophy* **VI**, University of Minnesota Press, Minneapolis 1981, 37–58.

extensional contexts. Parsons argues that a Frege-style semantic theory, when it is suitably formulated, yields a theory of truth satisfying Tarski's material adequacy condition for truth. Moreover, it is supposed to furnish a theory of meaning which satisfies what Parsons regards as the adequacy condition of meaning: it entails for every sentence S of a language L a sentence of the form "'S' means S". Surprisingly, he shows these results to be basically independent of Frege's doctrine that sense determines reference. That doctrine is also examined at some length.

Later in his paper, Parsons focuses on several hypotheses about the way terms behave in multiply embedded indirect contexts. In his mature period, Frege held that every reembedding of a term "t" in a new indirect context induces a new shift in what "t" expresses and refers to. A simpler view considered by Parsons is that reembedding a term causes no further semantic shifts. Other related theories rest on Carnap's proposal that both the extension and the intension of a term "t" should remain stable, no matter which context "t" occurs in. Parsons maintains that under certain conditions these theories are "outwardly equivalent" to one another, that is, they assign to each unembedded sentence the same sense and reference. Thus there is no apparent way in which direct evidence might lead one to prefer one over another.

Richard Mendelsohn's essay is closely linked to Parsons's paper. It tackles the problem of the infinite hierarchy of indirect senses which faces Frege's treatment of oblique contexts. An infinite hierarchy is here called *rigid* if expressions which agree in customary sense, agree at every level of indirect sense; otherwise it is called *nonrigid*. The challenge for the advocate of a nonrigid hierarchy is to explain the structural connection between indirect sense and customary sense, given that the former is not a function of the latter. Mendelsohn has deep misgivings about the nonrigid hierarchy, but his argument here is chiefly directed against Dummett's version of the rigid hierarchy. It is argued that Dummett's proposed revision of Frege's theory, according to which the indirect sense of a word is just its customary reference, must be considered inadequate. In Mendelsohn's view, Dummett's interpretation appears to require more drastic changes for Frege's semantic approach than Dummett himself acknowledges. It rejects not only the Fregean principle that sense determines reference, but also what Mendelsohn labels the NO SELF-REFERENCE PRINCIPLE: No sense can refer to itself. There is, indeed, considerable evidence that Frege endorses the second principle as well. Dummett's way out for Frege, so we are told, would work only if Frege relinquished the assumption that indirect reference is compositional in nature, which, Mendelsohn stresses, lies at the very heart of Frege's analysis of oblique contexts.

There are a number of related critical observations Mendelsohn makes on the way. One point made is that Dummett became vulnerable to Russell's argument in 'On Denoting', which attempts to undermine the sense/reference distinction. Another point is that sentences which require that we speak of the indirect reference of an expression in nonembedded contexts prove to be extremely problematic for Dummett's position. Mendelsohn concludes that if Frege's treatment of indirect reference is to be considered coherent and fruitful, it must be shown that the nonrigid hierarchy poses no insurmountable obstacle to semantic theory.

It is a characteristic feature of Frege's ontological distinction between function and object that it is based upon syntactic observations. If this conception is to be viable, if, in particular, his controversial thesis that numbers are objects is to be sustained, it must be possible to frame criteria by means of which expressions functioning as singular terms may be recognized and distinguished from expressions not so functioning. In his contribution, Bob Hale sets himself the task of formulating such criteria. Special attention is paid to a number of inferential tests suggested by Dummett as well as to a quite different criterion, stemming from Aristotle's dictum that substance has no contrary. Hale maintains that Dummett's inferential tests, if they are modified to deal with certain objections and difficulties, yield an appropriate means of discriminating singular terms within the broader class of substantival expressions. It is pointed out, however, that this approach cannot accomplish all that we should require. Hale accordingly tries to develop a workable version of the Aristotelian criterion. He finally argues that while this improved criterion cannot – any more than one resting exclusively on Dummett's inferential tests – serve on its own, there is a natural way to combine the two that solves the problems of both.[44]

[44] I am grateful to Dirk Greimann for carefully reading an earlier version of this introduction and to Richard Heck for his critical comments on a draft I had written about three years ago on § 10 of *Grundgesetze.* As far as I can see, Richard and I still disagree on several points. Thanks to Ulrike Ritter and Tobias Hürter.

Part I:
Logic and Philosophy of Mathematics

On Positing Mathematical Objects

Michael D. Resnik

Even the mathematician cannot create things at will, any more than the geographer can; he too can only discover what is there and give it a name.
Frege, *Foundations of Arithmetic*, section 96.

1. Introduction

Frege sharply condemned talk of creating or constructing mathematical objects through definitions and postulates or through mental acts of abstraction or invention. He directed much of his rebuke at textbook introductions to real and complex numbers systems, but he also targeted the great works of Cantor, Dedekind, and Hilbert. His differences with them were at bottom as much philosophical as methodological. Yet by focusing on a lack of precision and rigor in their formulations, Frege managed to lump their innovative writings with those of lesser lights and to ignore or overlook the importance of their mathematical contributions and the philosophical issues with which they were attempting to grapple.

I will begin by reviewing Frege's criticisms of creative definition and mental creation of mathematical objects. Then I will turn to the idea of recognizing preexisting mathematical objects through acts of postulation. It is plausible to think of postulationalism as Hilbert's way of sanitizing Dedekind's and Cantor's more psychological talk of creation and abstraction. Frege (and Russell echoing him) rejected postulational epistemologies – even those stripped of psychological overtones. (Recall Russell's quip in *Introduction to Mathematical Philosophy* that postulation has all the advantages of theft over honest toil.) Despite this, the subsequent history of mathematics favors a postulational epistemology over one, such as Frege's, based upon self-evident foundations. As I see it, the main problem with postulationalism is reconciling it with realism about mathematical objects, for at first glance it seems all too much like make-believe. I will to deal with this worry and the sense of realism appropriate for postulationists in the final part of this paper.

2. *Frege on creative definitions and the construction of mathematical objects*

In section 143 of the *Grundgesetze* Frege presents two passages from Otto Stolz as examples of creative definitions. The second is more interesting for our purposes, and I will quote it now:

> 1. *Definition.* "If in case (D_1) no magnitude of System (I) satisfies the equation *box=a*, then it shall be satisfied by *one and only one new thing not found in* (I); this may be symbolized by *aub*, since this symbol has not yet been used. We thus have
> *bo* (*aub*) = (*aub*) *ob* = *a*."
> Since the new objects possess no further properties, we can assign them properties arbitrarily, so long as these are not mutually inconsistent.

The obvious problem with this definition – its creative aspect – is that Stolz uses the phrase "it shall be satisfied" as if *his very words* can bring the requisite object into being.

Unlike Lesniewski, Frege never precisely specified what he meant by "creative definition". Yet from the examples he offers and his discussions of them, we may infer that a definition is *creative in the Fregean sense* if it defines an object as *the F* in the absence of previously established theorems showing that something does uniquely fall under *the concept F.* Of course, Frege rightly berated attempts, such as the one just quoted, to use definitions to introduce mathematical objects when one really needed to prove a theorem or to introduce a new axiom. He may also have been correct in thinking that the logical error committed here arises from failing to distinguish concepts and objects so that one actually defines a concept (e. g., the concept x+x=x) while being under the illusion of defining the object that uniquely falls under it (e. g., zero).

Yet, perhaps prompted by the very texts he cited, Frege often allowed his criticism to wander from methodological issues to more philosophical ones. For example, *Grundgesetze*, 143 is embedded in a more general discussion entitled "Construction of new Objects; Views of R. Dedekind, H. Hankel and O. Stolz" in which he criticizes Dedekind's thesis that we create irrational numbers corresponding to cuts as well as Stolz's claim, also advocated by Hilbert, that in mathematics we are free to postulate any objects we can consistently describe. These sections, by the way, are also reminiscent of the even wider ranging critique of Hankel's and Kossak's treatment of real and complex numbers Frege presented in sections 92–103 of the *Grundlagen*. There we also find him advocating strong anti-formalist and anti-psychologist themes.

Frege did not seem to distinguish clearly between the methodologically erroneous use of creative definitions and the more defensible appeal to mental constructions or acts of postulation. As a result he seemed unable to appreciate fully the work of Cantor, Dedekind and Hilbert. Thus he rejected the definitions of real numbers given by Cantor, Dedekind and Weierstrass as couched in terms which were too psychological, too physical or just too confused, and then he went on to develop a theory which, from a mathematical point of view, was not significantly different from theirs.[1] Similarly, even after he had found a way to clarify Hilbert's ideas on using axioms to define the primitives of geometry, he characterized Hilbert's work to Heinrich Liebmann as follows:

> Clever and inventive as it is in many points, I think that it is on the whole a failure and in any case that it can be used only after a thorough criticism. Hilbert, like many mathematicians, seems to lack a clear awareness of what a definition can do ... (*Correspondence*, p. 90)[2]

As far as I know Frege never reversed his unfortunate assessment of his great contemporaries. Of course, by his and our own standards Dedekind's and Cantor's views were more than a bit vague and philosophically naive. In his 1872 essay on continuity and the real numbers Dedekind writes as if individual mathematicians create individual irrational numbers (p. 15), while in his 1888 essay on the natural numbers he states that they are a free creation of "the human mind" obtained through mental abstraction. Frege did not seize upon these remarks; but if he

---

[1] Michael Dummett's *Frege: Philosophy of Mathematics*, which contains a very helpful discussion of these topics, explains that Frege's deeper reason for rejecting these definitions is that they fail to explain the possibility of applying the reals.

[2] It is interesting to juxtapose this passage with the next one taken from Frege's letter to Hilbert of 27 December 1899. Here he warns Hilbert that "there is already widespread confusion with regard to definitions in mathematics, and some seem to act according to the rule: If you can't prove a proposition, then treat it as a definition". (*Correspondence*, p. 36)

had, he would surely have ridiculed Dedekind's talk of abstraction and derided his psychologism just as he objected to the "impossible abstractions" and other psychological processes Cantor invoked. (Cf. Frege's review of Cantor's *Zur Lehre Vom Transfiniten*, pp. 164–5).

If neither definitions nor mental abstraction serve to bring mathematical objects into existence, why not introduce axioms that stipulate that they exist? Frege tells us why we cannot in *Grundlagen*, 102 where he writes

> It is common to proceed as if mere postulation were equivalent to its own fulfillment. We postulate that it shall be possible to in all cases carry out the operation of subtraction [There is a footnote here to Kossak], or of root extraction, and suppose that with that we have done enough. But why do we not postulate that through any three points it shall be possible to draw a straight line? Why do we not postulate that all the laws of addition and multiplication shall continue to hold for a three-dimensional number system just as they do for real numbers? Because this postulate contains a contradiction. Very well then, what we have to do first is to prove that these other postulates of ours do not contain any contradiction. Until we have done that, all our hard-won rigor is so much moonshine.

Clearly Frege was foreshadowing this passage when in the introduction he wrote:

> Yet it must still be borne in mind that the rigor of the proof remains an illusion, however flawless the chain of deductions, so long as the definitions are justified only as an afterthought, by our failing to come across any contradiction. By these methods we shall, at bottom, never have achieved more than an empirical certainty, and we must really face the possibility that we may still in the end encounter a contradiction which brings the whole edifice down in ruins. For this reason I have felt bound to go back rather further into the general logical foundations of our science than perhaps most mathematicians will consider necessary. (p. ix)

The Russell contradiction later made this passage ironical. However, Frege continued to reiterate these ideas even after learning that his own system contained a contradiction.

Section 103 of the *Grundlagen* is part of a discussion of proofs that use complex numbers to derive results restricted to the reals, and in both organization and content this discussion is remarkably similar to Frege's remarks in *Grundgesetze* 140–143 on the use of auxiliary objects in proofs. These passages make it quite plain that Frege believes that we are not free to posit new objects simply because doing so will enable us to complete a proof. I think not only we, but also Cantor,

Dedekind and Hilbert would readily agree to that. As will emerge when we discuss Hilbert, Frege thought that positing mathematical objects would be pointless if justified. I suspect this left Frege unable to see Hilbert's turn of the century approach to the foundations of mathematics as anything but a more sophisticated variant of the erroneous view of Kossak that he had already thoroughly criticized in the *Grundlagen*.

We can get a clearer picture of Hilbert's ideas and the philosophical and methodological issues they raise by comparing them with the views of Cantor and Dedekind. Let us turn to that now.

3. Dedekind, Cantor and Hilbert

Along with a number of mathematicians of their day, including some of Frege's favorite formalist targets, Cantor, Dedekind and Hilbert were attempting to determine how and when we can be justified in introducing new mathematical objects. The introduction of the complex numbers had already focussed the mathematical community on these questions, but Cantor's infinite numbers and sets surely made them all the more pressing. One way of responding to new and questionable entities is to show that they can be generated from previously accepted ones. The standard account of how the complex numbers gained acceptance, through being constructed geometrically, fits this approach. And it is certainly the approach that Frege used and insisted upon. Cantor also seems to embody this approach even when he writes of abstraction; for starting with a familiar set or sequence he has us abstract in thought in order to obtain a cardinal or ordinal number. I will not venture to say whether Cantor held that we can obtain all infinities in this way.

From the historical point of view, however, it makes more sense to see us as introducing most mathematical entities in order to complete or enrich known structures, whose deficiencies we may have discovered more or less by accident. This certainly seems to be the best account of how we came to recognize the irrational and imaginary numbers, and it probably works well for zero and the negative numbers too. In the following passage Dedekind comes close to affirming this historical thesis:

> ... every theorem of algebra and higher analysis ... can be expressed as a theorem about natural numbers ... But I see nothing meritorious ... in actually performing this wearisome circumlocution ... On the contrary, the greatest and most fruitful advances in mathematics and

> other sciences have invariably been made by the creation and introduction of new concepts, rendered necessary by the frequent recurrence of complex phenomena which could be controlled by the old notions only with difficulty. (1888, pp. 35–36)

Yet despite this endorsement of mathematical creativity, he also held that, however fruitful they may be, such acts of creation must be validated through a proper foundation.

> ... Just as negative and fractional rational numbers are formed by a new creation, and as the laws of operating with these numbers must and can be reduced to the laws of operating with positive integers, so we must endeavor completely to define irrational numbers by means of the rational numbers alone. (1872, p. 10)

But we should not let his talk here of defining the irrational numbers mislead us, for Dedekind was no reductionist. We do not identify the reals with cuts; rather for each cut we create a real number to associate uniquely with it (1872, p. 15). Similarly, we do not reduce the natural numbers to sets or systems but instead we create them by starting with a simply infinite system and abstracting from the particular features of its elements and generating relation (1888, p. 68).

This is surely a curious position. On the one hand, Dedekind allows us to create new systems of mathematical objects as the need arises; on the other hand, he insists that we establish that their structure is realized by some familiar system. (Moreover, not any system will do: "That such comparisons with non-arithmetic notions have furnished the immediate occasion for the extension of the number concept may, in a general way, be granted (though this was certainly not the case in the introduction of the complex numbers); but this surely is no sufficient ground for introducing these foreign notions into arithmetic, the science of numbers" (1872, p. 10)).

The next passage will help us reconcile these seemingly contradictory themes, for here Dedekind explains that numbers are "free creations of the human mind" is the sense of being abstracted from a previously given system:

> With reference to this freeing the elements from every other content (abstraction) we are justified in calling the numbers a free creation of the human mind. (1888, p. 68)

Now suppose that we combine this explanation of creation as abstraction with Dedekind's other thesis that the natural number system is a structure with the individual numbers being nothing but positions in that structure. Then we see that what we create when we create the numbers is a structure and that we do so by abstracting it from a previously given simply infinite system. But this system in turn is no structure composed of featureless positions. Rather it is an arrangement of full-blown objects – one whose existence Dedekind had previously attempted to demonstrate in his notorious proof that the possible objects of his thought constituted a simply infinite system. Generalizing from this case, it is plausible that Dedekind believed that we create number systems by creating new patterns or structures in which the numbers in question are nothing but positions, and that we must justify the creation of such structures by showing that they can be abstracted from an appropriate system of thought objects. (I am not sure whether he would hold that in all cases the original act of creation involves abstracting a structure from one of its instantiations. This would fit the complex numbers poorly – a point he seems to note in a passage I lately quoted.) In modern dress, this doctrine might declare that we may introduce a new mathematical structure by laying down a set of categorical axioms for describing the structure, *provided* that we are prepared to show that these axioms have a set theoretic model.[3]

This way of formulating Dedekind's theory reveals a defect that is obvious to us today: it merely transfers the problem of mathematical existence to that of sets. Of course, we could hardly expect Dedekind to have appreciated this problem in 1888. But when Hilbert turned to the question of mathematical existence in 1899 and 1900 he was already well aware of the difficulties with trying to ground mathematics in set theory or logic. Writing to Frege on 7 November 1903, he says

> ... Your example [the contradiction at the end of *Grundgesetze*, Vol. II] ... was known to us here;* I found other even more convincing contradictions as long as four or five years ago; they led me to the conviction that traditional logic is inadequate and that the theory of concept formation needs to be sharpened and refined. As I see it, the most important gap in the traditional structure of logic is the assumption made by all logicians and mathematicians up to now that a concept is already there if one can state of any object whether or not it falls under it. This does not seem adequate to me. What is decisive is the recognition that the axioms that define the concept are free from contradiction. (Footnote * I believe Dr Zermelo discovered it three or four years ago after I had communicated my examples to him.) (*Correspondence*, p. 51)

[3] I am merely using sets as a convenient substitute for Dedekind's less reputable thought objects. I do not mean to interpret him as committed to them or to make any detailed claim about his ontology.

Here Hilbert diverges from both Dedekind and Frege in rejecting the search for *a priori* foundations for mathematics. Furthermore, instead of thinking of axioms as describing a specific system of previously given mathematical entities, Hilbert took them as devices for specifying mathematical concepts. Unfortunately, the essays Hilbert wrote on the foundations of mathematics at this time are far from clear on points of logical and philosophical detail. As a consequence, Frege, rather than Hilbert, gave the clearest contemporary explanation of the sense in which Hilbert's axioms specify mathematical concepts; to wit, if we construe the non-logical predicates appearing in his axioms as placeholders for first-order concepts, then their conjunction defines a second-order relation. Construed in this way, Hilbert's geometrical axioms, for instance, define the concept of a Euclidean space.[4]

Hilbert's axiomatics resembles Dedekind's structuralism. It is tempting to read Dedekind 1888 as defining the concept of an omega sequence. Moreover, in the following passage from a letter to Frege dated 29 December 1899 Hilbert expresses structuralist ideas:

> ... You say that my concepts, e. g. 'point', 'between', are not unequivocally fixed; e. g. 'between' is understood differently on p. 20, and a point is there a pair of numbers. But it is surely obvious that every theory is only a scaffolding or schema of concepts together with their necessary relations to one another, and that the basic elements can be thought of in any way one likes. If in speaking of my points I think of some system of things, e. g. the system: love, law, chimney-sweep, ... and then assume all my axioms as relations between these things, then my propositions, e. g. Pythagoras' theorem, are also valid for these things. In other words: any theory can always be applied to infinitely many systems of basic elements ... (*Correspondence*, pp. 40–41)

Yet I would not want to make much of these similarities between Hilbert and Dedekind. Today structuralism is taken to comprise a variety of positions, which differ from each other in philosophically important ways. For this reason, I believe that a more thorough investigation would reveal significant discrepancies between their views.

With the preceding qualification in mind, it is profitable to view Hilbert as trying to replace Dedekind's *ontological* thesis that we *create* mathematical structures through acts of mental abstraction by the *epistemological* thesis that we can *recognize* new structures by postulating them through axioms. For us to see a real distinction between these

---

[4] For an account of the Frege-Hilbert controversy see my "The Frege-Hilbert Controversy".

views, we must appreciate that in the sciences postulation, though a form of stipulation, is a really a way of hypothesizing. If we read a bit more into the text than is justified, we can take the conjunction of Hilbert's axioms as asserting of itself that it is satisfied by some system of objects and relations. One way technically to clean up this interpretation is to build upon Frege's rendering of Hilbert: First we relativize the quantifiers in the conjoined axioms to a new one-place predicate representing the posited domain, second we replace all the non-logical predicates and functors by second-order variables, and finally we prefix the formula obtained thus far with second-order existential quantifiers binding its free second-order variables. To illustrate this, suppose that we start with the number theoretic axioms:

(x) (y) (z) (Sxy & Sxz → y=z)
(x) (Ey) Sxy
(x) (Zx ↔ -(Ey) Syx)
(F) {(x) (Zx → Fx) & (x) (y) (Fx & Sxy → Fy) → (x) (Fx)}.

Then we would relativize the individual quantifiers to a new second-order variable, say, "X" and replace "S" (successor) and "Z" (zero) by new second order variables, say, "Y" and "W", conjoin the results and existentially quantify this conjunction. This would yield:

(EX) (EY) (EW) [(x) (y) (z) (Xx & Xy & Xz → (Yxy & Yxz → y=z)) & (x) (Xx → (Ey) (Xy & Yxy)) & (x) (Xx → (Wx ↔ -(Ey) (Xy & Yyx))) & (F) {(x) (Xx → (Wx → Fx)) & (x) (y) (Xx & Xy → (Fx & Yxy → Fy)) → (x) (Xx → Fx)}].

(Since Hilbert's axioms for geometry use three sorts of variable, the first step would take three new predicates.)

In some ways this departure from Dedekind represents real progress. It shifts the debate from ontology to epistemology, thereby dispensing with the problem of how we could create actual non-denumerable infinities. On the other hand, it raises questions of control. What, for example, is to keep us from postulating non-existent or impossible structures. (Recall Frege's example of positing that any three distinct points determine a line.)

Hilbert believed that we can deal with this difficulty by proving the consistency of our postulates. In a 1900 paper on the real numbers Hilbert writes:

> ... But if we succeed in proving that the characteristic marks assigned to the concept can never lead to a contradiction when subjected to a

> finite number of logical inferences, then I say that this proves the mathematical existence of the concept of, e. g., a number or function satisfying certain requirements. In the present case, where we are dealing with the axioms for real numbers in arithmetic, the demonstration that the axioms are free from contradiction is at the same time a proof of the existence of what is contained in the concept of real numbers or in that of the continuum. (*Correspondence*, p. 50)

But how are we to prove the consistency of the axioms of arithmetic? Since Hilbert had used arithmetic models in his *Foundations of Geometry* to prove the consistency and independence of his geometric axioms, and since he rejected set theoretic foundations, he had hit model theoretic bedrock with the axioms of arithmetic. His reference in the passage to a finite number of logical inferences together with his claim a few lines earlier that he was convinced that one could find a "direct proof" of consistency by "carefully working through the known methods of inference in the theory of irrational numbers" suggest that he had in mind a proof theoretic approach of the sort he used some years afterwards.

Frege was highly skeptical. To him Hilbert's new version of postulationalism smacked of the version he criticized years ago in the *Grundlagen*:

> Strictly, of course, we can only establish that a concept is free from contradiction by first producing something that falls under it. The converse inference is a fallacy, and one into which Hankel falls. (Section 95)

He puts this point in the form of a dilemma in the *Grundgesetze*:

> ... How do we tell that properties are not mutually inconsistent? There seems to be no criterion for this except the occurrence of the properties in question in one and the same object. But the creative power with which many mathematicians credit themselves thus becomes practically worthless. For as it is they must certainly prove, before they perform a creative act, that there is no inconsistency between the properties they want to assign to the object that is to be, or has already been, constructed; and apparently they can do this only by proving that there is an object with all these properties together. But if they can do that, they need not first construct such an object. (Section 143)

Ironically, in the next sentence Frege poses the question "Or is there perhaps still another way of proving consistency?". But he continued with his critique instead of taking his query seriously. Had he done otherwise he might then have considered the possibility of using proof theoretic methods to demonstrate consistency. And maybe he actually did so later, for in writing to Hilbert on the 16th of September 1900, he indicates that he believed that Hilbert may have found a new way for proving consistency. He remained quite doubtful of its ultimate success, however. (*Correspondence*, pp. 49–50)[5]

Frege never attached any importance to consistency proofs, anyway; whereas for Hilbert they were absolutely crucial. The sharp contrast in their thinking emerges vividly in the following well known dialogue:

> Frege: ... I call axioms [of Euclidean geometry] propositions that are true but are not proved because our knowledge of them flows from a source very different from the logical source, a source which might be called spatial intuition. From the truth of the axioms it follows that they do not contradict one another. There is therefore no need for a further proof. (*Correspondence*, p. 37.)
> Hilbert: I found it very interesting to read this very sentence in your letter, for as long as I have been thinking, writing and lecturing on these things, I have been saying the exact reverse: if arbitrarily given axioms do not contradict one another with all their consequences, then they are true and the things defined by the axioms exist. This is for me the criterion of truth and existence. (*Correspondence*, pp. 39–40)

I find this passage is a bit difficult to interpret, for Hilbert seems to waiver between ontic and epistemic concerns. In speaking of consistency as the *criterion* of truth and existence he may have been suggesting the familiar idea that mathematical existence and truth *is nothing but* consistency[6], or he may have been making the epistemic proposal that we can *know* that some axioms are true by proving their consistency.

Now without having to resolve this exegetical question, we can see a strong epistemological consideration on Frege's side: On pain of an infinite regress, we must bring the chain of consistency proofs to an end with mathematical principles whose truth we do not attempt to establish by further (consistency) proofs. Even proof theoretic demonstrations of consistency must assume some kind of mathematical ontology and truths about it. I believe that Hilbert ultimately came to appreciate

---

5 Frege's reasoning is faulty too. We can establish the consistency of the unsatisfied condition "is a white unicorn" by showing that it has the same logical form as some satisfied condition, say, "is a white horse".

6 Taking this interpretation to heart, we might prefix our second-order versions of Hilbert's postulates with a primitive modal operator expressing logical consistency, converting Hilbert's view into a close relative of the modal-structuralism of Geoffrey Hellman's *Mathematics without Numbers*.

this point, which in turn prompted him to formulate his program for proving the consistency of mathematics by taking finitary mathematics as a starting point.

4. *Postulational epistemologies and realism*

I think that in abandoning a postulational epistemology Hilbert made a mistake. It was probably due to his accepting, along with Frege, the idea that good mathematical practice obliges us to justify any set of postulates we introduce by proving their consistency. Thanks to Gödel, we now know that in the most important cases this approach to warranting our axioms is misguided; in proving their consistency we must draw upon principles that are unlikely to be more secure than the ones we are trying to vindicate.

To be sure, we can and should insist that contradictory systems be rejected. However, just as we cannot derive infinite numbers or sets from a purely finite basis, so in general we must increase the strength of our axioms as we try to capture progressively complex mathematical structures. At each step there are a number of essentially equivalent ways to postulate the structures we want, but none can be justified through models or consistency proofs which appeal to the more elementary structures we have already accepted. We must simply try out new axioms and seek indirect evidence for them.

In the balance of this paper I argue that we can combine a realist, mathematical structuralism with a postulational epistemology to obtain a coherent philosophical view. Since realism has its sources in Frege, structuralism in Dedekind and postulationalism in Hilbert, it is wise to review the lessons we can draw from our examination of their views.

First, Frege was certainly right to object to the idea that we can create mathematical objects through postulating them. Even constructivists can agree that merely postulating that infinitely many twin primes exist is no more potent than wishing that they exist. Moreover, it boggles the mind to think, for example, that in positing choice functions (by adding the Axiom of Choice to ZF) we have literally created them. How could we individually create so many things of such complexity? And what sense can we make of creating all of them in one blow? Better to see our positing them as a way of acknowledging or recognizing something existing independently of our positing.

Thus if we are going to be postulationists, we should be epistemological postulationists rather than ontological ones. This means that we must regard positing mathematical objects as a way of coming to *recognize* something that already exists independently of our positing them. This in turn almost commits us to mathematical realism.

I say *almost* commits us to realism, because I think mathematical realism involves more. Specifically, it consists in three theses: first, that mathematical objects exist (in the same sense that anything else does); second, that our current (classical) mathematical theories are largely true; third, that the existence of mathematical objects and the truth our mathematical theories is independent of us, our beliefs, postulates, proofs and constructions, etc. The sense of truth I have in mind in the second and third theses is just immanent, disquotational truth. It is disquotational, because for each sentence *S* in the language of science and mathematics, I postulate the metalinguistic sentence obtained from the schema

S is true if and only if *p*,

by putting *S* itself in place of '*p*' and a name of *S* in place of '*S*'. It is immanent because I do not require the predicate "true" to apply beyond our own scientific and mathematical language. And I do not presuppose that our language somehow uniquely corresponds (or hooks onto) reality.

Notice that this version of realism counts as anti-realists those who deny the existence of mathematical objects, or deny classical mathematics, or hold that mathematics is true by convention or consists only of conditional assertions. Thus it is already strong enough for formulating the difference between realism and anti-realism in contemporary philosophy of mathematics.[7]

Frege was wrong, however, in thinking that the only way to prove the consistency of a set of axioms is to prove that they are true. Short of proving them true, we can prove them consistent *via* syntactic proofs or model theoretic ones using models in previously accepted domains. For example, the proofs that the axiom of choice is consistent with and independent of first-order ZF set theory presupposes no sets not already available in ZF without Choice. (These proofs can even be formulated as results in first-order number theory.)

This example also shows that Frege's dilemma is a false one. To prove that we can consistently posit a choice function for every set it is not necessary that we first prove that such functions exist.

---

[7] I expound this conception of truth and versions of realism based upon it in my "Immanent Truth".

From Dedekind I take structuralism, the view that mathematical objects are just positions in structures; that like mathematical points, they receive their identity and purely mathematical features through their relationships to other positions in the structure containing them. On this view, when we posit mathematical objects we are positing positions standing in certain relations to each other. In this sense, we are positing structures themselves. Yet, strictly speaking, we only posit the positions, for our quantifiers range over positions and not the structures in which they occur.[8]

My structuralism parallels Hilbert on mathematical existence and truth; for I take any consistent set of axioms to describe a structure. Hilbert was right in the sense that consistent axioms are true of some structure or other. Categorical axioms are true of just one structure while others may be true of infinitely many structures. Now this raises somewhat technical questions concerning axioms that have non-standard models, and their exact resolution depends upon the mathematical articulation of structuralism one adopts. I am not in a position to address them here, and I do not think we need do so.

However, since I have made much of disquotational truth in this paper, I want to digress a bit now to make some remarks about reconciling it with structuralism. Here is the problem (which Stephen Leeds pointed out to me): Consider ZF set theory, which in its strongest second-order form only manages to characterize a nested hierarchy of structures. (Geoffrey Hellman calls second-order ZF *quasi-categorical* in his *Mathematics without Numbers.* First-order ZF does not even approach this limited kind of categoricity.) The axioms that are independent of second-order ZF, such as the axiom postulating an inaccessible cardinal, fix the extent of this hierarchy. Should we count the axiom of inaccessibles as true? Or true in some structures and not others? As a structuralist I want to hold the latter. But excluded middle holds in ZF, so by applying disquotational truth to the language of ZF we may infer that the disjunction of the axiom of inaccessibles with its negation is true. If my truth-theory holds that a disjunction is true if and only if at least one of its disjuncts is (and why shouldn't it!), then we get bivalance: the axiom of inaccessibles is straightforwardly true or false, not true in some set theoretic structures and false in others. What can the structuralist say?

---

[8] See my "Mathematics as a Science of Patterns: Ontology and Reference".

Here are several ways of responding that have occurred to me: 1) Acknowledge that set-theory, like group theory, treats a multiplicity of structures.[9] 2) Hold that our current talk of sets *does* refer to a definite structure, a definite iterative hierarchy, but that we have not succeeded in fully describing it.[10] 3) Hold that our talk of sets does not determine a definite iterative hierarchy (perhaps, our set theoretic language is ambiguous or it is not determinate which language we speak), but that our set-theoretic practice does require us to seek to try to eliminate indeterminateness in our set talk when we find it. In other words, on our current conception of sets there is but one iterative hierarchy of sets; when we settle on what it is, we will say that this is what we would have meant all along, if we had known enough, and regard the other alternatives as close relatives that we did not intend, much as some might regard non-standard models of number theory.[11] 4) Restrict our truth theory, and therewith bi-valance to talk of those portions of the hierarchy that are currently settled. We could probably do this by restricting our truth-theory to so-called bounded set theoretic sentences whose bounds are less than a suitable ordinal.[12]

Alternative (1) implies that epsilon is a schematic letter; so bi-valence fails, the independent set theoretic sentences are not true or false *simpliciter* but true under some interpretations in iterative hierarchies, false in others. Structuralists can adopt this so long as they are willing to regard set theory on a par with group theory. Option (2) is compatible with bi-valence, but it commits us to a mysterious theory of reference that some how fixes our current reference to sets. We can save bi-valence while adopting (3) by maintaining that it holds for each of the indefinitely many languages that we might be speaking. The last option allows bi-valence where it is unproblematic, but denies it elsewhere. I currently waiver between (3) and (4).

Note, however, that although the problem Leeds has raised calls for a further elaboration of my structuralism, it does not show that the view is untenable. Nor does it refute my realism. I can still admit that a large variety of structures exist, including those corresponding to bifurcations in set theory. *The difficulties do not concern the nature of structures or even our ability to know that they exist*, but only our inability

---

[9] Maddy labels this position as "*no fact*" in her "Indispensability and Practice".

[10] This is Maddy's *fact*.

[11] This is like Maddy's *beginning of story no fact*. Cf. Shapiro's position on the incompleteness of second-order logic in his *Foundations without Foundationalism*.

[12] Cf. Hellman's treatment of set theory in his book.

to describe them categorically. The same difficulties would also arise if I formulated my structuralism using correspondence truth – except that then they would concern the uniqueness of correspondence relation.

Let us return to positing mathematical objects (positions in structures). I am suggesting that we can arrive at beliefs concerning structures by introducing new axioms that purport to extend structures already familiar to us by adding new positions to them (as our ancestors did in introducing, say, the imaginary numbers) or to introduce entirely new ones (as the founders of set theory did). However, I am not claiming that positing new mathematical objects automatically gives us knowledge of them (or the structure in which they occur) but only that positing can be a route to knowledge. Initially our axioms may function as fairly tentative hypotheses, and as we acquire more evidence for them, our belief in them can evolve into knowledge.

We have already seen that my postulational epistemology commits me to the independent existence of mathematical objects. So it cannot be that postulationalism is incompatible with realism. On the other hand, one might think that it is an implausible epistemology for realists to adopt, that positing smacks of make-believe. A world of posits seems a better companion for fictionalism or instrumentalism than for realism. Or if it must be realism, then it should at least be the sort of modal realism Hartry Field ("Is Mathematical Knowledge Just Logical Knowledge") and Geoffrey Hellman embrace in substituting the logical possibility of mathematics for its truth.

(Actually, we have no reason to think that the postulational approach will do better with modalism than with my supposedly less tempered realism. Modalists are realists about the truth of certain possibility statements, which means in part that their truth-values are independent of our postulating them to be true. Thus just as mathematical realists must face the problem of explaining how we obtain knowledge about an independent mathematical reality, so must modalists explain how we obtain knowledge about an independent "realm of possibilities". Hellman acknowledges as much when he concludes *Mathematics without Numbers* with this candid assessment: "... We see no way of explaining away [modal-existence claims] as linguistic conventions, or of otherwise reducing them to a level of observation, computation or formal manipulation. At best, the modal approach involves a trade-off *vis-a-vis* standard platonism ..."(pp. 143–144))

As to combining realism with postulationalism, I think that we need to address two questions: 1) What distinguishes mathematical posits epistemically from fictional characters? 2) What sense can we make of mathematical theories introduced through postulation being about an independent reality? I will tackle these in the order given.

In order to dramatize our first question, let me phrase it much as Terry Parsons did when I read an earlier version of this paper in Munich: Couldn't a literary mathematician write some mathematics that one could read as a piece of fiction? Conversely, couldn't a piece of fiction contain some significant mathematics? After all, our inability physically to detect the objects they introduce counts against neither mathematics or nor fiction. What then makes a piece of prose mathematics rather than fiction?[13] Now, of course, we can read mathematics as fiction and also read mathematics in fiction; just look at mathematical riddles, parables for popularizing mathematical ideas or the longer works by Lewis Carroll and Rudy Rucker. Generally, very obvious differences in style and vocabulary differentiate articles intended as mathematics from essays in fiction, but these exceptions prove the rule. The matter does not rest on syntax.

Nor on semantics: fiction does not have to be about unreal or unmathematical characters; it might even be true – accidently. In any case, having a recognizably real subject matter could not be the test for separating mathematics from fiction, since frequently when someone introduces us to new mathematical objects through positing them they are not recognizably real. Rather we mark something as a piece of mathematics by treating it and expecting it to be treated differently than we treat fiction. We expect it to meet different standards of accessibility, clarity, precision, rigor, coherence and thoroughness, to mention some criteria that come to mind, and to play different roles in our overall intellectual life. We applaud different reasons for positing new mathematical objects than we do for creating new fictional characters, and we treat mathematics as a much more integral part of our theoretical endeavors than we do fiction.

To be more specific, we rarely introduce mathematical posits lightly. Unlike fiction, it is not enough that they entertain. To be taken seriously they must answer to a clear mathematical need such as allowing us to answer questions that were independent of our previous mathematics or

---

[13] When I read an earlier version of this paper to the Department of Philosophy at the University of Colorado in Boulder George Bealer also raised these questions. Dale Jamieson showed me how to clarify my response.

to systematize and extend a body of previous results.[14] Furthermore, after introducing them we seek evidence for mathematical posits, which we do not seek for fictional characters. We want to know that our mathematical postulates characterize some structure or other, and preferably the one we thought we intended. They are hypotheses that we are prepared to modify or withdraw in the face of evidence that they are inconsistent, have unwanted models, fail to yield the consequences we seek or poorly fit our broader mathematical and scientific programs.

Towards increasing our confidence in our axioms we may conduct comparative studies, to provide measures of relative consistency proofs or expressive power, or devise alternative presentations of the same set of theorems or model them in different mathematical guises – for example, geometrically as opposed to analytically, or set theoretically as opposed to categorically. Experience counts too: If we have studied and developed a system, such as ZF, for a considerable time without encountering a contradiction, then we have grounds for believing it to be consistent and to describe a structure or structures. This sort of consideration favors the weaker and older systems. So does their relative simplicity, since the simpler a system the more likely we are to understanding its workings and to find our expectations concerning its theorems verified through our proofs. Our situation is not as bad as that Frege attributed to his contemporaries; we do not attempt to justify our postulates "only as an afterthought, by our failing to come across any contradiction", but in the end we can not achieve anything more than "empirical certainty" of their consistency. (Cf. *Grundlagen*, p. ix)

So far I have concentrated on how we might know that our postulates are consistent, that is, describe some structure or other. But there is also the question of how we might know that they are true (in the appropriate structure), that is, describe the structure that we intended them to describe. Now if we are trying to extend an old structure, then we might appeal to the sort of considerations Penelope Maddy surveyed in her "Believing the Axioms". We might, for example, show that axioms introduced to extend the old structure have previously accepted consequences in more elementary domains or that they connect previously disunited mathematical theories or that they conform to principles guiding the development of their domain.

---

[14] I do not mean to exclude the possibility of someone introducing a structure that so fascinates the mathematical community that it is taken seriously, although there was no prior practical or mathematical motivation for introducing it.

Furthermore, in some, albeit relatively rare, cases we can even test a mathematical hypotheses observationally. Thus a mathematician who conjectures that a certain equation has a solution within certain bounds might confirm it by *looking* at the output of a computer programmed to compute it. Or take computer programs, considered now as mathematical objects. Some of our best evidence – often our only evidence – concerning how programs behave comes from observing computer runs. Observationally confirmed mathematical hypotheses provide indirect evidence for axiom systems that imply them.

Seeing a computer print-out as confirming a mathematical hypothesis involves holding fixed a number of hypotheses concerning the reliability of the computer, the relationship between the mathematical function and the mathematical algorithm computing it and the implementation of the latter on the computer. But we must hold certain hypotheses fixed in confirming physical hypotheses too. Even when we speak of *observing* a physical particle, what we often perceive in the first instance is the behavior of instruments designed to detect it. And with increasing regularity even these are being supplanted by read-outs from computers linked to our instruments. To connect these to posited particles we need lengthy chains of theoretical development whose links include the physical properties and capacities we postulate posited particles to have.

By contrast, none of these evidential constraints apply to the stipulations through which authors premise their fictions. They are extraneous to the practice of fiction writing. Not only do some good stories mock science and commonsense, but in principle they might violate elementary logic as well.

In emphasizing analogies between positing in mathematics and positing in the empirical sciences I am assuming that the latter does not undermine realism about theoretical entities in science. Granted, some philosophers of science will chuckle at my basing a case for mathematical posits on realism about theoretical entities; but virtually all philosophers of mathematics presuppose scientific realism anyway, since they attribute the epistemic problems with mathematical objects to their non-physical nature.

On the other hand, there is a disanalogy between mathematical positing and scientific positing, which some may find disquieting. Consider the case of some physicists who propose a theory of a new kind of particle. Good physical practice requires them or their colleagues to develop experiments for confirming their theory, and some of these will involve detecting the particles the theory posits. I see no mathematical match for this case. When we introduce new mathematical theories, even those proposed with applications in mind, we do not try to detect the mathematical entities we posit.

But is not this as it should be? We posit our paradigm mathematical objects to delineate structures; we posit our paradigm physical objects to explain events. Consequently, we expect the latter but not the former to be detectable. That may be one reason why physicists seem more suspicious of the undetectable interior of black holes or virtual processes than mathematicians are of the natural numbers.

Doing fiction, mathematics, and physics may grade off into each other so that it could be unclear whether one discourse should be taken as fiction or mathematics (whether we should acknowledge its truth) or an other as mathematics or, say, theoretical economics (whether we should expect it to have empirical content). But this does not mean that in doing mathematics we produce fiction any more than it means that in doing theoretical science we refrain from attributing empirical import to our theories.

One might grant that the practice of mathematics is much more like the practice of science than the practice of fiction writing, but still wonder why we should believe that mathematical posits are real, much less as real as physical ones. We should not, if by "just as real" one means "just as detectable, just a much a part of the spatiotemporal, causal network". On the other hand, if one means "exist in the same sense of 'existence'", then we need only look at the applications of mathematics to see that mathematical and scientific existence assertions are treated as part of a seamless linguistic network. Even a cursory examination of the use of mathematics in science shows that scientists refer to mathematical objects in defining many of their most fundamental concepts. For instance, if the real numbers did not exist, then the concept of a velocity would be no more well-defined than that of the sum of a divergent infinite series and all generalizations about velocities would be vacuously true. It is also easy to see that scientists not only refer to mathematical objects but also freely appeal to their standard mathematical properties, such as the mathematical relationship between velocities and accelerations, used to calculate how a the path of an object will change when subjected to a force. Using mathematics in inferences requires them to take its existence assertions as true, otherwise they could not detach the conclusions they draw from them.[15]

---

[15] This is the point of the so-called indispensability argument. See my "Scientific vs. Mathematical Realism: The Indispensability Argument".

Colin McLarty suggested to me that divisions in mathematics/science network might be marked by different styles of variables – one style for mathematical objects, another for physical ones. The resulting picture of science would give it a many-sorted logic, and the mathematical existential quantifier might answer to a species of existence intermediate between the fictional and physical. However, if there is such a division between the fictional, the mathematical, and the physical, it does not correspond to one in the meaning of "exists". In so far as individual words have meanings, this word means the same thing in fiction, mathematics and empirical science. After all, don't we have to re-assure children that the story "Once upon a time there was an evil monster who feasted upon lost children ..." is only make-believe, and that monster never really existed. Whatever distinction there is here has to do with the larger discourse – the way it is too be taken, evaluated, etc. – not with the word "exists". In any case, we also apply mathematics within mathematics, say, by counting the number of numbers less than a given sum of numbers. This too requires us to acknowledge the truth of mathematical existence assertions, and it also skotches a thoroughgoing application of McLarty's many-sorted idea.

I hope that the foregoing has made it clear that in positing mathematical objects we are not writing fiction, and that positing does not detract from our justification in recognizing the existence of mathematical posits or truths. Let me turn now to the second question: What sense can we make of mathematical theories introduced through postulation being *about* an independent mathematical reality? We can appreciate the worry one might have by considering the example of a team of astronomers whose theorizing leads them to posit a galaxy for which they currently have no observational evidence. Suppose that they do subsequently observe a new galaxy. Obviously their positing it did not bring it into existence – it was already there. What, then, would make it the one to which they were previously referring? The answer given by the popular causal theory of reference is that it would be if they stood in the "appropriate" causal relationships with that galaxy at the time they posited.[16] Obviously the same account cannot work for mathematical objects, which by their very nature, cannot participate in causal chains. This is why many fans of the causal theory are nominalists.

[16] So far as I know causal theorists have never explained what "appropriate" means in a context like this.

I do think we should worry much about objections to mathematical objects based upon the causal theory. For it cannot make sense of much positing by scientists either. Our astronomers may have posited the new galaxy to make sense of other sorts of observations. However, in many cases purely theoretical, indeed mathematical, considerations rather than experimental ones lead scientists to posit new objects. Dirac was led to posit anti-matter because he want to make physical sense of the negative square roots in the equations of special relativity, and group theoretic reasoning has led contemporary particle physicists to posit certain bosons. It is hard to think of any causal processes involving anti-matter or these bosons that would appropriately connect them with the events that led physicists to refer to them. Moreover, physicists also refer to supposedly physical things that are not even physically detectable, such as particles within the interior of black holes and virtual processes. It is not clear that it makes sense to speak of our being causally connected to such entities. But if we are, they cannot affect our nervous system; so it is hard to understand how any event involving them would fix our reference to them.

Furthermore, the causal theory sees the reference relation as a sort of linguistic anchor by which our language hooks onto or corresponds to reality. While I do not deny that reference is a word-world relation or I that we have linguistic traffic with reality, I use a disquotational approach to reference, which is the proper companion for a disquotational theory of truth. For each name N and n-place predicate P in the language of science and mathematics, it should yield metalinguistic sentences obtained from the schemata

> N name-refers to x if and only if x is a
> P n-place-predicate-refers to $x_1, x_2, \ldots, x_n$ if and only if $x_1, x_2, \ldots, x_n$ stand in K,

by putting N (respectively P) itself in place of '*a*' (respectively '*K*') and a name of *N (P)* in place of '*N*' ('*P*'). But it need not presuppose that there is more to reference than these sentences or that our language somehow uniquely corresponds (or hooks onto) reality. Such a theory can be worked out using list-like definitions of primitive reference in the manner of Tarski. (Thus means that we use definitions like this: N name-refers to x $\leftrightarrow$ (N is 'Adam' & x is Adam) or N is 'Beth' & x is Beth or ... or N is 'Zeb' and x is Zeb.) On this approach, the principles of reference (and truth), instead of being determined by causal or other "empirical" relations, turn out to be mathematical consequences of these list-like definitions and the syntactic principles for our language.[17]

If we treat reference disquotationally, then there is no special problem referring to mathematical objects. The predicate "number", for instance, refers to an object if and only if it is a number – end of story. Taking our language at face value, we can safely affirm within it that "number" refers to numbers and not to quarks or rabbits. And we will have affirmed truly. For numbers exist, and they are neither quarks nor rabbits. What is more, we have referred to something existing independently of our constructions, proofs, etc., since our constructing a mathematical object or proving theorems about it is not necessary for its existence.

Disquotational reference does not undercut the independence of the reality to which we refer anymore than disquotational truth does. Whether or not the sentence, "there are infinitely many primes" is true depends upon whether or not there are infinitely many primes, not on whether we have proved it or believe it, and so on. And whether or not "the number of apostles" refers to a prime number depends upon whether the number of apostles, i. e., 12 is a prime number.

Thus we can refer wrongly. Moreover, even mathematicians positing new entities could mistakenly refer to something else – just as our astronomers could refer to the wrong galaxy. Suppose, for example, that some mathematicians introduce the term "finor" to refer to some new entities they intend be positions in a structure that is some kind of simple ordering. Also suppose that they subsequently lay down a consistent set of axioms to characterize the structure of finors, and these axioms have no models in simple orderings. Then the entities they posited would not be finors, and they would be mistaken in referring to them as such. Of course, if their axioms were inconsistent, then they would not be referring to anything at all.

When we posit new mathematical objects we typically introduce new terms for them. If our positing is successful, then our terms will refer to them in the disquotational sense. This is one of the ways they relate to reality. Although this relationship is not the rich notion of reference some philosophers seek, it is not utterly trivial. In positing we may fail to have captured the structure we intended – our axioms may turn out to have the wrong sorts of models – or we may fail to capture

---

[17] For more on this see my "Immanent Truth" and "Beliefs About Mathematical Objects".

any structure at all. In either case, we will not refer to what we think we are referring.

One of the arguments for the causal theory is that it can handle cases that descriptional theories cannot. Now I am not advocating a descriptional theory of reference. But we should remember that my concern is with referring to *positions in structures* not to isolated objects. For me the number 2 is not an object standing on its own but the second member of the natural numbers series, and we identify this series *via* its structure. Thus to inform someone that we are referring to the number 2 we could first communicate that we are referring to a member of the natural number series. One way to do this is describe this series by conditions whose models are omega sequences. After we established that the structure in question is the natural number series it is easy to communicate that we are referring to its third position. Thus, when it comes to structures, descriptional methods for fixing references might be more successful than they have been elsewhere.[18]

References

Dedekind, Richard. "Continuity and Irrational Numbers" (1872) and "The Nature and Meaning of Numbers" (1888) translated in Richard Dedekind, *Essays on the Theory of Numbers*. Wooster Woodruff Beman, translator. New York: Dover, 1963.

Dummett, Michael. *Frege: Philosophy of Mathematics*. Cambridge, Massachusetts: Harvard University Press, 1991.

Frege, G. *Die Grundlagen der Arithmetik* (1884). Reprinted and translated by J. L. Austin in *The Foundations of Arithmetic*. Oxford: Blackwell and Mott, 1950.

Frege, G. *Grundgesetze der Arithmetic*, Vol. 1 (1893), Vol. 2 (1903). Quotations taken from *Translations from the Philosophical Writings of Gottlob Frege*. Peter Geach and Max Black, editors. Oxford: Blackwell, 1952.

Frege, G. *Mathematical and Philosophical Correspondence*, B. McGuinness, editor, H. Kaal, translator. Oxford 1980.

Frege, G. "Rezension von: Georg Cantor, zur Lehre Vom Transfiniten. Gesammelte Abhandlungen aus der Zeitschrift für Philosophie und Philosophische Kritik" in G. Frege, *Kleine Schriften*. I. Angelelli (ed.). Hildesheim: Georg Olms, 1967.

[18] I would like to thank Mark Balaguer, Colin McLarty, Barbara Scholz, the audience in Boulder, especially, Dale Jamieson, Stephen Leeds, George Bealer and William Reinhardt, and the audience in Munich, especially, Crispin Wright, Bob Hale and Terence Parsons for useful comments and help with this paper..

Field, Hartry, "Is Mathematical Knowledge Just Logical Knowledge", *Philosophical Review* **93** (1984), pp. 509–52.

Hellman, Geoffrey. *Mathematics without Numbers*. Oxford: Clarendon Press, 1989.

Maddy, Penelope. "Believing the Axioms" I, II. *The Journal of Symbolic Logic*, vol. **53** (1988), pp. 481–511, 736–764.

Maddy, Penelope. "Indispensability and Practice", *Journal of Philosophy*, **89** (1992), pp. 275–89

Resnik, Michael D. "The Frege-Hilbert Controversy," *Philosophy and Phenomenological Research*, **XXXIV**, No. 3 (1974), pp. 386–403.

Resnik, Michael D. "Mathematics as a Science of Patterns: Ontology and Reference," *Nous*, **15** (1981) pp. 529–566.

Resnik, Michael D. "Beliefs about Mathematical Objects" in *Physicalism in Mathematics*, edited by Andrew Irvine, Dordrecht: Kluwer, 1989.

Resnik, Michael D. "Immanent Truth", *Mind* **99** (1990), 405–424.

Resnik, Michael D. "Scientific vs. Mathematical Realism: The Indispensability Argument" *Philosophia Mathematica*, Series III Vol. **3** (1995), pp. 166–174.

Shapiro, Stewart. *Foundations without Foundationalism*. Oxford: Clarendon Press, 1991.

Frege versus Cantor and Dedekind: On the Concept of Number

W. W. Tait*

There can be no doubt about the value of Frege's contributions to the philosophy of mathematics. First, he invented quantification theory and this was the first step toward making precise the notion of a purely logical deduction. Secondly, he was the first to publish a logical analysis of the ancestral R* of a relation R, which yields a definition of R* in second-order logic.[1] Only a narrow and arid conception of philosophy would exclude these two achievements. Thirdly and very importantly, the discussion in §§ 58–60 of the *Grundlagen* defends a conception of mathematical existence, to be found in Cantor (1883) and later in the writings of Dedekind and Hilbert, by basing it upon considerations about meaning which have *general* application, outside mathematics.[2]

Michael Dummett, in his book [Dummett (1991)][3] on Frege's philosophy of mathematics, is rather stronger in his evaluation. He writes "For all his mistakes and omissions, he was the greatest philosopher of

---

* This paper is in honor of my colleague and friend, Leonard Linsky, on the occasion of his retirement. I presented the earliest version in the Spring of 1992 to a reading group, the other members of which were Leonard Linsky, Steve Awodey, Andre Carus and Mike Price. I presented later versions in the autumn of 1992 to the philosophy colloquium at McGill University and in the autumn of 1993 to the philosophy colloquium at Carnegie-Mellon University. The discussions following these presentations were valuable to me and I would especially like to acknowledge Emily Carson (for comments on the earliest draft), Michael Hallett, Kenneth Manders, Stephen Menn, G. E. Reyes, Teddy Seidenfeld, and Wilfrid Sieg and the members of the reading group for helpful comments. But, most of all, I would like to thank Howard Stein and Richard Heck, who read the penultimate draft of the paper and made extensive comments and corrections. Naturally, none of these scholars, except possibly Howard Stein, is responsible for any remaining defects.

1 Frege (1879). Dedekind (1887) similarly analyzed the ancestral F* in the case of a one-to-one function F from a set into a proper subset. In the preface to the first edition, Dedekind stated that, in the years 1872–78, he had written a first draft, containing all the essential ideas of his monograph.

2 However, it was only in the hands of Wittgenstein, in *Philosophical Investigations*, that this critique of meaning was fully and convincingly elaborated.

3 All references to Dummett will be to this work, unless otherwise specified.

mathematics yet to have written" (P. 321). I think that one has to have a rather circumscribed view of what constitutes philosophy to subscribe to such a statement – or indeed to any ranking in philosophy of mathematics. If I had to choose, I would perhaps rank Plato first, on grounds of priority, since he was first, as far as we know, to conceive of the idea of a priori science, that is science based on primitive truths from which we reason purely deductively. But, if Plato seems too remote, then Frege still has some strong competitors even in the nineteenth century, for example Bolzano, Riemann, Weierstrass and, especially, Cantor and Dedekind. Contributing to Dummett's assessment is, I think, a tendency to make a sharp distinction between what is philosophical and what is technical and outside the domain of philosophy, a sharper distinction between philosophy and science than is historically justified or reasonable. Thus we read that Frege had answers (although not always the right ones) "to all the philosophical problems concerning the branches of mathematics with which he dealt. He had an account to offer of the applications of arithmetic; of the status of its objects; of the kind of necessity attaching to arithmetic truths; and of how to reconcile their a priori character with our attainment of new knowledge about arithmetic." (p. 292) The question of existence of mathematical objects, their 'status', certainly needed clarification; but, otherwise, are these the most important philosophical problems associated with the branches of mathematics with which he dealt? Surely the most important philosophical problem of Frege's time and ours, and one certainly connected with the investigation of the concept of number, is the clarification of the infinite, initiated by Bolzano and Cantor and seriously misunderstood by Frege. Likewise, the important distinction between cardinal and ordinal numbers, introduced by Cantor, and (especially in connection with the question of mathematical existence) the characterization of the system of finite numbers to within isomorphism as a simply infinite system, introduced by Dedekind, are of central importance in the philosophy of mathematics. Also, the issue of constructive versus non-constructive reasoning in mathematics, which Frege nowhere discussed, was very much alive by 1884, when he published his *Grundlagen*. Finally, although Frege took up the problem of the analysis of the continuum, his treatment of it appeared about thirty years after the work of Weierstrass, Cantor, Dedekind, Heine and Moray (the latter four in 1872) and, besides, was incomplete. What it lacked was, essentially, just what the earlier works supplied, a construction (at least up to isomorphism) of the complete ordered additive group of real numbers. Whether Frege had, as he thought, something to add to that construction in the definition of the real numbers is a question on which I shall briefly comment in § VII, where I discuss the analogous question of Frege's versus Dedekind's treatment of finite cardinal numbers. The issue here concerns the matter of applications. It is true that Frege offered an account of application of the natural numbers and the real numbers and that this account structured his treatment of the real numbers and possibly, as Dummett suggests, his treatment of the natural numbers. But there is some question as to whether his account of application should enhance his stature as a philosopher.

However, more important to me in this paper than the question of Frege's own importance in philosophy is the tendency in the literature on philosophy to contrast the superior clarity of thought and powers of conceptual analysis that Frege brought to bear on the foundations of arithmetic, especially in the *Grundlagen*, with the conceptual confusion of his predecessors and contemporaries on this topic. Thus, in Dummett (1991), p. 292: "In Frege's writings, by contrast [to those of Brouwer and Hilbert], everything is lucid and explicit: when there are mistakes, they are set out clearly for all to recognize." Aside from the contrast with Brouwer, I don't believe that this evaluation survives close examination. Frege's discussions of other writers are often characterized less by clarity than by misinterpretation and lack of charity, and, on many matters, both of criticism of other scholars and of substance, his analysis is defective. Dummett agrees with part of this assessment in so far as Volume II of the *Grundgesetze* (1903) is concerned. He writes

> The critical sections of *Grundlagen* follow one another in a logical sequence; each is devoted to a question concerning arithmetic and the natural numbers, and other writers are cited only when either some view they express or the refutation of their errors contributes positively to answering the question. In Part III.1 of *Grundgesetze*, the sections follow no logical sequence. Each after the first ... is devoted to a particular rival mathematician or group of mathematicians ... From their content, the reader cannot but think that Frege is anxious to direct at his competitors any criticism to which they lay themselves open, regardless of whether it advances his argument or not. He acknowledges no merit in the work of those he criticizes; nor, with the exception only of Newton and Gauss, is anyone quoted with approbation. The Frege who wrote Volume II of *Grundgesetze* was a very different man from the Frege who had written *Grundlagen*: an embittered man whose concern to give a convincing exposition of his theory of the foundations of analysis was repeatedly overpowered by

> his desire for revenge on those who had ignored or failed to understand his work. (pp. 242–43)

Concerning the relative coherence of the two works, Dummett is surely right. But I think that, in Frege's treatment of other scholars, we can very well recognize the later Frege in the earlier one. Establishing this purely negative fact about Frege would be, by itself, very small potatoes. But unfortunately, his assessment of his contemporaries in *Grundlagen* and elsewhere lives on in much of the philosophical literature, where respected mathematicians, such as Heine, Lipschitz, Schröder and Thomae, are regarded as utterly muddled about the concept of number and great philosophers, such as Cantor and Dedekind, are treated as philosophical naifs, however creative, whose work provides, at best, fodder for philosophical chewing. Not only have we inherited from Frege a poor regard for his contemporaries, but, taking the critical parts of his *Grundlagen* as a model, we in the Anglo-American tradition of analytic philosophy have inherited a poor vision of what philosophy is.

I

The conception of sets and of ordinal and cardinal numbers for which Cantor is perhaps best known first appeared in print in 1888 and represents a significant and, to my mind, unfortunate change in his position. He first introduced the concept of two arbitrary sets, finite or infinite, having the same power in Cantor (1878). In Cantor (1874) he had already in effect shown that there are at least two infinite powers (although he had not yet defined the general notion of equipollence). Prior to 1883, all of the sets that he had been considering were subsets of finite-dimensional Euclidean spaces, all of which he had shown to have the power of the continuum. New sets, the number classes, with successively higher powers, were introduced in Cantor (1883). So here, for the first time, he obtained sets which might have powers greater than that of the continuum. In this connection, it should be noted that, although he defined the concept of a well-ordered set and noted that the ordinal numbers corresponded to the order types of well-ordered sets,[4] the ordinal numbers themselves were defined *autonomously* and

[4] Cantor 1932, p. 168.

not *as* the order types of well-ordered sets.[5] Indeed, in general, the only well-ordered set of order type α available to him was the set of predecessors of α. In discussing what had been gained by his construction of the ordinals, the application to well-ordered sets is mentioned only *second*, after the founding of the theory of powers. For Cantor, at this time, the construction of the number classes was essential to the theory of powers. In speaking of their significance, he writes

> Our aforementioned number classes of determinately infinite real whole numbers [i. e. the ordinals] now show themselves to be the natural uniform representatives of the lawful sequence of ascending powers of well-defined sets. [Cantor (1932), p. 167]

Just prior to this he wrote that "Every well-defined set has a determinate power", so his view at that time was that every infinite well-defined set is equipollent to a number class.[6] In particular, he notes in (1883) that neither the totality of all ordinals nor the totality of all cardinals has a power. It follows then that neither is a well-defined set.

It was in "Mitteilungen zur Lehre vom Transfiniten" (1887–88) and, later, in "Beiträge zur Begründung der transfiniten Mengenlehre" (1895–97) that Cantor introduced the much-criticized abstractionist conception of the cardinals and ordinals. To quote from the "Beiträge":

---

[5] Frege obviously appreciated this point. In (1884) § 86 he wrote "I find special reason to welcome in Cantor's investigations an extension of the frontiers of science, because they have led to the construction of a purely arithmetical route to higher transfinite numbers (powers)."

[6] At the beginning of § 3 of (1883), Cantor explicitly states as a 'law of thought' that every set can be well-ordered. His assertion that the powers form an absolute infinity seems to imply that the construction of the number classes is to be continued beyond the finite number classes. He isn't explicit about how one proceeds to construct the αth-number class for limit ordinal α, but presumably, if its power is to be the next highest after those of all the β number classes for $\beta<\alpha$, we should take it to be the union of the number classes of smaller index. But *when* should we introduce the αth number class for limit α? If we require that α be already obtained in some earlier number class, then the only ordinals Cantor's scheme yields are those less than the least fixed point $\alpha=\aleph_\alpha$. But if, as seems perfectly consistent with Cantor's ideology, we require only that the cofinality of α be obtained in some earlier number class, then the ordinals that would be obtained are precisely those less than the least weakly inaccessible cardinal (i. e. the least *regular* fixed point of $\aleph$). After his proof in Cantor (1891–92) that the power of a set is strictly less than that of its power set, possibly higher powers are obtained. The analogous hierarchy of powers leads to the least strongly inaccessible cardinal. Not until Zermelo (1930) does it seem that anyone pursued this 'constructive' approach to set theory, to obtain ordinals beyond the least strongly inaccessible cardinal.

> *By the "power" or "cardinal number" of M we mean the general concept, which arises with the help of our active faculty of thought from the set M, in that we abstract from the nature of the particular elements of M and from the order in which they are presented. ...* Since every single element m [of M], if we abstract from its nature, becomes a 'unit', the cardinal number ... [of M] is a definite aggregate composed of units, and this number has existence in our mind as an intellectual image or projection of the given aggregate M. [Cantor (1932), p. 282–83.]

In the analogous way, he introduced the *order type* of a linearly ordered set M:

> By this we understand *the general concept which arises from M when we abstract only from the nature of the elements of M, retaining the order of precedence among them. ...* Thus, the order type ... is *itself an ordered set* whose elements are pure units ... . (p. 297)

In particular, ordinal numbers are identified in the "Mitteilungen" and the "Beiträge" with the order types of well-ordered sets. It would be interesting to conjecture about the reasons for the change from the point of view of (1883) to that of (1887–88); but I shall not go into that here, other than to register my regret.[7] Husserl (1890), who notes the change with approval, defends the later definition and, as does Cantor, argues that the essential principle that, if two sets are equipollent, then they have the same cardinal, is derivable from it. (Two sets are *equipollent* or, as Cantor expressed it, *equivalent* if they are in one-to-one correspondence.) But the argument is not entirely clear: Why should abstraction from two equipollent sets lead to the *same* set of 'pure units'? And the conception of the cardinal number as a set plays no other role in Cantor's theory. We shall discuss this further in § VIII.

II

In § 71 of "Was sind und was sollen die Zahlen?" (1888), Dedekind defines the notion of a 'simply infinite system' M with respect to a one-to-one function ϕ:M $\rightarrow$ M. Namely, there is an element e of M which is not in the range of ϕ and M is the least set containing e and closed

[7] The assimilation of the theory of ordinals to the more general theory of order types may be part of the explanation. The discovery that a set is strictly less in power than its power set does not seem to be part of the explanation, since the "Mitteilungen" preceeds that discovery.

under ϕ. ϕ is said to *order* M and e is called the *base element* of M (with respect to ϕ). He then goes on to introduce the simply infinite system of natural numbers:

> 73. Definition. If, in considering a simply infinite system [M] ordered by the mapping ϕ, we completely disregard the particular nature of the elements, retaining only their distinguishability and considering only those relationships in which they are placed to one another by the ordering map ϕ, then these elements are called *natural numbers* or *ordinal numbers* or simply *numbers*, and the base element [e] is called the *base element* of the *number series* [M]. In consideration of this freeing of the elements from every other content (abstraction) one can with justice call the numbers a free creation of the human intellect [des menschlichen Geistes].

When M is a simply infinite system with respect to the map ϕ and base element e, let us call the triple $\mathbf{M}=\langle M,\phi,e\rangle$ a *simply infinite set*. We may denote the system of numbers by $\mathbf{N}=\langle N,',1\rangle$. Dedekind goes on to cite his proof in § 134 that all simply infinite sets are isomorphic, in order to show that arithmetic depends only on the axioms of the (second order) theory of simply infinite sets and not on the choice of any particular such system.[8] The step from **M** to **N** is an instance of Cantor's abstraction in the case of ordered sets.

III

The response in the literature on philosophy of mathematics to Cantor's and Dedekind's abstractionist treatment of numbers has generally been negative.

An early direct attack on Cantor is contained in Frege (1891), a review of Cantor (1890) (which includes the "Mitteilungen").[9] In what must count as one of the more impertinent passages in the history of philosophy, he writes

> If Mr. Cantor had not only reviewed my *Grundlagen der Arithmetik* but also read it thoughtfully, he would have avoided many mistakes. I believe that I have done there already a long time ago what he is here

[8] The axioms in question are

$$\forall x(e\neq\phi(x))$$
$$\forall xy(\phi(x)=\phi(y)\rightarrow x=y)$$
$$\forall Z[e\in Z\wedge\forall x(x\in Z\rightarrow\phi(x)\in Z)\rightarrow\forall x(x\in Z)].$$

[9] A more virulent attack, which Frege chose not to publish, is contained in a partial draft of that review [Frege (1979), p. 68–71].

> trying in vain to do. Mr. Cantor repeats (p. 13) a definition he had given in his review of my book as his own intellectual property. It seemed to me at the time that it differed from mine, not in its essentials, but only in its wording ... I now see that the truths I enunciated in my book were not, after all, like coins dropped in the street which anybody could make his own simply by bending down. For Mr. Cantor goes on to give some other definitions (pp. 23 and 56) which show that he is still firmly ensconced in an antiquated position. He is asking for impossible abstractions and it is unclear to him what is to be understood by a 'set', even though he has an inkling of the correct answer, which comes out faintly when he says (p. 67 n.): 'A set is already completely delimited by the fact that everything that belongs to it is determined in itself and well distinguished from everything that does not belong to it.' This delimitation is, of course, achieved by characteristic marks and is nothing other than the definition of a concept. On this point compare my proposition (*Grundlagen*, § 46): '... the content of a statement of number is an assertion about a concept'. (1984, p. 179)

And shortly after:

> ... we once again encounter those unfortunate ones which are different even though there is nothing to distinguish them from one another. The author evidently did not have the slightest inkling of the presence of this difficulty, which I dealt with at length in §§ 34 to 54 of my *Grundlagen*.

The page references are to Cantor (1890). But the 'other definitions (pp. 23 and 56)' are the "Mitteilungen" versions of the above quoted definitions of power and order type, in Cantor (1932), p. 387 and p. 422, respectively. The definition given 'as his own intellectual property' in Cantor's review (1885) of Frege (1884) and which he 'repeats' is clearly the definition of the cardinal number or power of an aggregate (Inbegriff) or set as 'that general concept under which all and only those sets fall which are equivalent to the given set.' [Cantor (1932), p. 380]. Frege's own definition is that the number $N_xF(x)$ of F's, where F is a concept, is the extension of the second level concept 'is equipollent to F'. Frege had thought that it differs only in wording from his own, because he thought that it is inessential whether one speaks of a general concept here or of its extension and because he thought, incorrectly (see § XII below), that Cantor's notion of a set could be understood to mean 'extension of a concept' in his sense.[10]

---

[10] See the footnote at the end of § 68 of Frege (1884) and his reply in Frege (1984), p. 120, to Cantor's review [Cantor (1932), pp. 440–41].

In (1883), § 1, Cantor writes "Every well-defined set M has a power, such that two sets have the same power when they [are equipollent]" The passage from this, in response to the question "What *is* the cardinal of M?", to the definition of the cardinal of M as the general concept under which fall precisely those sets equipollent to M, surely owes nothing to Frege. In defining the cardinal number of M as a 'general concept', it seems clear that Cantor did not have in mind Frege's technical notion of a concept; rather, he was following the traditional view according to which, for example, the numeral '10' is a common name, under which fall all ten-element sets (for example, see Aristotle's *Physics*, 224a3–16). Moreover, as we shall see, Cantor had already pointed out, implicitly in his (1883) and explicitly in his review of Frege's (1884), that it is a mistake to take the notion of a set (i. e. of that which has a cardinal number) to simply mean the extension of a concept. Finally, the very definition of equinumerosity in terms of equipollence, applied to sets *in general*, is due to no one but Cantor – not to Hume, Kossak or Schröder, whom Frege cites but who were concerned entirely with finite sets (see § X below). As for Frege, who did intend the definition to apply also to the infinite, he was not only anticipated by Cantor by seven years, but even after that time and, presumably, after having read Cantor's (1883) (to which he refers in his (1884)), he was unaware of the difficulties involved in treating the infinite (see § XII below).

But Frege is right that Cantor crucially modifies his definition of the power of a set M when he goes on to define it, not merely as a concept under which fall all sets equipollent to M, but also as an equipollent set of pure units: the concept becomes a paradigm instance of itself. Frege's own conception of abstraction (although he disapproves of the term) is, as we shall see, in agreement with the view that abstracting from the particular nature of the elements of M would yield the concept under which fall all sets equipollent to M. His target was the idea that abstraction leads to the paradigm set of pure units. We have already noted a difficulty with this idea, at least with respect to the work to which Cantor wanted to put it. But Frege's own arguments against Cantor's conception, that he cites from §§ 34–54 of his (1884), are invalid (see § VIII below).

A very influential attack on Dedekind's theory occurs in Russell's *Principles of Mathematics* (1903):

> Moreover, it is impossible that the ordinals should be, as Dedekind suggests, nothing but the terms of such relations as constitute a pro-

> gression. If they are anything at all, they must be intrinsically something; they must differ from other entities as points from instants, or colours from sounds. (p. 249)

Russell's point is often expressed by saying that it is impossible that objects should 'have only structural properties'. The confusion that lies behind this objection is discussed in § VII. His criticism is echoed in Dummett (1991) In Chapter 5, comparing Frege's and Dedekind's treatment of the foundations of arithmetic, Dummett writes

> One of the mental operations most frequently credited with creative powers was that of abstracting from particular features of some object or system of objects, that is, ceasing to take any account of them. It was virtually an orthodoxy, subscribed to by many philosophers and mathematicians, including Husserl and Cantor, that the mind could, by this means, create an object or system of objects lacking the features abstracted from, but not possessing any others in their place. It was to this operation that Dedekind appealed in order to explain what the natural numbers are. His procedure differed from the usual one. Husserl ... supposed that each individual cardinal number was created by a special act of abstraction: starting with any arbitrary set having that number of elements, we abstract from all properties possessed by the individual members of the set, thus transforming them into featureless units; the set comprising these units was then the relevant cardinal number. Cantor's variation on this account was a trifle more complex: we start with an ordered set, and abstract from all the features of the individual members, but not from their ordering, and thus obtain their [sic] order type; next, we abstract from the ordering relation, and obtain the cardinal number as an unordered set of featureless units, as before. Frege devoted a lengthy section of *Grundlagen*, §§ 29–44, to a detailed and conclusive critique of this misbegotten theory; it was a bitter disappointment to him that it had not the slightest effect. (p. 50)

In questioning that "the mind could, by this means, create an object or system of objects lacking the features abstracted from, but not possessing any others in their place", Dummett seems to be merging Russell's criticism with another: Is it abstract objects to which we should object or is it their creation by the mind? The same double-barreled objection arises immediately after when, discussing Russell's reaction to (and misunderstanding of) Dedekind, Dummett writes that Dedekind "believed that the magical operation of abstraction can provide us with specific objects having only structural properties: Russell did not understand that belief because, very rightly, he had no faith in abstraction thus understood." (p. 52) Dummett is taking 'abstraction' here to be a psychological term.

> Dedekind's philosophy of mathematics was that mathematical objects are 'free creations of the human mind', as he says in the Preface. The idea, widely shared by his contemporaries, was that abstract objects are actually created by operations of our minds. This would seem to lead to a solipsistic conception of mathematics; but it is implicit in this conception that each subject is entitled to feel assured that what he creates by means of his own mental operations will coincide, at least in its properties, with what others have created by means of analogous operations. For Frege, such an assurance would be without foundation: for him, the contents of our minds are wholly subjective; since there is no means of comparing them, I cannot know whether my idea is the same as yours. (p. 49)

So Dummett believes that Cantor's and Dedekind's operation of abstraction is psychologistic.

Frege himself does not criticize Dedekind's treatment of the number concept on grounds of psychologism when he discusses it in the introduction to the *Grundgesetze I* (1893). And, in his review of Cantor (1890), he makes the criticism rather mildly when he writes: "Besides, the verb 'abstract' is a psychological expression and, as such, ought to be avoided in mathematics." (p. 181) But this is by no means his principal attack on Cantor's conception of cardinal and ordinal numbers. *His* reference to his refutation of what Dummett calls the 'misbegotten theory' of Cantor is not to §§ 29–44 in his (1884), but to §§ 34–54, in which psychologism is not the issue.

IV

Actually, Dedekind did not say in the Preface that mathematical objects are free creations of the human mind. He did say this of the natural numbers and there is little doubt that he would have said it also of the real numbers; but it is too hasty to reduce his 'philosophy of mathematics' to a psychologistic reading of this metaphor. Indeed, this tendency to attack forms of expression rather than attempting to appreciate what is actually being said is one of the more unfortunate habits that analytic philosophy inherited from Frege. If one reads § 73, quoted above, the metaphor of 'free creation' is justified by the fact that we arrive at the system of numbers by abstraction, by freeing the elements from every other content. Therefore, it is reasonable to conclude that Dedekind's conception is psychologistic only if that is the only way to understand the abstraction that is involved. And we shall see that it is not.

The difficulty with abstraction as a psychological operation would be that what is abstracted is mental, that what I abstract is mine and what you abstract is yours (see Frege (1884), §§ 26–27). We are no more communicating when I say "0<1", meaning that my 0 is (my) less than my 1, and you say, "No, 1<0", meaning that your 1 is (your) less than your 0, than when I say "I am shorter than Jones" and you say, "No, I am taller than Jones". Of course, this does not mean that you and I cannot argue objectively about your, my or some third party's mental states as an empirical question. But when we discuss the nature of arithmetic truth, that is not what is going on: on whatever grounds I might be seduced into thinking that I am expressing something about *my* mental states when I assert that 0<1, you would, on the same grounds, be seduced into thinking that you are expressing something about *your* mental states when you assert that 1<0. Frege's point is that the objectivity of mathematics demands that we both resist this seduction.

So Frege's argument against psychologism in the context of abstraction is not that the source of judgement about the abstracted objects is not in some sense to be found in the common human psyche, but rather that the objects abstracted should not be found in the individual psyche.[11] For example, he himself ultimately traces the source of logical principles to our logical disposition (Frege (1979), p. 269). So the objectivity of logic rests, for him, upon the fact that we are disposed to agree in logical judgement.[12] The same point is illustrated by his defense of Kant's conception of geometry as objective in (1884), § 26.

> Space, according to Kant, belongs to appearance. For other rational beings it might take some form quite different from that in which we know it. Indeed, we cannot even know whether it appears the same to one man as to another; for we cannot, in order to compare them, lay one man's intuition of space beside another's. Nevertheless, there is something objective in space all the same; everyone recognizes the same geometrical axioms, even if only by his behavior, and must do so if he is to find his way about in the world. What is objective in it is

[11] There is of course another side to psychologism, which Frege discusses and opposes in (1884), p. x and § 60. There the issue is the confusion of the logical with the psychological, e. g. of the senses of words or sentences with our (psychological ideas). But it would seem that Dummett's charge against Dedekind is not this, but that he is taking the *reference* of number words to be ideas.

[12] It should be noted that Frege does not observe, as Wittgenstein later did in the *Investigations*, that objectivity requires not only agreement in judgement concerning the laws of logic, but also agreement in drawing consequences, i. e. in moving from premises to conclusion (computing according to a rule).

> what is subject to laws, what can be conceived and judged, what is expressible in words. What is purely intuitable is not communicable.

The abstractionism of neither Cantor nor Dedekind is subject to the criticism that it is psychologistic: For neither of them are numbers psychological objects nor are the laws of number to be understood in any way as subjective.

V

Concerning the notion of abstraction, Frege writes

> For suppose that we do, as Thomae demands, "abstract from the peculiarities of the individual members of a set of things", or "disregard, in considering separate things, those characteristics which serve to distinguish them". In that event we are not left, as Lipschitz maintains, with "the concept of the number of the things considered"; what we get is rather a general concept under which the things considered fall. The things themselves do not in the process lose any of their special characteristics. For example, if I, in considering a white cat and a black cat, disregard the properties which serve to distinguish them, then I get presumably the concept "cat". [(1884), § 34]

There is a serious misunderstanding of both Thomae and Lipschitz in this passage, which we shall take up later. But the point I want to make here is that Frege is not really correct about abstraction resulting always in concepts, at least not if he is referring to the traditional meaning of the term 'abstraction'. For example, although it is true that Aristotle regards attributes such as 'white' (the concept 'x is white', for Frege) to be obtained by abstraction, he also regards geometric objects such as lines and surfaces to be obtained by abstraction from sensible things (*Metaphysics* 1061a29).

But, of course, there is a sense in which no object is really created by this latter kind of abstraction. The geometric magnitude, e. g. the line segment or the plane figure, was for Aristotle just the sensible substance; but in geometry we regard it, not *qua* sensible object, but only with respect to those properties it has in virtue of its extension. On this view, truths about a geometric object are simply truths about the sensible object, but restricted to the language of geometry. Analogous to this is the so-called 'forgetful functor' by means of which we pass, say, from a ring $\langle M,+,\times\rangle$ to the corresponding group $\langle M,+\rangle$. But these cases are somewhat different from the case of the abstraction of the power $|A|$ from the set A. For if A is equipollent to the set B, then $|A|=|B|$. But this does not imply that A=B. Hence, we cannot regard $|A|$ as really being just A, taken in abstraction, unless 'taking in abstraction' has the power to identify distinct sets. Or rather, since Frege is certainly right that the process cannot literally identify distinct things, the process of abstraction in this case must be understood to create new objects. Nevertheless, there is something common to Aristotle's conception of geometry, right or wrong, and the abstraction of cardinal numbers. Namely, sentences about the abstract objects have a canonical, truth-preserving, translation into sentences about the objects from which they are abstracted. In the case of cardinals, the objects from which the powers are abstracted are sets. The relation $\sim$ of equipollence is clearly an equivalence relation among sets and it respects the relation

$$X \preccurlyeq Y$$

of X being equipollent to a subset of Y as well as the operations of *sum* or *disjoint union*

$$\Sigma_{i \in I} X_i$$

(i. e. the set $\{(i,a) | \, i \in I \;\&\; a \in X_i\}$) and *cartesian product*

$$\Pi_{i \in I} X_i$$

of a family $\langle X | \, i \in I \rangle$ of sets. In view of this fact, the order relation $\leq$ among the cardinals and the arithmetical operations of addition and multiplication on cardinals may be defined by

$$|X| \leq |Y| \Leftrightarrow X \preccurlyeq Y$$
$$\Sigma_i |X_i| = |\Sigma_i X_i|$$
$$\Pi_i |X_i| = |\Pi_i X_i|.$$

It follows that any proposition about the arithmetic and ordering of cardinal numbers translates into a proposition about sets, providing only that '=' and '$\leq$' are translated as '$\sim$' and '$\preccurlyeq$', respectively, and 'Σ' and 'Π' are interpreted as the corresponding operations on families of sets. The one difficulty with this translation is that we are passing from the cardinal $|X|$ to the set X, and it could happen that X is itself a set containing cardinals or whose transitive closure contains cardinals;[13] and so the translation does not entirely eliminate reference to cardinals. But, assuming the well-ordering principle, we can always take the representative X of the cardinal $|X|$ to be a pure set, i. e. one, such as the corresponding initial von Neumann ordinal, whose transitive closure contains only sets.

---

[13] The transitive closure of a set X is the least transitive set which includes X. A set X is transitive iff $Y \in X$ and $b \in Y$ imply that $b \in X$.

Of course I am not accurately presenting Cantor's abstractionist conception of the cardinals here, since for him they are sets of pure units. But I will leave this aside for the moment, noting only that it is entirely inessential to the reduction.

In any case, what seems to me to be essential to this kind of abstraction is this: the propositions about the abstract objects translate into propositions about the things from which they are abstracted and, in particular, the truth of the former is founded upon the truth of the latter. So the abstraction in question has a strong claim to the title *logical abstraction*: the sense of a proposition about the abstract domain is given in terms of the sense of the corresponding proposition about the (relatively) concrete domain.[14]

Dedekind's treatment of the finite ordinals is also a case of logical abstraction, providing that we assume given some simply infinite set **M**=⟨M,ϕ,e⟩, such as the system of finite von Neumann ordinals. We introduce the simply infinite set **N**=⟨N,',1⟩ of finite ordinals by stipulating that **M**≅**N**. As we noted, the isomorphism is unique. In terms of this isomorphism, any arithmetical proposition, i. e. proposition about **N**, translates into a proposition about **M**. Moreover, because all simply infinite sets are isomorphic, the truth value of the arithmetical proposition does not depend upon the particular simply infinite set **M**.[15]

Of course, this treatment depends upon having a simply infinite set **M** to begin with, from which to abstract **N**. Dedekind in fact showed that it suffices to have a so-called Dedekind infinite set ⟨S,ϕ⟩, i. e. a set S with a one-to-one function ϕ from S into a proper subset of S. As Frege had essentially done previously, Dedekind notes that, if e is an element of S which is not a value of ϕ and M is the intersection of all subsets of S which contain e and are closed under ϕ, then ⟨M,ϕ,e⟩ is a simply infinite set.[16]

---

14 Dummett (1991), pp. 167–68, uses the term 'logical abstraction' for the construction of the abstract objects as equivalence classes. But it is not clear why we should call this construction 'logical'.

15 For further discussion of Dedekind's conception, see Parsons (1990) and Tait (1986) and (1993).

16 Dedekind in fact 'constructs' such a Dedekind infinite set, where S is the domain of all objects of thought and ϕ(x) is the thought of x. He argues that his ego is in S but is not a thought; and so ⟨S,ϕ⟩ is a Dedekind infinite set. Frege begins, in effect, with the Dedekind infinite set ⟨S,ϕ⟩ in which S is the totality of all objects or at least all cardinals and ϕ(m)=n means that n is the cardinal of some concept F and, for some b in the extension of F, m is the cardinal of 'Fx & x≠b'. (See Frege (1884), § 76.)

One may ask: What is the point of logical abstraction? That is, instead of abstracting the simply infinite set of numbers from an already given simply infinite set **M**, why did Dedekind not simply take the system of numbers to *be* this latter system? Similarly, noting that every ordinal or cardinal is the order type or cardinal of a unique von Neumann ordinal, why not take this pure set to *be* the ordinal or cardinal? Dedekind discussed this question, not in connection with his monograph on the natural numbers, but in a letter to Weber about his earlier monograph, Dedekind (1872), on the irrational numbers.[17] He explains why he takes an irrational number to be *represented* by the corresponding Dedekind cut rather than defining it to *be* that cut. His argument is that to identify the real numbers with cuts – or with the objects in any other representation of them – is to endow them with properties which have nothing to do with them *qua* numbers but only to do with a particular and arbitrary representation of them.[18] Dummett seems to entirely misunderstand Dedekind's point here when he writes of the latter's refusal to identify the real numbers with the cuts: "Dedekind's resort to construction was not a means of avoiding labor. It was due solely to his philosophical orientation, according to which mathematical entities are to be displayed as creations of the human mind." (p. 250).

An objection often leveled at the abstractions of Cantor and Dedekind is that the abstractions do no work – they play no role in proofs. Of course Cantor intended his conception of cardinals as sets of pure units to do work, namely to yield the equivalence of the equinumerosity of two sets with their equipollence; but this is not convincing. But it would seem that logical abstraction, as it is described here, does play a role, not in proofs, but in that it fixes grammar, the domain of meaningful propositions, concerning the objects in question, and so determines the appropriate subject matter of proofs. For example, proving the categoricity of the axioms of simply ordered sets fixes the sense of all propositions in the pure theory of numbers; but it would not do so if numbers were sets, since the sense of $0 \in 1$ is not fixed.

---

[17] *Werke*, vol. 3, pp. 489–90. This letter was brought to my attention by Howard Stein. Cf. Stein (1988).

[18] Dedekind's argument is close to Benacerraf's in his paper "What numbers could not be": The latter argues that there is no one representation of the numbers by sets and so nothing intrinsic to the notion of number itself which decides the answer to the question "Is $0 \in 2$?", for example. Hence *no* representation of the numbers as sets should be regarded as *defining* the numbers.

In Dedekind's final judgement of the matter, it is not clear that his foundation of arithmetic (as opposed to the foundation of the theory of real numbers) should be regarded as abstractionist. In the well-known letter to Keferstein in 1890,[19] in which he explained the argument of "Was sind und was sollen die Zahlen?", Dedekind casts a somewhat different light upon his foundation of arithmetic and, in particular, his construction of the simply infinite set from the Dedekind infinite set of all objects of thought. He writes

> ... the question arose: does such a system *exist* at all in the realm of our ideas? Without a logical proof of existence it would always remain doubtful whether the notion of such a system might not perhaps contain internal contradictions. Hence the need for such proofs. (van Heijenoort, p. 101.)

What is stressed here is not the abstractionist reduction of arithmetic to something else, but rather the question of the *internal* consistency of arithmetic itself. In this respect, Dedekind was the precursor of Hilbert's view that mathematical existence is established when one has proved completeness and consistency.[20] In view of this, there is all the more reason to question Dummett's psychologistic reading of Dedekind – in particular when he writes

> For Dedekind, however, the process of creation involved the operation of psychological abstraction, which needed a non-abstract system from which to begin; so it was for him a necessity, for the foundation of the mathematical theory, that there be such systems. That is why he included in his foundation for arithmetic a proof of the existence of a simply infinite system, which had, of necessity, to be a non-mathematical one. (p. 296)

VI

There is another objection to Dedekind's foundation of arithmetic raised by Dummett: the characterization of N as a simple infinity does not tell us whether it is the system of numbers beginning with zero or beginning with one (or beginning with some other number) – in other

---

[19] Dedekind (1932), p. 490.

[20] Of course, when second order logic is involved, completeness and consistency are not purely formal notions, because of the incompleteness of formal higher-order logic. But categoricity and existence of a model suffice to establish these properties, both in Dedekind's case and in Hilbert's case of Euclidean geometry.

words, it does not tell us whether 1 is really zero or one. But why hasn't Dedekind eliminated that ambiguity by *telling* us that 1 is the number one? Another way to state Dummett's objection is this: Dedekind proves (§ 126) the general principle of definition by primitive recursion according to which we may define unique functions F and G on N such that

$$F(m,1) = m \qquad G(m,1) = m'$$
$$F(m,n') = F(m,n)' \qquad G(m,n') = G(m,n)'.$$

The ambiguity in question then concerns whether it is F or G that is to be called 'addition', and Dedekind himself opts for the latter. Dummett suggests that we could eliminate the ambiguity by identifying the system of numbers with, say, the structure $\mathbf{N}^{+}=\langle N,{}',1,G\rangle$ rather than with **N**. But that is surely inadequate *unless we then specify that G is to denote addition*! After all, no matter what number we take "1" to denote, the function G is well-defined on N. But, if we must specify that G is addition in $\mathbf{N}^{+}$, then we might as well stick with the structure **N** and specify that 1 is the number one. Dummett introduces this topic in the context, not of Dedekind's theory, but of Benacerraf's 'neo-Dedekindian' thesis "that structure is all that matters, since we can specify a mathematical object only in terms of the structure to which it belongs". He objects to eliminating the ambiguity by passing from **N** to $\mathbf{N}^{+}$, not because it doesn't work, but because he believes that it betrays Benacerraf's thesis. **N** is already characterizable to within isomorphism as a simple infinity and G is definable in the structure **N**. Therefore it is contrary to Benacerraf's thesis to consider the structure $\mathbf{N}^{+}$ instead of **N** as giving the structure of the numbers. So Benacerraf's thesis is false. (P. 53) But, having noted that $\mathbf{N}^{+}$ is simply a definitional expansion of N, how can Dummett believe that substituting the former for the latter would determine which number 1 denotes?

But putting this aside, it is clear that, for Dummett, it is the fact that Dedekind's definition of the numbers does not 'intrinsically' determine for the number $1''$, say, whether it is the cardinal of two-element sets or the cardinal of three-element sets that is the defect in Dedekind's treatment of number. He writes

> Frege and Dedekind were at odds over two interconnected questions: whether or not the use of natural numbers to give the cardinality of finite totalities is one of their distinguishing characteristics, which ought therefore to figure in their definition; and whether it is possible, not merely to characterize the abstract structure of the system of natural numbers, but to identify the natural numbers solely in terms

> of that structure. Unlike Frege's, Dedekind's natural numbers have no properties other than their positions in the ordering determined by their generating operation, and those derivable from them; the question is whether such a conception is coherent. (p. 51)[21]

In consequence of the above alleged ambiguity in the meaning of "1", he writes of Benacerraf's thesis

> The thesis is false, and the example Benacerraf chose to illustrate it is the very one that most clearly illustrates its falsity. The identity of a mathematical object may sometimes be fixed by its relation to what lies outside the structure to which it belongs; what is constitutive of the number 3 is not its position in any progression whatever, or even in some particular progression, nor yet the result of adding 3 to another number, or of multiplying it by 3, but something more fundamental than any of these: the fact that, if certain objects are counted 'One two, three', or, equally, 'Naught, one, two', then there are three of them. The point is so simple that it needs a sophisticated intellect to overlook it; and it shows Frege to have been right, as against Dedekind, to have made the use of the natural numbers as finite cardinals intrinsic to their characterization. (p. 53)

There are several difficulties with Dummett's assessment in this connection. One concerns the dominance in his argument *here* of the role of numbers as cardinals: Why should we single out one kind of application of the natural numbers as being of their essence?[22] We have already noted that Dedekind focused on a different one; namely their role as ordinals or counting numbers. Thus, Dummett notes (p. 51) as a point of criticism of Dedekind's account that whereas he defines the addition of natural numbers by the recursion equations for G, i. e. as ordinal addition, Frege defines it as cardinal addition. But of course it is precisely what one would expect from someone who is analyzing the notion of finite *ordinal* or counting number that he would define addition as ordinal addition. What is surprising is that Dummett should find, in the case of finite numbers, that 'giving cardinality' is a more 'distinguishing characteristic' than serving as counting number. He even writes about Frege that "[H]e assumed, as virtually everyone else at the

---

21 Of course Dummett does not mean that Frege and Dedekind were at odds over whether the structure of the system of numbers could be characterized: this was not a question that Frege even considered in (1884). In (1893), he formalizes (without citation) Dedekind's proof of categoricity (after having given a new proof of Dedekind's principle of definition by primitive recursion, again without citation) in his system. (Cf. Heck.(1993))

22 Stein (1988) raises this question as a mild criticism even of the "und was sollen" part of the title of Dedekind's monograph.

time would have done, that the most general application of the natural numbers is to give the cardinality of finite sets." (p. 293), although he then goes on to point out that Cantor took the ordinal numbers to be primary: in his generalization of the cardinals and ordinals into the transfinite, it is the ordinals that he called 'numbers'. With an apparent reversal of judgement, he also suggests *here* that the notion of ordinal is the more fundamental one.[23] One must also put Dedekind on the side of the ordinal numbers. Kronecker also, in his "Über den Zahlbegriff" (1887), writes "I find the natural starting point for the development of the number concept in the *ordinal numbers.*" But anyway, when we are speaking of applications, what about the role of the natural numbers in the foundation of analysis, e. g. in the foundation of the theory of rational and real numbers? Even if we attempt to go the route of Frege (1903) and construct the real numbers as ratios, the natural numbers must function as exterior multipliers on any system of magnitudes. That is, there must be the operation $n \cdot x$ defined for numbers n and quantities x by $1 \cdot x = x$ and $n' \cdot x = n \cdot x + x$ (when $n > 0$)[24]. As Frege himself noted in (1884), § 19, in his criticism of Newton's definition of numbers in terms of ratios, the definition of the relation of 'having the same ratio' between pairs of like magnitudes presupposes the operation $n \cdot x$.[25]

What makes one of these applications of the natural numbers privileged, so that it, rather than others, should be one of their 'distinguishing characteristics'?

VII

So far as the application of the numbers as cardinals is concerned, what is wrong with Dedekind's definition (§ 161) (also given in

---

[23] One must question this judgement. For example, the proof in affine geometry that all lines contain the same number of points is not a counting argument.

[24] Actually, it suffices to define the notion that a pair (x,y) of magnitudes are equimultiples of the pair (a,b). We could do this, using Frege's (and Dedekind's) analysis of the ancestral F^* of a function F, without introducing the natural numbers. Namely (x,y) is an equimultiple of (a,b) iff $(a,b)F^*(x,y)$, where $F(c,d)=(c+a,d+b)$. But then we have essentially introduced the (positive) natural numbers; namely, they are the ratios x:a, where (x,x) is an equimultiple of (a,a).

[25] Dummett (1991), p. 73, writes that Frege "flounders somewhat, and fails to make the simple point as cleanly as he ought", the simple point being that Newton's definition is of the positive real numbers whereas Frege is concerned with the natural numbers. But, of course, the positive real numbers, with the operation +1, form a Dedekind infinity in terms of which a simple infinity may be constructed. So in fact it is essential to Frege's point that Eudoxos' definition already involves the numbers as multipliers.

Kronecker's paper) of 'The set X has cardinal n' as meaning that X is equipollent with the set {1,...,n}? It should be noted that, if this analysis is acceptable, then a certain argument that Frege repeats over and over again fails. The argument appears in "On formal theories of arithmetic" [Frege (1984), pp. 112-21] as an *a priori* argument for logicism:

> ... the basic propositions on which arithmetic is based cannot apply merely to a limited area whose peculiarities they express in the way in which the axioms of geometry express the peculiarities of what is spatial; rather, these basic propositions must extend to everything that can be thought. And surely we are justified in ascribing such extremely general propositions to logic. (p. 112.)

Frege's point appears to be endorsed in the foreword to the first edition of Dedekind (1887):

> In speaking of arithmetic (algebra, analysis) as part of logic I mean to imply that I consider the number concept entirely independent of the notions or intuitions of space and time, that I consider it an immediate result from the laws of thought.

But there is a difference in their arguments. Dedekind's point, echoing Bolzano (who, in turn, quotes Aristotle)[26], is in effect an expression of unwillingness to admit, in reasoning about numbers, any principles drawn from alien sciences. Frege's argument, on the other hand, is bound up with the idea that the definite description 'the number of x such that $\phi(x)$' should apply to any concept $\phi(x)$ at all and not just to those concerning which we have some special source of knowledge. (We can count anything.) We find this argument repeated many times in Frege (1884). In § 14, discussing Kant's conception that arithmetic is founded on intuition, Frege contrasts arithmetic and geometry:

> The fact that this is possible shows that the axioms of geometry are independent of one another and of the primitive laws of logic, and are consequently synthetic. Can the same be said of the fundamental propositions of the science of number? Here, we have only to try denying any one of them, and complete confusion ensues. Even to think at all seems no longer possible. The basis of arithmetic lies deeper, it seems, than that of any of the empirical sciences, and even than that of geometry. The truths of arithmetic govern all that is numerable. This is the widest domain of all; for to it belongs not only the existent (Wirkliche) not only the intuitable, but everything thinkable. Should not the laws of number, then, be connected very intimately with the laws of thought?

---

[26] Bolzano (1817), Preface. (p. 160 in the translation Russ (1980)); *Posterior Analytics* 75ª39.

In § 19, in the discussion of Newton's definition of number as a ratio between like magnitudes, Frege again presents the argument for logicism from the universal applicability of number. We have already mentioned one objection that he raised to Newton's definition. But his second objection is this:

> [W]e should still remain in doubt as to how the number defined geometrically in this way is related to the number of ordinary life, which would then be entirely cut off from science. Yet surely we are entitled to demand of arithmetic that its numbers should be adopted for use in every application made of number, even although that application is not itself the business of arithmetic. Even in our everyday sums, we must be able to rely on the science of arithmetic to provide the basis for the methods we use. And moreover, the question arises whether arithmetic itself can make do with a geometric concept of number, when we think of some of the concepts in it, such as the number of roots of an equation or of the numbers prime to and smaller than a given number. On the other hand, the number which gives the answer to the question *how many*? can answer among other things how many units are in a length.

In § 40, where he is discussing the problem of how one could understand cardinal numbers as sets of 'pure units', he considers the possibility of the units being points in space/time. He again writes

> The first doubt that strikes us about any such view is that then nothing would be numerable except what is spatial and temporal.

But, of course, on Dedekind's analysis, this line of argument is fallacious: *Whatever* simple infinity we take the system of numbers to be, the question of whether the extension of a concept is equipollent to {1,...,n} makes sense. In § 42 Frege continues

> Another way out is to invoke instead of spatial or temporal order a more generalized concept of series, but this too fails in its object; for their position in the series cannot be the basis on which we distinguish the objects, since they must have already been distinguished somehow or other, for us to have been able to arrange them in a series.

But this is a strange idea. Why can't we have a series all of whose members are identical? Frege is confusing the notion of a series with that of a linearly ordered set ⟨A,<⟩ (where, for x and y in A, x<y implies that x and y are distinct). Why does Frege preclude the series (a,a,...,a)? His response is:

> When Hankel speaks of our thinking or putting a thing one or twice or three times, this too seems to be an attempt to combine in the things to be numbered distinguishability with identity. But it is ob-

> vious at once that it is not successful; for his ideas or intuitions of the same object must, if they are not to coalesce into one, be different in some way or other. Moreover we are, I imagine, fully entitled to speak of 45 million Germans without having first to have thought or put an average German 45 million times, which might be somewhat tedious.

But to regard the number n as an n-element sequence whose members are identical is not to say that the things numbered must be identical. Frege's second point surely betrays his confusion. To say that there are 45 million Germans is to say that there is a set of Germans which is equipollent to {1,...,45,000,000} – and, again, *this is quite independent of how the numbers are defined.*

Schröder's proposal to define the numbers as expressions, i. e. sequences of atomic symbols, is essentially the same as Hankel's proposal, except that Schröder decides against the expressions 11...1 because, for example, three, identified with 111, might be confused with one hundred and eleven. so he wishes to use the expressions 1+1+...+1 instead. This may be silly, but it is not open to Frege's objection in § 43 that "This passage shows that for Schröder number is a *symbol.* What the symbol expresses, which is what I have been calling number, is taken, with the words 'how many of these units are present', as already known." Why shouldn't numbers be symbols? 111 is, in any case, a different symbol than 3. Why not take the latter to denote the former? In truth, there is an objection, namely the objection that numbers are numbers and symbols are symbols and that the grammar of the one is different from the grammar of the other. In essence, this is the objection of Dedekind to any reductionist account of the natural numbers or the real numbers. But it cannot be Frege's objection, since he will make numbers be extensions of concepts!

Presumably Dedekind's analysis of "The set X has n elements" won't do for Frege, as Dummett understands it, because it does not make "the use of natural numbers to give cardinality of finite sets ... one of their distinguishing characteristics". But we are owed a definition of 'distinguishing characteristic', as well as of 'intrinsic property'. Since Dedekind's treatment is being contrasted with Frege's in this connection, we should presumably look to Frege to find cases of distinguishing characteristics and intrinsic properties. The two cases with which this presents us are the cardinal numbers and the real numbers. In both instances, the numbers are defined as equivalence classes: A set has the cardinal number n iff it is an element of n and a pair of like magnitudes has the real number r iff it is an element of r. Of course, one problem in the case of cardinal numbers is that this definition is inconsistent for n>0, since there is no set consisting of all sets of power n. But let us restrict the equipollence relation to subsets of some infinite set A, so that n is an equipollence class of the subsets of A. To be sure, for sets X which are not subsets of A, the statement that X has cardinal n will no longer be equivalent to the assertion that it is an element of n; and so it will not be a 'defining characteristic' of n that X has cardinal n. But, even for subsets X of A, why does its being an element of n make the fact that it has cardinal n a defining characteristic or intrinsic property of n? The relation $\in$ of set membership (or 'being in the extension') is being given a distinguished role without any indication of why this should be so. For example, suppose that we instead define the cardinal numbers to be the equipollence classes of non-empty subsets of A and define 'X has cardinal n' to mean that $X \cup \{X\} \in n$ (assuming for this that A contains all of its finite subsets). Then $\{\emptyset\}$ will no longer be a cardinal of any subset of A, the set of all unit subsets of A will be the cardinal of the null set and, in general, the cardinal of an n-element subset of A will be the equipollence class of all n+1-element subsets of A. Why is the relation $X \in n$ between X and n to be preferred to the relation $X \cup \{X\} \in n$?

The same considerations apply to Frege's rejection of earlier treatments of the real numbers on the grounds that they do not adequately account for applications. Given a (linearly ordered) system of magnitudes, any ratio a:b of magnitudes from that domain is equal to a ratio r:1 in any connected ordered field, and so can be assigned the 'real number' r. It has yet to be made clear why it would be preferable to treat the real number of the pair (a,b) *as* the ratio, i. e. as the equivalence class of all pairs (c,d) (from any system of magnitudes) such that $a:b \approx c:d$ – leaving aside the question of the consistency of doing so.

Finally, as used by Dummett, the term 'structural property' is misleading.

> *There is no absolute notion of 'structural property'. It is only relative to a specific structure, e. g. $\langle M,e,\phi\rangle$, on M that we may speak of the structural properties of the objects of M, namely the properties definable in terms of the structure (in the example, in terms of e and ϕ, with individual quantifiers ranging over M, second-order quantifiers over subsets of M, etc.)*

Thus, when Dummett speaks of "objects or systems of objects lacking the features abstracted from, but not possessing any others in their place" or of Dedekind's natural numbers as having "no properties other than their positions in the ordering determined by their generating operation, and those derivable from them" or of "specific objects having only structural properties", he is guilty of confusion. What is true of the numbers on Dedekind's account is that it is possible to specify a structure on them, e. g. $\mathbf{N}$ or $\langle N,' \rangle$ or $\langle N,< \rangle$, in terms of which they are characterizable to within isomorphism by finitely many axioms. But *this* can hardly be the basis of a metaphysical objection to them.

From this it is clear that the objection of Russell and Dummett cannot reasonably be about the paucity or kind of properties that Dedekind's numbers have nor about the fact that they can be categorically determined in terms of a structure which distinguishes some of those properties. Perhaps the objection is, rather, that they are *given* to us only in terms of that structure. In Dummett's words, the question is "whether it is possible, not merely to characterize the abstract structure of the system of natural numbers, but to identify the natural numbers solely in terms of that structure". But what more is required to 'identify' the numbers, even on Frege's own grounds? The sense of every arithmetical proposition A is fixed: A is true just in case it can be derived from the axioms for a simple infinity. Moreover, we lack no account of the application of the numbers on this foundation. The idea that the numbers can be identified or, perhaps, further identified in terms of some particular application of them is, as we have seen, neither a very clear idea nor a desirable one.

At the end of the day, I think that, for Russell and for Dummett, the objection to Dedekind's treatment of the natural numbers is that, for Dedekind, they are just numbers and not something else as well. For Dedekind, the question "What are the numbers?" could only be answered by exhibiting their structure. For many writers since Frege, the question has rather meant: "What *besides* numbers are the numbers?" This becomes clear in the case of Dummett in his discussion of 'structuralism', when he writes

> On the stronger interpretation, structuralism is the doctrine that mathematics in general is solely concerned with structures in the abstract sense, that is, with systems left no further specified than as exemplifying the structure in question. This doctrine has, again, two versions. According to the more mystical of these, mathematics relates to *abstract structures*, distinguished by the fact that their elements have no non-structural properties. The abstract four-element Boolean algebra is, on this view, a specific system, with specific el-

ements; but, for example, the zero of the algebra has no other properties than those which follow from its being the zero of that Boolean algebra – it is not a set, or a number, or anything else whose nature is extrinsic to that algebra. This may be regarded as Dedekind's version of structuralism: for him the natural numbers are specific objects; but they are objects that have no properties save those that derive from their position in 'the' abstract simply infinite system (sequence of order type ω.)

VIII

There is an aspect of Cantor's abstractionist conception of ordinal and cardinal numbers that I have so far ignored and which was the main target of Frege's attack. Namely, the cardinal number of a set is to be itself a set, equipollent to the given set, and its elements are to be 'pure units'; and the order type of an ordered set is to be itself an ordered set of 'pure units', isomorphic to the given ordered set.

Consider just the case of cardinals. Clearly the idea of a cardinal as a set of pure units – call it a *cardinal set* – is inessential to the foundation of the theory of cardinals on logical abstraction, at least as that kind of abstraction is described in § V. But what did Cantor actually mean by speaking of pure units? There are two ways in which we might try to understand his idea of the cardinal set corresponding to the set M as a set of 'pure units'. One is that the units, the elements of a given cardinal set, should be obtained by abstracting from the particular properties that distinguish the elements of M from one another and thus should be indistinguishable in some sense from one another. That is the way in which Frege understands Cantor. The other interpretation is that the abstraction concerns, not the individuating properties of the elements relative to one another, but rather the individuating property of the set itself, for example the concept of which it is the extension. On this interpretation, the cardinal set C corresponding to a set M is to be constituted of unique elements, specified in no way other than that they are the elements of C and that C is equipollent to M. Thus, the cardinal sets are not sets of points in Euclidean space or of numbers or of sets, or of apples or etc. Once we decide in the first place that the cardinal number of M should be a set equipollent to M, then the argument for Cantor's conception of the cardinal set, on this interpretation, is exactly Dedekind's argument that the system of numbers is a 'creation of the human spirit': The cardinal set corresponding to M should not be a set of points or of numbers or of apples or of sets (as in the case of the initial von Neumann ordinals). The things that we may say about these other kinds of sets would be ungrammatical when speaking of cardinal numbers. Now, the role that Cantor would have his doctrine that cardinal numbers are sets of pure units play in the theory of cardinals is to infer the equivalence of Card(M)=Card(N) and the equipollence of M and N, and only this. It is clear from this that it is the *second* interpretation of his doctrine that is correct, and that Frege has entirely misunderstood him when he referred to his *Grundlagen* for a refutation of Cantor's doctrine.

As a matter of fact, Cantor's motivation for the conception of cardinals as cardinal sets is weak: one is introducing one cardinal set by 'abstraction' corresponding to each equipollence class, but one is not analyzing the notion that M is equipollent to N *in terms of* the notion of cardinal number. Abstraction, as Frege says, is strong lye. To abstract the cardinal set corresponding to M from M, we must specify what it is from which we are abstracting. The only possible answer is that we are abstracting from all properties and relationships of M except those which respect the equipollence relationships of M. So we do not derive the equipollence of M and N from Card(M)=Card(N); rather, the former notion is built into the latter. (Frege makes essentially the same point in criticizing Husserl's formulation of Cantor's doctrine.) Moreover, Dedekind's argument from grammar applies to Cantor's doctrine of cardinals as sets, itself: it is ungrammatical to ask whether a particular object is an element of a particular cardinal number. So I think that Cantor's view that cardinal numbers are sets of pure units is ill-conceived.

But Frege's claim was that it is *incoherent*. But certainly, on the second interpretation, which I think that we must accept as the correct one, the theory of cardinal-sets is perfectly coherent. We may take the pure units in a cardinal-set to be atoms (i. e. non-sets or *urelements*). If κ and λ are distinct von Neumann cardinals (i. e. initial von Neumann ordinals), we may even assume that the corresponding cardinal-sets C_κ and C_λ are disjoint sets of atoms. So the question is whether or not we can coherently assume that, corresponding to every von Neumann cardinal κ, there is a set C_κ of atoms which is equipollent to κ such that, for distinct von Neumann cardinals κ and λ, $C_\kappa \cap C_\lambda = \emptyset$. But we can construct a standard model **M** of this assumption together with the axioms of second-order set theory with urelements from a standard model of second-order set theory with a strongly inaccessible cardinal.[27] Moreover, any permutation of the elements of the cardinal set C_κ induces an automorphism of **M** in an obvious way. For example, all of the cardinal sets C_λ and all pure sets (i. e. whose transitive closures contain no atoms) are fixed points of this automorphism. So, in any reasonable sense, even on Frege's mistaken interpretation of Cantor's doctrine, it is perfectly coherent. Of course there is a property that distinguishes the element b of C_κ from all of the other elements, namely the property x=b. But, however Frege was reading Cantor, surely it would have been unreasonable of him to suppose that Cantor either overlooked this or would have denied it.

The suggestion of incoherence of the notion of the cardinals being sets of pure units seems to arise for Frege from two sources. One source is Leibniz's Principle of the Identity of Indiscernibles: If nothing distinguishes between two 'pure units' in the cardinal-set C, then they must be the same; and so every cardinal number must be 0 or 1. For example, recall that, in his review of Cantor, Frege writes of "those unfortunate Ones which are different even though there is nothing to distinguish them one from another" (p. 270). He goes on to say: "The author evidently did not have the slightest inkling of the presence of this difficulty, which I deal with at length in §§ 34–54 of my *Grundlagen.*" Unfortunately, the discussion in these sections is interwoven with a discussion, beginning at § 29, of quite another notion, that of a 'unit', as well as with Frege's view of abstraction. But, concerning the issue at hand, he writes in § 35

> We cannot succeed in making different things identical simply by dint of operations with concepts. But even if we did, we should then no longer have things in the plural, but only one thing; for, as Descartes says, the number (or better, the plurality) in things arises from their distinction.

The first sentence of this passage refers to his argument in § 34 that abstraction cannot produce new objects ('make distinct things identical'). His correction of Descartes, choosing the term 'plurality' over 'number', illustrates another source of confusion in his discussion of other authors. By a number, Descartes meant essentially a set and, as we shall see below, that was a common usage of the term in earlier times.

[27] Take the domain to be $V_{\kappa+\kappa}$, where κ is the least inaccessible cardinal. The sets of rank $\leq\kappa$ are taken to be atoms, except for the null set, which is taken to be the null set. The sets in $V_{\kappa+\kappa}$ of rank $>\kappa$ thus become the non-null sets of rank $<\kappa$ over the set of atoms. The elements of C_μ, for μ a cardinal $< \kappa$, are the atoms $\mu\times\{\mu\}$.

Moreover, the passage cited in Descartes' *Principia* (Part I, § 60) in no way supports Frege's point. 'Distinct' there means only 'not identical', not 'distinguished by some property', and Descartes is distinguishing three different kinds of 'distinction': real, modal and rational. Nowhere in this discussion (§§ 60–62) does he imply that non-identity of P and Q requires there to be some property possessed by one of them and not the other. But there can be no doubt that Frege himself subscribed to this principle: In *Grundlagen* § 65 he takes as his own definition of identity (being the same object) Leibniz's:[28]

> Things are the same as each other, of which one can be substituted for the other without loss of truth.

But, of course, this principle doesn't tell us anything until we know in which propositional contexts F(x) the substitution may occur. For example, any two distinct objects will be distinguished by the context '$x \in M$', where M is a set which contains one but not the other.

Frege himself might defend against this by insisting that '$x \in M$' is not a primitive concept and must be replaced by the concept F(x) of which M is the extension. And, presumably, F(x) cannot itself be defined in terms of the identity relation, since otherwise $x=b$ distinguishes the object b from all other objects and Leibniz's principle is trivialized. I have already expressed the view that it would have been entirely unreasonable for Frege to have supposed Cantor to have rejected this trivial form of Leibniz's principle or to have overlooked it in formulating his doctrine of cardinal numbers as sets of pure units. So Frege's objection would then be that there is no such concept F(x), not already involving the identity relation, which distinguishes between two 'pure units' of the cardinal-set.

But, of course, this whole line of argument presupposes the validity of the non-trivial form of Leibniz's principle, and, in Frege's case, it seems hard to defend. Certainly no point in Euclidean space is distinguished from any other by a concept, unless that concept itself is defined by reference to specific points, since the space is homogeneous. But there is a difficulty with individuating points by means of concepts which themselves refer to points. For example, the points p and q may be distinguished by the concept 'x is between r and s', which is satisfied by p and not by q. But then q satisfies a corresponding concept 'x lies

---

[28] He later rejected this as a *definition* (cf. Frege (1984) p. 200), but he continued to affirm its validity.

between t and u, e. g. where (p,r,s) is congruent to (q,t,u). The problem is that our grounds for calling these two concepts distinct is only that (r,s) and (t,u) are distinct (i. e. non-identical) pairs of points: the individuation of such concepts presupposes the individuation of pairs of points and so, ultimately, of points. Hence, there is a circle.[29] Of course, since Frege believed that Euclidean geometry is the science of physical space, he may have believed that any two points are distinguished by empirical, non-geometric, concepts, e. g. by physical scalars. But it would be hard to accept the non-trivial form of Leibniz's principle as a metaphysical principle based upon such a belief.

The other source of the appearance of incoherence in the notion of a cardinal number as a set of pure units is the argument that, if cardinal numbers are sets at all, then, for example, 1+1 must equal 1. In § 35 Frege writes

> Jevons goes on: "Whenever I use the symbol 5 I really mean
>
> $1 + 1 + 1 + 1 + 1$
>
> and it is perfectly understood that each of these units is distinct from each other. If requisite I might mark them thus
>
> $1' + 1'' + 1''' + 1'''' + 1'''''$."
>
> Certainly it is requisite to mark them differently, if they are different: otherwise the utmost confusion must result. ... ... The symbols
>
> $1', 1'', 1'''$
>
> tell the tale of our embarrassment. We must have identity – hence the 1; but we must have difference – hence the indices; only unfortunately, the latter undo the work of the former.

Frege and Jevons collaborate in a confusion here. Jevons problem is that he is thinking of 5 as a set of five elements on the one hand and as 1+ ... +1 on the other. If the different occurrences of '1' do not denote different unit sets, then Jevons worries that 1+ ... +1 will not yield a set of five elements. *But that is because he thinks that, if cardinals are sets, then their addition is just their union.* But even if we take cardinal numbers to be sets, m+n does not denote union of m and n; it denotes the cardinal of their disjoint union. Frege (§ 38) somewhat misses the point here: He thinks that Jevon's problem arises from confusing 'unit' with 'one' and of treating numbers as 'agglomerations'. He is right that one

[29] Leibniz did not face this difficulty. For him, the principle of identity of indiscernibles applies to substances (which have no real relations), and not to ideal things such as points in space.

cannot take the units in 5 to be the unit in 1, since there are five of the former and only one of the latter. But Jevons' equation 5=1+1+1+1+1 leads to this identification, not because he thought that numbers are sets (after all, the von Neuman cardinals are sets), but because he confused the addition of cardinals with the union of sets. Frege's argument, were it valid, would not simply be an argument against cardinals being sets of pure units, *it would be an argument against the cardinal of a set M being a set equipollent to M at all.* I am persuaded by Dedekind's grammatical argument that numbers are not sets; but Frege's line of argument would exclude even the *representation* of cardinals by initial von Neumann ordinals.

Of course, Frege might have made the argument that, since the identification of cardinals with sets does not admit the identification of cardinal addition with set-theoretic union, then there is no point in regarding cardinals as sets at all. But this is not the argument that he gave.

IX

One problem with reading the literature on the number concept prior to Cantor, Dedekind and Frege is that, aside from Bolzano, the authors generally have not fully distinguished the notion of a set and tended to subsume it under a more general notion of a 'multitude' or 'plurality'. Anything with proper parts was regarded as a plurality – a line segment, Socrates, a heap of stones, a flock of sheep. It was understood from the time of Plato that number does not unambiguously apply to pluralities. Socrates is one but has many parts, the flock of ten sheep also includes a plurality of twenty sheeps' eyes and a plurality of forty sheeps' legs, etc. Aristotle explicitly understood that assigning number to pluralities in this sense requires a prior choice of the parts to be numbered.[30] He referred to this as a choice of 'unit': To assign numbers to line segments, for example, we must first choose a particular line segment as the unit of measurement. This conception, that what can be numbered, the ἀριθμός, is some object (in a generalized sense that admits flocks of sheep, the aggregate of planets, etc.) relative to a partition – a choice of unit –, survived even into the late nineteenth century in the form of the rejection of the null set: no object can be partitioned into zero parts. There was also a surviving conceptual difficulty with

[30] Aristotle, *Metaphysics*, Book X, Ch. 1 and 2.

unit sets, which is reflected in the sometime rejection of the number 1 by the classical Greeks: if what can be numbered is an object X relative to a choice of unit, the unit set of X would be X *qua* part of X, which was indistinguishable from X.[31]

Indeed, Frege reveals some of this confusion about the concept of a set in *Grundlagen*, § 28, where he is discussing the notion that a number is simply a set. He writes

> Some writers define the number as a set or multitude or plurality. All of these views suffer from the drawback that the concept will not then cover the numbers 0 and 1. Moreover, these terms are utterly vague: sometimes they approximate in meaning to "heap" or "group" or "aggregate" ["*Aggregat*"], referring to a juxtaposition in space, sometimes they are so used as to be practically equivalent to "number", only vaguer.

In *Paradoxien des Unendlichen* (1851), Bolzano had already made a distinction between what he called an '*Inbegriff*' and a set:

> There exist aggregates [*Inbegriffe*] which agree in containing the selfsame members, and nevertheless present themselves as *different* when seen under different aspects or under different conceptions, and this kind of difference we call 'essential'. For example: an unbroken tumbler and a tumbler broken into pieces, considered as a drinking vessel. We call the ground of distinction between two such aggregates their *mode of combination* or their *arrangement*. An aggregate whose basic conception renders the arrangement of its members a matter of indifference, and whose permutation therefore produces no essential difference, I call a *set* ... (§ 4)

Bolzano's 'Inbegriff' seems to be Frege's 'Aggregat' – a plurality with the unit chosen. But it is fair to say that even Bolzano had not entirely liberated the notion of a set from that of an aggregate and, in particular, could not accommodate unit sets, since in § 3 he writes "For if A were identical with B, it would of course be absurd to speak of an aggregate composed of the objects A and B." Moreover Bolzano uses the term 'Teil' to refer to elements of a set, suggesting, too, that the element/set relation is not entirely distinguished from the part/whole relation. Notice also that Cantor did not clearly distinguish between Bolzano's notions of Inbegriff and Menge in the quote above from (1895–97), p. 282, in as much as he speaks of obtaining the power of the set (Menge) M by abstracting both from the nature of its elements and

---

[31] *Metaphysics* 1052^{b}31. Also see the second quote below from Bolzano's *Paradoxien des Unendlichen*.

from the order in which they are given.[32] Yet, when he speaks of a set as "already completely delimited by the fact that everything that belongs to it is determined in itself and well distinguished from everything that does not belong to it", it would seem that he intends the notion of a set to be independent of the order of its elements.

A further difficulty with reading the pre-twentieth century literature on the number concept is manifested in Frege's "[these terms] are so used as to be practically equivalent to 'number', only vaguer". The fact is that *the term "number", in the sense of the whole numbers, often really did just mean a (finite) set* – in the somewhat confused sense of 'set' that we have just discussed. So a 'number' is given only relative to the choice of unit. This meaning goes back to the Greek meaning of ἀριθμός which came to be translated as "number". Thus at the beginning of Book VII of the *Elements*, Euclid has these definitions

1. A unit is that with respect to which each of the things that exist is called one.
2. A number is a multitude composed of units.

In (1884), § 29, Frege writes that Euclid "seems to mean by the word "μονάς" sometimes an object to be numbered, sometimes a property of such an object and sometimes the number one." But the meaning is clear. In Definition 1 Euclid says that numbering begins with the choice of unit, of what is to be counted – what is to be called one. Definition 2 then defines a 'number' (an ἀριθμός) to be a set of such units. It is true that 'unit' in Definition 1 refers to a property and in Definition 2 to objects having that property. But that is a frivolous objection. We use the word 'man' sometimes to refer to a property and sometimes to men. So things may be called 'one' after the choice of unit in the same way that men may be called 'man'. It is because Frege does not understand that 'one' is a *common* name, applying to all units (once the unit has been chosen), that he misunderstands Euclid. Similarly, also in § 29, he misunderstands Schröder (1873), p. 5, when he quotes "Each of the things to be numbered is called a unit" and then goes on to object:

> "We may well wonder why we must first conceive of the things as units, instead of simply defining number right away as a "set of things", which would bring us back once again to the view just discussed.

[32] Of course, Cantor had been concerned with point sets in Euclidean space, and this is what he had in mind in this passage.

Here he is referring to his discussion in § 28, partly quoted above. But he can't have it both ways: if he believes that the term 'set' can only refer to heaps and the like, then he should also agree that number can be assigned only after the choice of unit.

Hume also uses the term 'number' to mean a set. In (1884), § 63, discussing how the notion of equinumerosity should be defined, Frege writes

> Hume long ago mentioned such a means: "When two numbers are so combined as that the one has always an unit answering to every unit of the other, we pronounce them as equal." This opinion, that equality of numbers must be defined in terms of one-to-one correspondence, seems in recent years to have gained widespread acceptance among mathematicians.

Frege goes on to say that the definition of numerical equality in terms of one-to-one correspondence raises certain "logical doubts and difficulties." In particular, he writes "It is not only among numbers that the relation of equality (Gleichheit) is to be found. From which it seems to follow that we ought not to define it specially for the case of numbers." The ambiguity of the German "Gleichheit" as between "equality" and "identity" helps (along with Austin's translation) to hide a confusion here: It is quite clear, even just from the passage in Hume's *Treatise* that is quoted (Bk. I, Part III, Sect. I), that by 'number' Hume is referring to finite sets and that, when he speaks of equality of numbers, he is not referring to the identity relation but to the relation of equinumerosity, *which indeed is to be defined specially for the case of 'numbers', i. e. sets.* The 'logical doubts and difficulties' were created by Frege's incorrect reading, not by Hume's conception.

Bolzano (1851) also used the term 'number' for sets. In § 8 he attempts to characterize "*finite or countable multitudes,* or quite boldly: *numbers*".

Certainly the term 'number' did not *always* refer to a set. Thus people spoke of 'the number ten' and of the set (and number) of prime numbers less than n, etc. But, on the one hand, there seems to be no obstacle to understanding number in these contexts as refering to species of finite sets – just as one might speak of the animal, man, or of the number of species in a certain genus. Into the nineteenth century, every theorem of number theory could be understood as a statement about an arbitrary finite set, free from any assumption of the existence of an infinite set. Moreover, there was always the resource (exploited in analytic number theory) of regarding the natural numbers as embedded in the system of real numbers. The move towards treating the natural numbers as forming an autonomous system of objects dates from later in the century. It required the explicit admission of the actual infinite into mathematics, and the motivation would seem to be the arithmetical foundation of the system of real numbers. Rational numbers are constructed from natural numbers and real numbers are, for example, sets of rationals. If the natural numbers are not to form an autonomous system of objects, then it is hard to make sense of this construction.[33]

X

In (1884), § 62, Frege raises the question of the meaning of "the number which belongs to the concept F is the same as that which belongs to the concept G". We have already quoted his attribution to Hume of the definition of this concept as meaning equipollence. He goes on to cite Schröder, Kossak and Cantor's 1883 paper as well. Dummett writes "By the time that Frege wrote *Grundlagen*, the definition had already become a piece of mathematical orthodoxy, though Frege undoubtedly gave it its most exact formulation and its most acute philosophical defense." (pp. 142–43) But Frege's citation of earlier authors is misleading. We have already noted his misunderstanding of Hume. But there is another and more compelling respect in which his citation is misleading and, in particular, slights Cantor's contribution. I do not mean the fact that Frege refers to Cantor's 1883 paper rather than to his earliest published definition of equality of power, Cantor (1878).[34] Rather, I refer to the fact that *the other authors cited were concerned only with finite 'numbers'*, i. e. sets, and not with the general notion of cardinal number which applies to infinite sets. Frege and Cantor, on the other hand were concerned with the general notion of cardi-

[33] Dummett's explanation (pp. 133–34) of why *Frege* required that the numbers be objects is unconvincing. His explanation is that 0,...,n have to be objects in order to prove that n has a successor, namely the number of {0,...,n}. In other words, we need a set (i. e. extension of a concept) with n+1 elements. But, whether or not we count the object, call it n*, which Frege takes to be the number n *as* the number n, it is an object in his system and so the set {0*,...,n*} still exists. Indeed, eliminating the middle man and making no commitment at all about the nature of the numbers, we may replace k by the von Neuman ordinal $N_k=\{N_0,\dots,N_{k-1}\}$ in Frege's argument.

[34] Not, as Dummett cites, (1874). In this paper, Cantor does not define the general concept of equipollence, although he does produce the first significant result concerning infinite powers.

nal. The extension to this general notion was not a trivial matter. As late as 1851, in his monograph just cited, Bolzano, who did understand the notion of set to include infinite sets, had argued that the characterization of numerical equality in terms of one-to-one correspondence, though correct for finite sets, could not be applied to infinite sets. His reason was the traditional obstacle to a coherent theory of infinite numbers: It would then happen that infinite sets could be numerically equal to proper subsets of themselves. (§ 23) *It was Cantor, in (1878), who took this bull by the horns and forever separated the notion of proper subset from that of proper numerical inequality.* [See Cantor (1932), p. 119] Quite clearly the *general* analysis of the concept of having the same cardinal number, finite or infinite, can be attributed only to Cantor.

XI

One must wonder why Dummett wrote that Frege gave the definition of equinumerosity in terms of one-to-one correspondence of sets "its most exact formulation and its most acute philosophical defense". (pp. 142–43) The definition is quite clear in the above quote from Hume once one understands that by 'number' he means the thing to be numbered, the set. But anyway it is stated with admirable clarity by both Bolzano and Cantor, even though the former did not accept it as the definition of equinumerosity in the case of infinite sets. As for a defense, who was attacking it and why was defense needed, philosophical or otherwise? Certainly one attack that needed to be answered was Bolzano's, which we have already mentioned. Cantor responded to this and Frege made no mention of it at all.[35] Another attack came from Cantor himself, who in (1883), § 2, distinguished the two conceptions of number, cardinal and ordinal, which essentially coincide in the finite case but not in the infinite. On the ordinal conception, it is not abstract sets but well-ordered sets to which number applies. Again, this analysis was undertaken by Cantor, not by Frege.[36]

[35] Although both Cantor and Dedekind cite Bolzano's fundamental work on the nature of the infinite, I have found no reference at all to him in the published works of Frege.

[36] Indeed, from the discussion in Frege (1884), § 86, it seems likely that he did not even understand Cantors concept of a well-ordered set at that time.

What *did* Frege contribute to this question? The problem as he saw it is formulated in (1884), § 39, as a dilemma:

> If we try to produce a number by putting together different distinct objects, the result is an agglomeration [Anhäufung] in which the objects contained remain still in possession of precisely those properties which serve to distinguish them from one another; and that is not a number. But if we try to do it in the other way, by putting together identicals [Gleiches], the result runs perpetually together into one and we never reach a plurality.

But distinguish two questions which, in his discussion of earlier authors, Frege tended to confuse: What are the things to which number applies? And, what are numbers? The first horn of his dilemma concerns the first question. And the things to be numbered are not 'agglomerations' but sets, which indeed arise by 'putting together' different distinct objects. These sets *were* called 'numbers' by some of these authors, and this is one source of Frege's confusion. The second horn of Frege's dilemma can concern only the view that the numbers themselves are sets of pure units. The argument is that, if the units are not distinguished by their properties, then they will be identical. We have already discussed this view and concluded that the notion of cardinal number as a set of pure units, though unattractive, is by no means incoherent. Distinct units are indeed distinguished by their properties; but when from a set of two cats, one white and one black, we 'abstract' the number two as a set of pure units, the units are not white and black, respectively, and they are not cats.

But also, unlike Frege, I find no sign of the conception of number as set of pure units in the quotes from other authors that he offers in (1884) (in contrast to Cantor in his later writings and to Husserl's book). His reading seems to me to have been misdirected by two related things: his interpretation of "*gleich*" to mean identity and his failure to understand the historical use of the term "number" to mean what is numbered. For example, in § 34, where he begins the discussion of whether the units are *gleich* he refers to the *very same page* (p. 5) of Schröder (1873) from which the quote "Each of the things to be numbered is called a unit" (cited in § 28) came. He now paraphrases Schröder as giving as the reason we call things "units" that it ascribes "to the items that are to be numbered the necessary identity (Gleichheit)". Clearly Schröder could not mean here that the things are being identified with each other. He means rather that they are being identified *as* the things to be counted. Indeed, in § 54, Frege quotes p. 7 of the same work on the notion of a unit: "This generic name or concept will be called the denomination of the number formed by the method given, and constitutes, in effect, what is meant by its unit." As further evidence of Frege's confused reading of other authors, let me repeat part of the quote cited above from (1884), § 34:

> For suppose that we do, as Thomae demands, "abstract from the peculiarities of the individual members of a set of things", or "disregard, in considering separate things, those characteristics which serve to distinguish them: In that event we are not left, as Lipschitz maintains, with "the concept of the number of the things considered"; what we get is rather a general concept under which the things considered fall. The things themselves do not in the process lose any of their special characteristics.

If we put together Thomae's "abstract ... set of things" with Lipschitz's "the concept of the number ...", then it seems that we are abstracting from the particular nature of the elements of a *set* to obtain the number of the set. But if we read only what Lipschitz wrote, then we are, in considering separate things, disregarding those characteristics which serve to distinguish them and are left with the concept of the number (Anzahl) of the things in question. But by "Anzahl" here Lipschitz means the set. So Frege is in complete agreement with Lipschitz: The set (or concept whose extension is the set) is obtained by abstracting from the differences among the elements of the set.

The citation of Thomae is equally misleading. *His* ultimate concern is with the theory of analytic functions and so with the complex numbers. He is sketching the construction of these numbers from the natural numbers. His account is 'formalistic', in the sense that he treats the natural numbers as signs and the real numbers as infinite sequences of signs. This account is by no means as defective as Frege makes it out to be (e. g. in (1903), pp. 96–139);[37] but that is not my point here. The numbers, being for Thomae signs, are certainly not sets of pure units. The context of the above quote from Thomae is this:

> We assume that one can count, i. e. that one is in the position to abstract from the peculiarities of the individuals in a set of objects and assign successive distinct names to distinct such sets of objects. Each individual of the set is called a unit, and, as a consequence of the required abstraction from all distinctive peculiarities, one may replace any unit by any other. *The units are equal to one another.* [Thomae (1880), p. 1.]

---

[37] And Frege's treatment of the 'formalism'of Heine (1872) in (1903) is totally unjustified.

The units are *equal*, not identical. The sense in which they are equal is explained: one can be substituted for any other without altering the name assigned, i. e. the number.

Frege's proposed solution to his dilemma is to be found in §§ 46–48 and in § 54: The things to which number applies are concepts or extensions of concepts. This is consistent with his view that Cantor's notion of a set can only be understood as the extension of a concept. Concerning the notion of a unit, Frege points out (§ 54) that the term was used with two senses: For the concept and for the objects that fall under the concept. Indeed, we saw that this was true of Euclid. But Frege seems to have thought that this was a source of confusion, whereas one would think that it would have been a commonplace observation: "Man" is sometimes used to denote a particular man and sometimes to denote the property of being a man. The difference between 'man' and 'unit' in this respect is that the meaning of the former is fixed and the meaning of the latter is relative and must be specified in any context – as meaning 'man in the room' in one context, 'sheep in the flock' in another, etc. It is to this relativity that Schröder refers when he speaks of it as a *generic* name or concept.

XII

Ultimately, Frege's contribution with respect to the definition of equinumerosity was to replace Cantor's sets as the objects of number attributions by concepts. Indeed, his proposal in his review of Cantor is that we should understand the term 'set' to refer to extensions of concepts. Perhaps it is this idea that Dummett thinks renders the definition more precise. *In the case of finite numbers*, Frege's proposal would indeed have clarified the discussion of number among his contemporaries: the meaning of choosing a unit and of all units being equal is well analyzed in terms of choosing a concept, in something like Frege's sense of concept. The number of people in the room is an attribute of the concept 'x is in the room', where x ranges over people. It is only a pity that he needed to so discredit other authors, from Euclid to Thomae, in order to make his contribution.

But notice that, in the above example, we do not really have Frege's notion of concept, since x ranges over people and, for Frege, the variable ranges over all objects. In the example given, the appropriate concept for Frege would be 'x is a person in the room'. But, when we admit Fregean concepts in general, that is concepts of the form $\phi(x)$, where x ranges over 'all objects', then Frege's idea that number is attributable to concepts goes wrong. In the case of *infinite* numbers, the fact is that Cantor had already noted in his (1883) that there are concepts, for example the concept that x is an ordinal (or cardinal), which do not have a power. In his note to § 4 he writes

> [E]ach of the number classes, and hence each of the powers, is associated with an entirely determinate number of the absolutely infinite totality of numbers, ... Thus the different powers also form an absolutely infinite sequence.

The 'absolute infinite' is contrasted with a determinate infinite, which has a power. His much maligned review of Frege's *Grundlagen* in 1885 may also be read as spotting immediately what is wrong with Frege's conception. He writes

> The author comes upon the unfortunate idea – and it appears that he is following in this respect a suggestion of Überweg in his *System der Logic*, § 53 – of taking what is called in Scholastic logic [Schullogik] the 'extension of a concept' as the foundation of the number concept. He entirely overlooks the fact that the 'extension of a concept' in general may be quantitatively completely indeterminate. Only in certain cases is the 'extension of a concept' quantitatively determinate: Then it has of course, if it is finite, a definite number or, in the case it is infinite, a definite power. For such a quantitative determination of an 'extension of a concept' the concepts of 'number' and 'power' must already be given from another source, and it is a *reversal of direction* if one undertakes to found the latter concepts on that of 'extension of a concept'. (Cantor (1932), p. 440)

Although neither Frege nor Cantor's editor, Zermelo,[38] understood him in this way, it is reasonable to believe that the concepts he had in mind, whose extensions are quantitatively indeterminate, are 'x is an ordinal' and 'x is a cardinal'. One might question this on the grounds that he does not explicitly name these concepts; but, since he doesn't give *any* examples, that would be an objection to *any* attempt to interpret his remark that there are quantitatively indeterminate extensions of concepts. Moreover, he had recently published his (1883) and Frege (1884) referred to it; so it would have been not unreasonable to assume that the reader of the review and, in particular, Frege would know that the totalities of ordinals and powers have no power. But most importantly,

[38] Cantor (1932), p. 441.

other interpretations – such as that he is referring to non-sortal concepts, such as 'water', or to concepts whose extensions are indefinite, such as 'living creature' – make no sense of Cantor's subsequent remark about reversal of direction. The point of that remark, it seems to me, must be that the number classes are in place to serve as a measure for an infinite totality being a determinate infinity, i. e. having a power. The number classes would play no conceivable role in distinguishing sortals from non-sortals or well-defined concepts from ill-defined ones. Again, Cantor's reference to Überweg's *System der Logik*, § 53, although it does not seem to be particularly apt, lends some support for my interpretation of his review. Überweg is speaking, not of concepts there, but of representations (Vorstellungen). Stripped of Überweg's psychologistic conception of them, they come closest to what Frege means by a 'concept'. Concepts for Überweg, on the other hand, are of a restricted kind of representation (§ 56). The extension of a representation (§ 53) consists of other representations. Thus the relation between representation and elements of the representation conforms not at all to Frege's sharp distinction between concept and object. An individual is in an extension by way of having its individual representation in the extension. Concerning number, Überweg writes

> The formal relation of the subordination of many representations under the same higher one leads to the concept [Begriff] of number which, in its original sense [as Anzahl], is the determination, by means of a unit, of the plurality [Vielheit] of individuals of the extension. (§ 53)

I do not understand the role of the unit in this conception. Shouldn't the plurality of individuals in the extension be determined already by the representation itself? But anyway, in an earlier passage, and perhaps closer to Frege's conception, Überweg writes

> Only on the basis of concept formation can *numerals* be understood; for they presuppose the subsumption of similar objects under the same concept. (§ 47)

I can't find any grounds in this for thinking that Cantor understood Frege's concepts to include non-sortals or concepts whose extensions are not well-defined.

It is easy to misunderstand Cantor's review because, for many, the primary question is to be formulated by asking whether a given totality is a set. If it is, then it has a cardinal number. Indeed, in his later papers, in which he seems to have abandoned his autonomous theory of ordinals and powers, this question is primary. But in the 1883 paper and in his review of Frege, Cantor's understanding was that the theory of the number classes is primary and with it, the theory of infinite cardinals *as the powers of the number classes*. Our question of whether a totality is a well-defined set *is* precisely his question of whether it is 'numerically determinate' – in other words, whether it is finite or equipollent to one of the number classes. It is for this reason that Cantor speaks of a reversal of direction. One cannot define the cardinals simply to be the extensions of second-order concepts of the form 'equipollent to F', since not every concept F has a cardinal.

Frege's reply [(1984), p. 122] to Cantor's review shows that he thinks that Cantor's remark is aimed at his definition of N_xFx, the number of F's' as the extension of the concept 'equipollent with F', and is pointing out that *this* concept might or might not be numerically determinate. Frege responds that this is of no consequence, since this concept does not have to have a number in order to *be* the number N_xFx. But Cantor means that F might itself be quantitatively indeterminate, so that N_xFx does not exist. On explanation for Frege's interpretation of Cantor's criticism, and one which lends some support to it, is Cantor's assertion that Frege took the notion of the extension of a concept as the foundation of the number concept. My suggested reading of this requires that we understand him to be saying that, for Frege, it is the extension of concepts which have number; but Frege's actual definition in (1884) attributes number to the concept and it is the number itself which is an extension of a concept. On the other hand, although Frege does not explicitly speak of higher level concepts in (1884), his official definition of $N_xF(x)$ makes it the extension of a second level concept, so that its elements are themselves first level concepts. It is not at all unreasonable to suppose that Cantor would understand this by taking the elements of $N_xF(x)$ to be the corresponding extensions of the concepts.

Of course, even with the misunderstanding, Frege should have taken heed. If there is *any* concept whose extension does not have a cardinal, then Frege's analysis fails. *Even if it were concepts to which number should be attributable, one would need to distinguish, on some other grounds, as Cantor insisted, those concepts which do have number from those which don't.* Frege might have responded that the concept 'equipollent with F' is a second-level concept, and that it is only first-level concepts to which number is attributable. But in (1893), which implicitly postulates as an object an extension of each first-level concept, this response fails; for every second-level concept induces the first-level concept of the corresponding extensions. In particular, the second-level concept 'equipollent with F' yields the first-level concept 'equipollent with the extension of F'.

There tends to be a picture of Frege as a tragic victim of fate: by his very virtue, namely his insistence on precision, he committed himself explicitly to a contradiction that was already implicit in mathematical thought. But in fact his assumption in the *Grundgesetze* that every concept has an extension was an act of recklessness, forewarned against by Cantor already in 1883 and again, explicitly, in his review in 1885.

References

Bolzano, B. (1817) "Rein analytischer Beweis des Lehrsatzes, dass zwischen je zwey Werthen, die ein entgegengesetztes Resultat gewähren, wenigstens eine reelle Wurzel der Gleichen liege." Prague. Translation by S. B. Russ: "A translation of Bolzano's paper on the Intermediate Value Theorem", *Historia Mathematica* 7 (1980), pp. 156–185.

– (1851) *Paradoxien des Unendlichen*. Leipzig. (Translated by D. A. Steele as *Paradoxes of the Infinite*, London: Routledge and Kegan Paul, 1959).

Cantor, G. (1874) "Über eine Eigenschaft des Inbegriffes aller reellen algebraischen Zahlen." *Journal für die reine und angewandte Mathematik* 77, 123–32. [Cantor (1932), pp. 115–18.]

– (1878) "Ein Beitrag zur Mannigfaltigkeitslehre." *Journal für reine und angewandte Mathematik* 84, 242–58. [Cantor (1932), pp. 119–33.]

– (1883) "Über unendliche, lineare Punktmannigfaltigkeiten, 5." *Mathematische Annalen* **21**, 545–86. [Cantor (1932), pp. 165–209] A separate edition of this, with a preface, was printed under the title *Grundlagen einer allgemeinen Mannigfaltigkeitslehre. Ein mathematisch-philosophischer Versuch in der Lehre des Unendlichen*. Leipzig: Teubner.

– (1887–88) "Mitteilungen zur Lehre vom Transfiniten I, II." *Zeitschrift für Philosophie und philosophische Kritik* **91**, 81–125, 252–70; **92**, 250–65. [Cantor (1932), pp. 378–439.]

– (1890–91) "Über eine elementare Frage der Mannigfaltigkeitslehre." *Jahresbericht der deutschen Mathematiker-Vereinigung* **1**, 75–78. [Cantor (932), pp. 278–80.]

– (1895–97) "Beiträge zur Begründung der transfiniten Mengenlehre, I, II." *Mathematische Annalen* **46**, 481–512; **49**, 207–46. [Cantor (1932), pp. 282–56.]

– (1932) *Gesammelte Abhandlungen mathematischen und philosophischen Inhalts* (ed. E. Zermelo). Berlin: Springer.

Dedekind, R. (1872) *Stetigkeit und irrationale Zahlen*. In Dedekind (1932). Republished in 1969 by Vieweg, Braunschweig; English translation by W. W. Berman in 1963 under the title *Essays in the Theory of Numbers*. New York: Dover.

– (1888) *Was sind und was sollen die Zahlen?* In Dedekind (1932). Republished in 1969 by Vieweg, Braunschweig; English translation by W. W. Ber-

man in 1963 under the title *Essays in the Theory of Numbers*. New York: Dover.
– (1932) *Gesammelte Werke*, vol. 3 (ed. R. Fricke, E. Noether, and O. Ore). Braunschweig: F. Vieweg.
Dummett, M. (1973) *Frege: Philosophy of Language*. New York: Harper & Row.
– (1991) *Frege: Philosophy of Mathematics*. Cambridge, MA: Harvard University.
Frege, G. (1879) *Begriffsschrift, eine der arithmetischen nachgebildete Formelsprache des reinen Denkens*. Halle: L. Nebert.
– (1884) *Die Grundlagen der Arithmetik. Eine logisch-mathematische Untersuchung über den Begriff der Zahl*. Breslau: W. Koebner. Translated in bilingual edition by J. L. Austin in 1950, Oxford: Basil Blackwell.
– (1893) *Grundgesetze der Arithmetik, Begriffsschriftlich abgeleitet*, Band I. Jena: H. Pohle.
– (1903) *Grundgesetze der Arithmetik, Begriffsschriftlich abgeleitet*, Band II. Jena: H. Pohle.
– (1984) *Collected Papers* (ed. B. McGuinness). Oxford: Basil Blackwell.
– (1979) *Posthumous Writings* (ed. H. Hermes, F. Kambartel, F. Kaulbach). Chicago: University of Chicago.
Heck, R. (1993) "The development of arithmetic in Frege's *Grundgesetze der Arithmetik*", *Journal of Symbolic Logic* **58**, pp. 579–601.
Heine, E. (1872) "Die Elemente der Functionenlehre", *Journal für die reine und angewandte Mathematik* **74**, pp. 172–88.
Kronecker, L. (1887) "Über den Zahlbegriff". Reworked and expanded version in *Werke* **3**, pp. 251–74 (ed. K. Hensel) Leipzig: Teubner.
Parsons, C. (1990) "The structuralist view of mathematical objects", *Synthese* **84**, pp. 303–346.
Russell, B (1903) *Principles of mathematics*. New York: Norton.
Schröder, E (1873) *Lehrbuch der Arithmetik und Algebra für Lehrer und Studirende* Band I. Leipzig: Teubner.
Stein, H. (1988) "Logos, logic and *logistiké*", *History and Philosophy of Modern Mathematics* (ed. W. Asprey and Ph. Kitcher). Minneapolis: University of Minnesota, pp. 238–59.
Tait, W. W. (1986) "Critical review of Charles Parsons' Mathematics in Philosophy". *Philosophy of Science* **53**, pp. 588–606.
– (1993) "Some recent essays in the history of the philosophy of mathematics: a critical review", *Synthese* **96**, pp. 293–331.
Thomae, J. (1880) *Elementare Theorie der analytischen Functionen einer complexen Veränderlichen*. Halle: Nebert.
Zermelo, E. (1930) "Über Grenzzahlen und Mengenbereiche: neue Untersuchungen über die Grundlagen der Mengenlehre." *Fundamenta Mathematicae* **14**, 339–44.

On Frege's Introduction of Cardinal Numbers as Logical Objects*

MATTHIAS SCHIRN

In his book *The Foundations of Arithmetic* of 1884, Frege set himself the task of making plausible the claim that the fundamental laws of pure number theory are analytic and consequently a priori. At the outset of his enquiry into the concept of cardinal number, he defines an analytic truth as one that can be derived exclusively from general logical laws (i. e., from primitive truths of logic neither needing nor admitting proof relative to a particular system) and definitions. The *Begriffsschrift* of 1879 had supplied the essential techniques required for laying the logical foundations of arithmetic. In the *Foundations*, Frege intended to make the decisive step in his project of reducing number theory to logic, namely to set up a formally correct and systematically fruitful definition of number *(Anzahl)* in purely logical terms. The book was designed not only to yield a profound *mathematical* analysis of the concept of cardinal number, but also to provide a *philosophical* justification of the claim that the truths of arithmetic are analytic and thus to illuminate the epistemological nature of arithmetic. The latter applies to both the critical and the constructive part of the book. It is worth emphasizing that Frege argued forcefully from the universal applicability of arithmetical laws to their purely logical nature.

Frege's principal scientific concern was to lay the logical foundations of number theory and analysis. He probably believed that for this purpose it was mandatory to establish not only the purely logical nature of the natural and the real numbers, but also their objectual status. It is a striking feature of Frege's conception of numbers as objects that it is almost entirely based on syntactic considerations. This applies notably to the strategy he pursues in the *Foundations*.

* References are to both the English translations of Frege's works and to the German originals. For the most part, I have modified the existing translations. The quotation taken from GGA II, § 164 is my own translation.

This article consists of three parts. In the first part, I characterize briefly some essential features of Frege's foundational programme. In the second part, I deal with his conception of numbers as objects by placing particular emphasis on his proposed analysis of numerical statements in the *Foundations*. In the third and final part, I scrutinize Frege's three attempts to define number and discuss, with respect to numerical terms, what I shall call *Frege's indeterminacy problem*.

1. The nature of the numbers and our cognitive access to them

In Frege's view, an adequate characterization of the nature of numbers must satisfy at least two basic, intimately related requirements. It must take account of both the objectivity and the unrestricted applicability of arithmetical laws.[1] By *objectivity* he means independence of our sensation, intuition and imagination, but not independence of our reason (cf. FA, GLA, 36). The advocate of a physicalistic conception of arithmetic may ignore Frege's second requirement, whereas the proponent of a psychologistic one disregards the first requirement and possibly also the second. Frege thinks that both advocates of what he considers to be rival theories of arithmetic fail to subject numerical statements to logical analysis. In particular, they fall short of what he presents as the most important result of his own investigation of the logical form of numerical statements, namely the insight that an ascription of number such as "The number 9 belongs to the concept *planet*" or "There are nine planets" contains a predication of a concept.[2] I term this insight *Frege's principle of numerical predication*, PNP for short.

Frege holds that if numbers were something perceptible (or intuitable), they could not be ascribed to anything non-perceptible (or non-intuitable).[3] Now, according to PNP, numbers are in fact ascribed to

---

[1] Frege's emphatic remark towards the end of the *Foundations* that there is nothing more objective than the laws of arithmetic comes immediately to mind.

[2] This is my preferred rendering of Frege's dictum "Die Zahlangabe enthält eine Aussage von einem Begriffe". Frege definitely uses the word "Aussage" to mean "predication" when he states this dictum. Dummett proposes to render it as follows: The content of an ascription of number consists in predicating something of a concept (1991b, 88).

[3] I do not endorse this view. According to Benacerraf's argument in 'What Numbers Could Not Be', the natural numbers could be any progression of objects satisfying the Peano axioms. They could be construed, for instance, as points of space, even as physical objects. As such the natural numbers could well be ascribed to concepts qua abstract entities. I am grateful to Charles Chihara for pointing this out to me.

something non-perceptible, namely to *concepts*, under which we subsume both the spatial and temporal and the non-spatial and non-temporal. To take the concept as the bearer of number is intended to preclude the possibility of regarding numbers as perceptible or intuitable objects and, at the same time, to yield a plausible explanation for their universal applicability. On the basis of these and related results, numbers are then to be characterized as follows: they are neither accessible to our sensation, intuition, or imagination, nor capable of involvement in causal interaction, nor subjective like ideas, but are non-spatial, non-temporal (i. e., causally inert, or in Frege's terminology: non-real), though still objective. Frege thinks we can only steer safely between the Scylla of physicalism and the Charybdis of psychologism by adhering strictly to his methodological maxim "Never ask for the meaning of a word in isolation, but only in the context of a sentence". Thus he paves the way for his envisaged introduction of numbers as *logical* objects.

1.1 How do we conceive numbers qua abstract objects?

In Section 62 of the *Foundations*, Frege raises the following epistemological question: "How, then, are numbers to be given to us, if we cannot have any ideas or intuitions of them?" Somewhat surprisingly, he replies by stating his famous contextual dictum "Only in the context of a sentence do words have meaning", and concludes that we have to define the sense of a sentence in which a number-word occurs. In modern terms, his question would be this. How do we succeed in referring to numbers qua abstract objects? How do we have cognitive access to them? We encounter essentially the same question in more general terms in the appendix to the second Volume of the *Basic Laws*. It is now highlighted as the prime problem of arithmetic. "How do we conceive logical objects, in particular numbers? What justifies our recognizing numbers as objects?" (BLA, 143, GGA II, 265). Evidently, Frege is reacting here to Russell's discovery of a contradiction in the logical system of the *Basic Laws*. Not only the logical nature of numbers appears to be questioned, but also their objectual status.[4]

[4] Obviously, should it turn out that the numbers are not objects, or not purely logical, or neither one nor the other, the question concerning the foundations of number theory and analysis would have to be formulated in a significantly different way. If numbers are not objects, then, according to Frege's syntactically based ontological dichotomy of *function* and *object*, they could only be functions.

In the spirit of Frege's view, though not in his own words, the answer both to the question "How do we apprehend numbers?" and to its more general version "How do we conceive logical objects?" would be this. We conceive them through *abstraction*, i. e., by transforming an equivalence relation into an identity. Still, in Frege's sense, we could also say: we conceive logical objects by applying a specific criterion of identity to them. As is well known, Frege uses the word "abstraction" for the most part in a pejorative, psychological sense, especially when severely taking to task certain rival theories of arithmetic.[5] Unlike Frege, I use it here to denote a special *logical* faculty which he is imputing to us.[6] Anticipating my discussion of his introduction of cardinal numbers, there is a more definite answer at hand to the first question. According to the *Foundations*, we apprehend cardinal numbers by converting an equinumerosity holding between first-level concepts into a numerical identity, relying on the identification of cardinal numbers with equivalence classes of equinumerosity.[7] The answer to the more general question along the lines suggested by Frege is this (cf. TF, 181; GGA II, 147). We conceive logical objects by making the transition from an equality holding generally between the values of two functions to an identity of courses-of-values.[8] In

[5] In Section 44 of the *Foundations*, Frege concedes that we may attain concepts by abstraction; e. g., by taking the moon as the starting-point of abstraction, we may attain the concepts *satellite of the Earth, satellite of a planet, non-self-luminous heavenly body, body, object.* Frege categorically denies, however, that one arrives at a number by abstraction from an object. In the case of 0, we have no object at all from which to start the process of abstraction. It is rather the concept attained by means of abstraction in which we then discover the number (FA, GLA, § 48). Strictly speaking, the kind of abstraction which Frege allows in the *Foundations* has no bearing whatsoever on his philosophy of arithmetic.

[6] We could appropriately speak of *logical abstraction* in contrast to *psychological abstraction*. It is the latter which Frege combats vigorously throughout his academic career.

[7] This is but a tentative answer. As to the relation between the transformation of an equinumerosity into an identity and Frege's final explicit definition of the number belonging to the concept F as an equivalence class of equinumerosity, see part 3 of this article.

[8] Since in the *Basic Laws* all numbers are defined as special courses-of-values, this answer applies also to the case of numbers. In his letter to Russell dated 28. 7. 1902, Frege writes: "But the question is, How do we apprehend logical objects? And I have found no other answer to it than this, We apprehend them as extensions of concepts, or more generally, as courses-of-values of functions" (PMC, 141; WB, 223). It is plain that the axioms of the formal theory of the *Basic Laws* require the existence not only of courses-of-values, but also of the two truth-values, the True and the False. I take it that Frege considered the truth-values to be primitive objects of logic. It is true, however, that in Section 10 of the *Basic Laws* he identified them with their own unit classes.

doing this, we are recognizing something in common to the two functions, namely their course-of-values. Thus, our grasp of the sense of the new sentence, i. e., the identity-statement, is mediated by our recognition of its equivalence with the old one, i. e., the equivalence statement. If Frege had been confronted with the postulation of a special faculty of mathematical intuition à la Gödel,[9] which is supposed to allow us direct cognitive access to the remote realm of abstract objects, he would probably have stigmatized it as a devastating "irruption of psychology into logic" (cf. CP, 209; KS, 192).

A few remarks on Benacerraf's epistemological challenge to platonism may be in order here. The challenge is that we must be able to combine the standard platonistic account of mathematical truth with a causal account of mathematical knowledge without offending against the plausible presupposition that we do have some mathematical knowledge. According to the platonist, mathematics describes a realm of abstract entities. Now if one endorses a causal theory of knowledge, one should hold that when a person knows something about an object, there must be some causal link between the object known and the knower. But how can abstract, causally inert entities enter into any physical interaction? Given that we possess some mathematical knowledge, it follows that either our best theory of mathematical truth at hand – i. e., the standard platonistic view – or the best theory of knowledge – and here Benacerraf favours the causal theory – comes to naught. Now, the platonist could align himself with those who claim to have good reasons for disavowing the causal theory of knowledge. Yet, even if he succeeded in rejecting the epistemological challenge just laid out, it would still remain for him to explain how we come to know anything about the subject-matter of mathematics. The platonist would have various options to attempt to do this; he might postulate a special faculty of mathematical intuition, for instance. Or he might argue that some familiar, more pedestrian source of knowledge such as sense-perception enables us to have the required access. And I think there are other options as well, and perhaps more promising ones, as I shall try to show in a planned paper on mathematical platonism.

[9] See Gödel 1947, 483 ff.

1.2 Arithmetical platonism

At least in one obvious respect, Frege's advocacy of logicism goes hand in hand with his endorsement of a number-theoretic version of ontological platonism: Numbers are logical or abstract objects existing independently of the human mind. The common view that Frege was a realist with respect to abstract objects has been challenged unsuccessfully, I think, by various renowned Frege scholars such as Sluga, Currie and Resnik. Although they restrict their assessment to the period of the *Foundations*, the arguments which they advance are either far-fetched, or awkward, or both.[10] The clearest expression of Frege's arithmetical platonism can be encountered in the context of his repudiation of Hankel's formal arithmetic: "even the mathematician cannot create things at will, any more than the geographer can; he too can only discover what is there and give it a name" (FA, GLA, 107 f.). Now, on the face of it, this dictum, stressing the non-creativity of mathematical definitions, seems to be at odds both with Frege's notion of objectivity and with his somewhat cryptic explanation in Section 105 of the *Foundations* of what he understands by the "rational character" of arithmetic. There he maintains that in arithmetic we are concerned with objects given directly to our reason and which, belonging to our reason, are utterly transparent to it. On closer reflection, however, it emerges that the conflict is only apparent. From Frege's point of view, we *conceive* numbers by using our reason, that is, we "lay hold of them" by means of our logical faculties without mediation of a non-logical source of knowledge – sense perception or intuition – but we do not create them. In that respect, he differs clearly from Dedekind, his fellow combatant for logicism.

The *epistemological* thesis that we derive our knowledge of numbers exclusively from the logical source of knowledge qua reason, obviously does not imply the *ontological* claim that they do not exist independently of human reason, let alone the stronger claim that they are creations of human reason. In Frege's opinion, the number 2, for instance, exists independently of our conceiving it through our reason, as the capacity to grasp thoughts and to make judgments, i. e., to recognize thoughts as true. At the same time, the number 2 *is what it is* – the sum of 1 and 1, the smallest prime number, the immediate successor of 1 in the natural sequence of numbers, etc. – independent of our arithmetical

---

[10] Cf. Sluga 1980, Currie 1982, Resnik 1980.

judgments about it. *To answer the question* independently of human reason of what the number 2 is, however, would be to judge without judging. Furthermore, several fragments written by Frege during the period of the *Foundations* make it clear that in his view judgeable contents in general and those expressed by the laws of arithmetic in particular are not the product of any mental process. On the contrary, the principles of arithmetic are not only true independently of our recognizing them to be so, but they are also independent of our thinking as such. Yet if this is so, then the numbers for which they hold good must likewise exist independently of human reason.

2. *The Mathematical-Object Picture*

The conception of numbers as objects or the Mathematical-Object Picture, as it has been termed by Hilary Putnam 1967 and Harold Hodes 1984, seems to be deeply rooted in our ordinary mathematical reasoning and discourse. Frege is certainly one of the foremost advocates of this position in the history of mathematics. How, then, is it to be explained that the Mathematical-Object Picture seems to force itself upon us? Following in Frege's footsteps, the answer is very simple. (a) Numerical expressions like "4", "the number of planets", "6 + 2", "the smallest prime number between 5 and 18", etc., behave logically like paradigmatic singular terms, such as proper names or definite descriptions of persons, cities, places, etc.; (b) expressions like "is prime", "is greater than", etc., function syntactically like authentic first-level predicates, and (c) expressions like "some number" and "all numbers" resemble ordinary first-order quantifier phrases. The step from these syntactic observations to the semantics of arithmetical discourse seems natural. (a') Numerical singular terms genuinely refer to numbers qua objects; (b') numerical predicates designate first-level concepts under which numbers do or do not fall, and first-level relations which do or do not hold between numbers; (c') number-theoretic quantifiers range over numbers qua objects; they designate second-level concepts. In a nutshell: numbers are what numerical singular terms refer to, what numerical first-level predicates apply to and what first-order number-theoretic quantifiers range over.

2.1 Syntactic criteria

It is generally agreed that the distinction between function and object lies at the very heart of Frege's philosophy of logic. One of its key elements is the primacy of the syntactic categories *function-name* and *proper name* over the ontological ones *function* and *object*. The latter are only derivative. Frege never abandoned this view, which fully crystallized during his mature period after 1891. Quite in the spirit of his approach, one might say that an object is anything that can be referred to by a singular term. A revealing example is his introduction of the two truth-values in 'Function and Concept': "An object is anything that is not a function, so that an expression for it does not contain any empty place. A declarative sentence contains no empty place, and therefore we must regard its reference [Bedeutung] as an object. Thus, the two truth-values are objects" (TF, 147; KS, 134).[11]

In the *Foundations*, Frege feels entitled to regard cardinal numbers as objects on purely syntactic grounds, thus subscribing to the Mathematical-Object Picture sketched above. He is, in fact, well aware that the kinship between ascriptions of number and existential statements[12] and, in particular, his number-theoretic key principle PNP might encourage the view that cardinal numbers are to be identified with second-level *concepts.*[13] He rejects such a view precisely by (a) appealing to certain syntactic criteria and (b) analyzing the logical form of an ascription of number. Here are his syntactic criteria.

---

[11] Here Frege is tacitly assuming that an expression containing one or more empty places cannot designate an object. Furthermore, he is not arguing for the claim that the reference of a declarative sentence is a truth-value. See in this connection his essay 'On Sense and Reference' (TF, 62 ff.; KS, 148 ff.). In 'Introduction to Logic' (PW, 194; NS, 211), Frege mentions the complete or saturated character of a thought as an additional reason for construing the True and the False as objects.

[12] In Section 53 of the *Foundations*, Frege points out that affirmation of existence is in fact nothing but denial of the number nought. It is likely that he considers, e. g., "There exist lions" and "The number of lions is not nought" to have the same judgeable content. Likewise he probably regards, e. g., the two sentences "$N_x^0(x \neq x)$" ("The number 0 belongs to the concept $x \neq x$") and "$\neg\exists x(x \neq x)$" as expressing the same content or thought (cf. FA, GLA, § 55).

[13] Clearly, if Frege had defended the view that numbers are to be identified with second-level concepts, he would have been obliged to interpret identity between numbers as a third-level relation and predicates like "is prime", "is greater than" etc. as third-level predicates.

(1) We characteristically apply the definite article to numerals and other numerical expressions.

(2) Numerical terms like "9", "the number of planets", etc., are typically used on both sides of the identity-sign.

(3) In the "language of everyday life", number-words also occur in attributive or adjectival constructions as, for example, in the statement "Jupiter has four moons". Such a statement can be transformed, however, into the equation "The number of Jupiter's moons = 4" without altering the sense or content (see also CP, 120; KS, 110).

(4) The words "all", "any", "no" and "some" are prefixed to concept-words. Expressions like "a two", "some twos" and "all twos" are inadmissible. A numeral or other numerical term can never stand in the logical place of a predicate.[14]

I am sure that Frege would unhesitatingly have accepted an additional criterion, at least as far as his Begriffsschrift is concerned. The latter was designed to be a logically perfect language, where all expressions wear their logical form on their sleeves. The additional criterion is this.

(5) Numerical terms like "9", the number of planets, etc., contain no argument-place, neither explicitly as symbols of a formal language, nor implicitly as expressions of a natural language.

Now, even if we are prepared to endorse Frege's account in giving priority to syntactic over ontological criteria, it is highly dubious whether these criteria provide a conclusive justification for recognizing numbers as self-subsistent objects.[15] Especially criterion (1) strikes me as somewhat parochial, though not every counterexample that has been brought forward is compelling. Surely a statement like "The whale is a mammal" is only an apparent counterexample. In this statement, the expression "the whale" can legitimately be replaced with "all whales" ("All whales are mammals"), but it would be nonsensical to substitute "all twos" for "the two" in a statement like "The two is a prime". To canvass Frege's criteria for singular termhood would be a major task in its own right and take me too far afield from my principal concern. Suffice it to say that in their entirety they have at least some weight and

[14] See also CP, 120; KS, 110.

[15] See Wright's critical examination of Frege's criteria in Wright 1983, 53 ff.

thus ought not to be jettisoned as completely worthless. As pointed out by Crispin Wright, the major weakness of Frege's syntactic criteria is that they hardly escape circularity. In order to classify, with their help, a given expression as a singular term, we must already invoke the notions of singular term, reference, or object.

2.2 Numerical statements analyzed

Let us now look at how Frege analyzes a statement such as "The number 4 belongs to the concept *Jupiter's moons*". In this statement, he emphasizes, the numeral "4" is only an element in the predicate (taking the concept *Jupiter's moons* to be the real subject). He adds: "For this reason I have avoided calling a number such as 0 or 1 or 2 a *property* of a concept. Precisely because it forms only an element in what is predicated, the individual number shows itself for what it is, a self-subsistent object" (FA, GLA, 68).[16] Here Frege is appealing to the *grammatical* categories of *subject* and *predicate*, yet in opposition to one of his guiding principles in *Begriffsschrift*.[17] We may achieve clarity about what he really wants to convey, and see further difficulties lurking in Section 57 of the *Foundations*, if we compare the following numerical statements or their respective formalizations. "F(x)" is to abbreviate the predicate "Jupiter's moons".

(1) The number 4 belongs to the concept Jupiter's moons.
(1*) $N_x^4 F(x)$.
(2) There are four moons of Jupiter.
(2*) $\exists_4 x F(x)$.
(3) The number belonging to the concept *Jupiter's moons* = 4.
(3*) $N_x F(x) = 4$.

In order to put matters into perspective, I should like to direct attention to four syntactic-semantic theses held by Frege.

(A) Different sentences can express the same thought.[18]

---

[16] Dummett's translation "The individual number, as being a self-subsistent object, appears precisely as a mere part of the predicate" (1991b, 108) is misleading, if not false. The individual number appears as a self-subsistent object *for the very reason* that it forms only a part of what is predicated.

[17] Cf. CN, BS, § 3; PW, 120, 141, 143; NS, 130, 153, 155; PMC, 68, 100 f.; WB, 103, 164.

[18] In Dummett 1973, 378 f. Dummett writes: "To say that the sense of a sentence is composed out of the senses of its constituent words is to say [...] that we can grasp that sense only as the sense of a complex which is composed out of parts in exactly

(B) A sentence and the thought which it expresses may be analyzed (or decomposed) in distinct ways (cf. CP, 188; KS, 173; PW, 187, 201 f.; NS, 203, 218).

(C) A thought is built up out of parts which correspond to the parts out of which the sentence expressing the thought is built up. (Frege formulates the homomorphism between the construction of a sentence out of sentence-constituents and the construction of the corresponding thought out of thought-constituents even more concisely: The structure of the sentence can serve as a picture of the structure of the thought.)[19]

(D) We start out from judgments and their contents, and not from concepts. We only allow the formation of concepts to proceed from judgments (PW, 16 f.; NS, 17 f.; cf. PMC, 101; WB, 164).

Sentences (1)–(3) and (1*)–(3*) confirm thesis (A); according to Frege, they all express the same thought.[20]

Consider sentence (1*). It can be split up into the first-level predicate "F" and the second-level predicate (α) "$N^4_x\varphi(x)$" ["the number 4

---

that way; only a sentence which had exactly that structure, and whose primitive constituents corresponded in sense pointwise with those of the original sentence, could possibly express the same sense. (Frege's notion of the senses of complex expressions thus tallies closely with Carnap's intensional isomorphism.)" These remarks are incompatible with thesis (A) if it applies, as I consider it does, to sentences which, under their standard interpretation, have significantly different logical structures such as our sample sentences (2) and (3). If thesis (A) did not hold, then, as Frege stresses, logic would be paralyzed; for its "task can hardly be performed without trying to recognize the thought in its manifold guises" (TF, 46 footnote *; KS, 170 footnote 7). Dummett's exegesis stands, moreover, in sharp contrast to thesis (B). One may compare, for example, Frege's observations in 'On Concept and Object', TF, 49 f.; KS, 173 f. Thesis (A) provides support for thesis (B) as relating to the process of decomposing a thought. It is in complete accordance with (A) and (B) to maintain that one and the same thought can be seen as put together out of parts in different ways (cf. PW, 201 f.; NS, 218). Contrary to what Dummett claims, Frege's notion of the sense of a complex expression has little if anything in common with Carnap's notion of intensional isomorphism. I should perhaps mention that I no longer endorse Dummett's distinction between two kinds of analysis; see Chapter 15 of Dummett 1981 and the discussion of it in Schirn 1983d. At any rate, Frege's writings fail to provide evidence that he would have accepted Dummett's distinction between *analysis* and *decomposition*.

[19] Cf. PW, 225, 243, 255; NS, 243, 262, 275; CP, 309; KS, 378; PMC, 79; WB, 127. Frege emphasizes that we talk figuratively when we transfer the relation of whole and part to thoughts (CP, 390; KS, 378).

[20] I have no doubt that Frege would consider (1) and (2) to be synonymous sentences. Since he regards (1) and (3) as expressing the same content, he would also maintain the synonymy of (2) and (3). But there is a difficulty with these claims, which I shall discuss in a moment.

belongs to the concept φ"].[21] Thus analyzed, (1*) expresses that the first-level concept F falls under the second-level concept $N_x^4\varphi(x)$. Analogous remarks apply to (2*). From the fact that the numeral "4" forms only an element in (α), Frege concludes (a) that the number 4 cannot be a property of the concept F and (b) that it shows itself for what it is, a self-subsistent object.[22] This conclusion is arguably unsound.

Let us focus on (b) first. When Frege claims that the number 4 appears as a self-subsistent object he seems to suggest that the numeral "4", as it occurs in the predicate "the number 4 belongs to", functions as a proper name of the number 4.[23] The latter, however, need not be the case. And it would be pointless to defend a substantival interpretation of "4" by appealing to the fact that in (1) it is combined with the definite article. Surely, the occurrence of "4" in both (1) and (1*) is syncategorematic, playing no referential role. If instead of Frege's contrived and tendentious formulation (1) we choose the more natural one (2), it is equally true that the numeral "4" forms only an element of a second-level predicate. Yet in sentence (2) we encounter not even a trace of the alleged objecthood of the number 4 or the purported singular termhood of "4". In (2) and (2*), "4" appears adjectivally as an undetachable element of a numerically definite quantifier. It does not even function syntactically as a singular term because $\exists_4$ is a single syntactic unit.

I now turn to (a). When Frege says that he has avoided calling an individual number a property of a concept he presupposes what he still has to prove, namely that numbers are to be construed as objects. Surely, conceived of as an object, a number cannot be a property of a first-level concept. Furthermore, from the fact that in (2) or (2*) "4" is only an element of the predicate (β) "$\exists_4 x\varphi(x)$" it does not follow that the number 4 cannot be considered a second-level concept. Frege tacitly assumes that a number cannot be designated by a predicate containing a numeral as a constituent such as (α) or (β). Plainly, only if he were able

21 The letter "φ" is to mark an argument-place of type 2, i. e., an argument-place which is appropriate to admit first-level concept-expressions (more generally: monadic first-level function-names).

22 Compare what he says in two late fragments concerning this issue, PW, 256 f., 276 footnote; NS, 276 f., 296 footnote.

23 If Frege did not suggest this, it would be difficult to make sense of (b). (b) in no way follows from the fact that "4" occurs only as a part in "$N_x^4\varphi(x)$". When we come to canvass both Frege's treatment of adjectival uses of number-words and his first attempt to define number, we shall see that his view about the role of a numeral as an element in an ascription of number lacks coherence.

to demonstrate that "4" as an element of either predicate denotes the number 4 would it follow that the whole predicate cannot denote the number 4. And if Frege wished to argue that the predicate (α) "$N^4_x\varphi(x)$", by virtue of its unsaturated nature, fails to designate the number 4, he could do so only by invoking his still unfounded thesis that numbers are objects. What a charitable reading of the opening remarks in Section 57 positively reveals is only that an ascription of number like (1) or (1*) expresses that the concept F falls under the concept $N^4_x\varphi(x)$.

In contrast to (1*) and (2*), sentence (3*) does not, under its standard interpretation, express that the unequal-leveled relation of *falling under* holds between a first-level and a second-level concept.[24] Rather, (3*) expresses that the (first-level) relation of identity holds between $N_xF(x)$ and 4. But clearly, there is no need to interpret (3*) in accordance with what we may regard as its surface structure. Sentence (3*) and the thought it expresses are, in effect, capable of being analyzed in a number of distinct ways, as is suggested by thesis (B), and can thus be seen as composed in different manners (cf. e. g., PW, 201 f.; NS, 218). We may, for instance, split up (3*) into the predicates "F" and (γ) "$N_x\varphi(x) = 4$". In doing this, we decompose the corresponding thought in a "non-standard" fashion. According to this analysis, (3*) expresses that the first-level concept F falls under the second-level concept $N_x\varphi(x) = 4$.

2.3 The twofold role of a numeral

I begin with an explanation. By a *Fregean* I mean someone basically subscribing to Frege's doctrines or at least sympathizing with them. For present purposes, we can ignore the fact that some of Frege's doctrines have undergone significant changes. If I am right, then none of these modifications amounts to a sea change, however.

The foregoing syntactic-semantic considerations yield a first interesting result for the Fregean. But they also entangle him in a difficulty. More specifically, they cast doubt upon the intuitively plausible view that if (3) qua "first-level" sentence is synonymous with (1) and (2),

[24] This relation is akin to the unequal-leveled relation of *falling under* holding between an object and a first-level concept. In Frege's formal language, the latter relation is designated by "$—\varphi(\xi)$".

then this must likewise hold for (3) qua "second-level" sentence.[25] The first result is that our sample sentences (1) and (2) do not involve reference to or quantification over numbers conceived of as objects. At least for (1) and (2) we can thus adequately explain what it is to grasp the thought expressed by either sentence, without having to explicate the mechanism of identifying reference to numbers qua abstract objects. And (3) qua identity-statement is regarded as expressing the same thought as (1) and (2). Moreover, if Frege were to endow (3) with a non-standard interpretation, displaying it as being of "second level", he would be completely justified in applying to (3) what PNP says about an ascription of number such as (1) or (2). So interpreted, (3) does contain a predication of a concept. Seen from this perspective, then, the Mathematical-Object Picture may appear less compelling even for the Fregean, though this is not to say that it loses its initial attractiveness.

As you may have anticipated, the difficulty I allude to rests on the different role which the numeral "4" plays in (1*) or (2*) on the one hand and in (3*) qua "second-level" sentence on the other.[26] I have been underscoring the fact that in (1*) and (2*) "4" occurs as a syncategorematic element of a second-level predicate. Not so in (3*), construed as a predication of a concept. As a constituent of (γ) "$N_x\varphi(x) = 4$", the numeral "4" figures as a singular term referring to the number 4. Unlike (α) "$N_x^4\varphi(x)$" and (β) "$\exists_4 x\varphi(x)$", (γ) is a complex predicate; a grasp of the sense of (γ) involves, in the light of Frege's later theory of sense, a grasp of the senses of the constituents "=", "$N_x\varphi(x)$" and "4".[27] And according to this theory, the sense of a complex expression is determined by the senses of its relevant parts and the manner in which these are put together to form the complex expression. Hence, if (3*) qua "second-level" sentence is to express the same thought as (1*) and (2*), then the sense of (γ) must coincide with the sense of (α) and (β). But

---

[25] The distinction I draw between "first-level" and "second-level" sentences is, of course, not to be taken in an absolute sense. That is why I use quotation marks. Considered a sentence in which something is asserted about an object, (3) is of first level; regarded as a sentence, in which something is asserted about a first-level concept, it is of second level.

[26] Brian Loar drew my attention to this point.

[27] Here I am invoking Frege's theory of sense only for the purpose of facilitating my line of argument, though, of course, I am aware that Frege had not yet formulated this theory at the time of writing the *Foundations*. Resnik (1980, 197) claims that in ascriptions of number numbers [it should probably read: numerals] are not treated as singular terms, but as elements of complex predicates. In contrast to Resnik, I maintain that a predicate such as "$\exists_4 x\varphi(x)$", for instance, is semantically simple precisely because the occurrence of "4" in that predicate is syncategorematic.

how can the Fregean endorse this conclusion, *if* he is prepared to grant that in (γ) "4" features syntactically and semantically as a singular term, while the occurrence of "4" in (α) and (β) is syncategorematic?

The Fregean might consider a number of options in response to this predicament.

(i) One option that may immediately come to mind is to gainsay that (3*) can properly be analyzed as containing a predication of a first-level concept. It is undeniable, however, that such a move would prove to be irreconcilable with one of the basic tenets to which Frege firmly and constantly adhered; I mean, of course, thesis (B): A sentence and the thought which it expresses may be analyzed in distinct ways.

(ii) A second option would be to deny that (3*), no matter how it is decomposed, expresses the same thought as (1*) or (2*). In the light of Frege's approach in the *Foundations*, I am almost certain that he would reject this possibility out of hand.

(iii) Likewise Frege would, I think, be quite against the proposal of treating (γ) as a simple predicate containing "4" as a syncategorematic element, which would certainly be a rather artificial ploy.

(iv) As I indicated under (ii), we can assume that the Fregean would be unwilling to retract the claim that (3*) qua equation expresses the same sense as (1*). Suppose now that he refuses to accept our previous conclusion that the sense of (γ) must coincide with the sense of (α), on the grounds that the premise is unsound. Evidently, in that case he would be obliged to adduce a cogent argument which entitles him to disclaim the synonymy of (3*) qua "second-level" sentence with (1*), while leaving the synonymy of (3*) qua identity-statement with (1*) intact. How the Fregean could redeem that obligation, I am at a loss to say.

(v) In order to state a final option, let us disregard (2*) as before. Now, Frege might wish to advocate the view that "4" functions as a singular term not only in (3*), no matter how it is construed, but also in (1*). Recall that in Section 57 of the *Foundations* he *seems* to be toying with the idea of espousing such a view. Plainly enough, if the view could be sustained, it might even be seen as buttressing Frege's advocacy of the Mathematical-Object Picture. We have observed, however, that the view is indefensible; in particular, the occurrence of the definite article in (1) is merely spurious, not genuine. And I imagine that Frege knew this.

So it appears, after all, that the Fregean is debarred from embracing candidly any of these options. I, for my part, fail to see an easy way out of the quandary. But we have not yet reached the end of our discussion of Frege's strategy to establish the Mathematical-Object Picture. We may attain a more complete idea of it if we subject his way of handling number-words to further scrutiny.

2.4 Substantival versus adjectival use of number-words

Frege bases his syntactic analysis of number-words on examples drawn from both number theory and natural language, taking advantage of the fact that in number-theoretic sentences their substantival use predominates. He does not, however, give any convincing reason why the adjectival use of number-words need not play a significant role for determining the ontological nature of cardinal numbers. He only points out that what really matters here is to understand the concept of number in such a way that it is useful in science. But why should an adjectival conception of numbers be scientifically inappropriate or even useless? Because we would then be unable to explain the occurrence of numerical expressions in equations whose form, Frege thinks, prevails in arithmetic? Suppose it could be shown that arithmetical statements in which number-words, as far as surface grammar is concerned, function syntactically as singular terms, could be transformed into statements in which number-words occur in the adjectival mode. Then every argument that Frege adduces in Sections 56–57 in favour of his avowed conception of numbers as objects would be off target. In fact, he takes a wavering attitude when he comes to characterize the ontological status of cardinal numbers by appeal to syntactic observations. On the one hand, he relies heavily on certain substantival uses of number-words in both number-theory and ordinary language, taking numerical terms at face-value. On the other hand, he attaches no particular importance to the fact that there are equally natural adjectival uses of number-words in numerical statements. He makes short work of them, trying to explain them away.

Let us dwell a little longer on Frege's somewhat opaque treatment of number-adjectives. In Section 57 of the *Foundations*, he contends that the adjectival use of number-words in natural language can always be avoided. And in Section 60 he emphasizes that the self-subsistence which he is claiming for number is meant to exclude the use of a number-word as a predicate or attribute. Frege indeed suggests an analysis of ascriptions of number which appears to accord well with an adjectival view of numerical expressions: they contain a predication of a concept. In the same breath, he apparently tries to conceal from us what clearly is an adjectival use of a numeral – for instance, the use of "4" in (1) – by smuggling in the use of the definite article in an ascription of number. Frege is skating on thin ice here. Not only is it not the true method of philosophy to hide what one knows to be really there, but such a manoeuvre is also pointless if one is prepared to advocate an objectual view of the numbers. Frege's concluding remarks in Section 56 make it clear that he is perfectly aware that a numeral figuring in a statement such as (1) forms an inseparable part of a second-level predicate and as such is not performing any referential role. Admittedly, this does not square with what he seems to be hinting at the beginning of Section 57 as regards his use of the definite article in an ascription of number. But let us set aside this discrepancy, be it genuine or only apparent, and turn to another difficulty looming in Section 57.

2.5 Ascriptions of number and numerical equations

The issue which I am going to raise now has a number of interwoven facets. The heading to Section 57 is as follows: An ascription of number should be regarded as an equation between numbers. Frege mentions the statement "Jupiter has four moons" as one that can be turned into a numerical equation in order to avoid an adjectival use of a numeral. At the same time, by way of analyzing an ascription of number like "$N_x^0 F(x)$" in accordance with PNP, and by appealing to the fact that the numeral "0" forms only an element of the predicate "$N_x^0 \varphi(x)$", he purports to have established that in such an ascription the individual number appears as an autonomous object. So, why and with what justification does he nevertheless insist that a sentence like (1) is to be considered a numerical equation?

My view of the matter is this. Even if Frege did not claim to have shown that in ascriptions of number an individual number appears as a self-subsistent object, he could certainly have accepted them as numerical statements in their own right. PNP is chiefly intended to pinpoint the logical structure of arithmetical sentences of a certain kind. On what grounds, then, could Frege attach so much weight to PNP when he thought he had to dispense precisely with that type of numerical statements to which PNP is meant to apply? In other words: If in the *Foundations* he believed that in order to defend an objectual view of cardinal numbers it was mandatory to explain away ascriptions of number by representing them through numerical equations, then in his foundational project PNP would have proved to be redundant. The latter, however, would run counter to available evidence. Even after having relinquished his logicist thesis, Frege expressly endorsed PNP (cf. PW, 278; NS, 298). In his late fragment 'A New Attempt at a Foundation of Arithmetic', he repeats two earlier assertions of his which he still regards as true. One of these enshrined insights is PNP; he refers to the first Volume of the *Basic Laws*, page 3: "In this way, then, it will finally be acknowledged that an ascription of number contains a predication of a concept." If I am right, then in the *Foundations* Frege is convinced that adherence to PNP need not undermine his conception of numbers as objects. At any rate, to construe a statement such as (1) or (2) as one containing a predication of a first-level concept does not commit him to regard cardinal numbers as second-level concepts. So once again: why should and how could he refuse to acknowledge ascriptions of number as numerical statements in their own right?

Dummett (1991b, 109) contends that Frege's solution of the problem of relating the adjectival and substantival uses of number-words to one another is apparently this: to treat the adjectival uses as disguised forms of the substantival ones. According to the analysis which Frege suggests in Section 57, the surface structure of a sentence like (2) is misleading. When its deep structure is uncovered, it can be recognized as being a numerical equation. Although Frege seems to encourage Dummett's assessment, I am not sure whether it is right. We have seen that Frege handles the surface grammar of numerical sentences in an unbalanced fashion. On the one hand, he takes it to be a reliable guide for determining the ontological status of the cardinal numbers; on the other hand, he dismisses the surface structure of certain numerical sentences as inessential for that purpose. In this respect, I basically agree with Dummett. I doubt, however, that Frege, if he were forced to put all his cards on the table, would describe the surface grammar of ascriptions of number as misleading, in the sense that they contain a disguised reference to a number qua object.

To begin with, I think that Frege would accept (1*) and (2*) as matching the real logical form of (1) and (2) respectively. That form does not emerge as the result of unearthing some concealed deep structure. At least sentence (2) appears to be free from the blemish of logical imperfection; it need not be transformed into a numerical equation in order to be scientifically respectable. A passage in Frege's 'Notes for

Ludwig Darmstaedter' from 1919 confirms that a correct symbolic representation of an ascription of number in his formalism would be along the lines of PNP. "Since an ascription of number based on counting contains a predication of a concept, in a logically perfect language a sentence used to make such an ascription must contain two parts, first a sign for the concept of which the ascription is made, and secondly a sign for a second-level concept" (PW, 256; NS, 277). If Frege were to regard the surface grammar of (1) or (2) as highly misleading, would he not then be giving the lie to PNP?

Let us take a closer look at what seems to be involved in Dummett's proposal. If Frege held the view that Dummett attributes to him, then the process of laying bare the hidden deep structure of an ascription of number would amount to this: the real logical form of what at the surface appears to be a "second-level" sentence such as (1) or (2) turns out to be the overt surface structure of a "first-level" sentence such as (3). The surface form of (3), unlike that of (1) or (2), is recognized as being irreproachable. According to this view, the surface structure and the deep structure of (3) would merge, unless Frege, as seems rather unlikely, were to maintain that the non-standard analysis of (3) exhibits its deep structure. As I already remarked, the distinction between "first-level" and "second-level" sentences is not to be construed in an absolute sense. While (1) and (2) admit no syntactic analysis which would reveal either sentence as being of "first-level", (3) can be split up in such a way that it emerges as a "second-level" sentence. Hence, if we were to interpret (3) in such a non-standard fashion, the transition from (1) or (2) to (3) would not involve different levels. In fact, what PNP says would also apply to sentence (3), whose surface form, as Dummett assumes, Frege takes to show the real face of (1) and (2).

Dummett does not specify in great detail how we are to understand the process of revealing the deep structure of an ascription of number. It is not sufficient to say that it consists in construing (1) or (2) as a disguised form of (3). Now, whatever the proposed transition from (1) or (2) to (3) by uncovering the deep structure of (1) or (2) may eventually amount to, it cannot have anything to do with Frege's method of decomposition which he endorsed throughout his academic career;[28] I mean the process (a) of removing some or all occurrences of an expression from a sentence (or, in the more general case, from a complex

[28] Recall thesis (D). I shall discuss it in more detail when I focus on the apparent incompatibility of (C) and (D).

term), and (b) of marking the resulting gap(s) as an argument-place of the appropriate type, leaving a function-name (concept-expression or relation-expression). This process is described, in precise terms, by the three rules governing the correct formation of function-names which Frege states in Section 26 of the *Basic Laws*; I label these rules *principles of gap formation*. Their application results in decomposing a sentence (or complex term) into component expressions, namely into the expression we remove and the function-name we arrive at. For Frege, the syntactic operation of analysis or decomposition goes hand in hand with the process of dividing a complex sense into simpler sense-components. If the analysis of a complex expression proceeds from a sentence, we have, in the realm of sense, a dissection of a thought into thought-components. It is worth emphasizing that the process of dividing a thought into parts is, in effect, inextricably intertwined with the operation of splitting up a corresponding sentence into parts. The inverse does not hold necessarily, as can be seen from the possibility of decomposing sentences of yet uninterpreted artificial languages.

Let me now return to our sample sentences. We can neither extract from (1) or (2) the numerical operator, nor can we discern in either sentence the equality sign. The numeral "4" occurs adjectivally as an inseparable element of a second-level predicate, that is, we are not entitled to remove "4" from (1) or (2) by applying Frege's first principle of gap formation.[29] Hence, neither sentence admits logical analysis as a numerical equation.

At first sight, one might try to interpret the transformation of (1) into (3) along the lines of the transition from (A1) "$a \parallel b$" to (A2) "$D(a) = D(b)$" or from (B1) "$E_x(F(x), G(x))$" to (B2) "$N_xF(x) = N_xG(x)$" which Frege suggests in Section 64 of the *Foundations*. There he maintains that one and the same (judgeable) content can be carved up in distinct ways and so emerge in sentences of significantly different logical form. Frege describes this process as follows: "We [...] replace the symbol $\parallel$ by the more general symbol =, by distributing the particular content of the former symbol to *a* and to *b*. We carve up the content in a way different from the original way, and thereby obtain a

[29] According to this principle, we remove from a complex object-name a proper name that forms a part of it or coincides with it, at some or all of the places where it occurs, and render recognizable the gap(s) as a single argument-place of type 1, i. e., appropriate to admit proper names. We thereby obtain a one-place function-name of first level.

new concept" (FA, GLA, 74 f.).[30] I find this hard to follow. What we are certainly given here is the transition from one mode of speaking to another, yielding the direction operator. Both modes of speaking are said to have the same content. It is equally obvious that it cannot be claimed that the transition from (A1) to (A2), as described by Frege, is a purely linguistic operation. It is rather a mixed operation, carried out both on the level of signs and on the level of content. In particular, the transition cannot be regarded as embodying any process of genuine decomposition as specified above.[31] Independently of the vagueness I see in Frege's way of characterizing the transition from (A1) to (A2), this characterization fails to provide a clue as to how we are to understand the transformation of (1) into (3). For the capacity to be carved up in terms of an operation by which a symbol is replaced and a particular content distributed, seems to be a privilege which only very special judgeable contents enjoy, namely those expressible by an equivalence statement. And to be sure, Frege denies that (A1) and (B1) can be construed as disguised ways of expressing (A2) and (B2). Yet Dummett claims precisely that Frege regards (1) or (2) as disguised forms of (3). So, there is ample evidence that the transition from (A1) to (A2) in no way serves as a model for disclosing the nature of the transformation of (1) into (3).

---

[30] Applied to the transition from (B1) to (B2), a sign for a second-level relation would be replaced by a sign for a first-level relation.

[31] This would be more conspicuous if one were to interpret the transition both from (A1) to (A2) and from (B1) to (B2) as one yielding new abstract *objects*. Frege comes close to putting it this way when he comments on the transition from an equivalence relation to an equality embodied in Axiom V. He would, however, refuse to describe the process of logical abstraction as one *creating* new logical objects; see TF, 179 ff.; GGA II, 147 ff. Some questions concerning the transformation from (A1) into (A2) remain. What is it to mean precisely that the particular content of "=" is distributed between a and b? What is the *original* way of splitting up the content of (A1) and (A2) supposed to be? As to the second question, the content or thought as expressed by the original sentence (A1) could be divided into the senses of "a", "b" and "||" by applying the method of decomposition as described above. It would be awkward, however, to see this operation as bringing it about that the thought in question emerges in (A1). To repeat: there simply is no way of decomposing a thought independently of analyzing a sentence which expresses it. In particular, the decomposition of the thought expressed by (A1) and (A2) into the senses of "a", "b" and "||" is only possible by analyzing (A1) into just these three expressions. So even if we were to accept the mixed operation by means of which Frege tries to explain the transition from (A1) to (A2), we could only say that the content of the original sentence (A1) can be carved up in such a manner that it emerges in a sentence of different logical form, in (A2) that is.

To conclude our prolonged discussion of the intricate issues contained in Section 57: Could Frege have adopted a position like this? Ascriptions of number are to be acknowledged as numerical statements in their own right. In such statements a numeral occurs adjectivally as an undetachable part of a second-level predicate and as such has no referential role. To admit predications of concepts such as (1) and (2) as numerical statements in their own right does not threaten, let alone undermine, the conception of numbers as objects. Now, my answer to our question is clearly "yes". Frege could have accepted a position like this and, indeed, without residual uneasiness about the adjectival occurrence of a numeral.

In Section 3.3 we shall see that Frege defines the number belonging to the concept F, after two failed attempts at definition, as the extension of the concept *equinumerous with the concept F*. While in his very last writings he jettisoned extensions of concepts altogether and turned, albeit in a rather fragmentary fashion, to a geometrical foundation of arithmetic, he never abandoned his conception of numbers as objects (cf. PW, 269, 278 ff.; NS, 288 f., 298 ff.). It is true enough that Frege expresses some doubts about the objecthood of numbers, but these seem to be mainly motivated by his growing scepticism about the viability of his logicism, ultimately to be totally rejected, not by new, clear-cut insights into the logical form of numerical terms and statements (cf. PW, 256 f., 263; NS, 277, 282). It therefore comes as no surprise that in his fragment 'Number' of September 1924 he reiterates an insight that had guided him in the *Foundations*. By paying due attention to the use we make of numerals and number-words, we may seek to discover something about the nature of numbers themselves. Since number-words and numerals are used, like names of objects, as proper names, by a number we are to understand an object (that cannot be perceived by the senses) (cf. PW, 265; NS, 248 f.).

2.6 The context principle and two other theses held by Frege

Generally, the Mathematical-Object Theory cannot claim to be sacrosanct, especially when we scrutinize its methodological underpinnings. It is easily conceivable that the true logical form of number-theoretic sentences is to be revealed under the cloak of their possibly misleading surface structure. In his stimulating article 'Logicism and the Ontological Commitments of Arithmetic' (1984), Harold Hodes argues in this vein.[32] There he launches a vigorous attack on the alleged legitimacy of the Mathematical-Object Theory. The fact that we cannot provide any causal account of reference for the case of numbers qua abstract objects constitutes a challenge for the platonist or the mathematical-object theorist to furnish an alternative account. Hodes seems to believe that Frege's method of conferring a definite reference, e. g., upon a numerical singular term, by fixing the sense or the truth-conditions of all relevant sentences in which it figures, is after all the only account to which the platonist may appeal in order to meet the challenge. Yet, as Hodes maintains, this account falls short of explaining the "microstructure" of reference, for example, to cardinal numbers. That is, it does not explain what our ability to refer to such objects consists in (cf. Hodes 1984, 126 ff.).

In this paper, I cannot examine Hodes' argument in detail. Whatever its purported strength may be, it rests partially on an outright misinterpretation of Frege's context principle.[33] Now, in order to show that Hodes misconstrues the context principle in an important respect, I should begin by making some brief remarks about both its nature and its role in the *Foundations*.[34] These remarks will also prove to be useful when we come to examine Frege's tentative contextual definition of the term "$N_xF(x)$" in part 2.

During the period of the *Foundations*, Frege had not yet drawn a terminologically strict distinction between reference *(Bedeutung)* and sense *(Sinn)*. There is, however, considerable evidence that his main concern was to apply the context principle to the *reference* of words. Taken as a thesis concerning reference, it states the general condition that a word (singular term or predicate) must satisfy in order to have a

---

[32] In his 1990 paper, Hodes develops, in a model-theoretic framework, the constructive ideas sketched in his 1984 paper. According to his theory, numbers are, roughly speaking, "fictions created to encode cardinality quantifiers, thereby clothing a certain higher-order logic in the attractive garments of lower-order logic" (350). Spelled out more fully: numerical singular terms, which appear to perform the semantic function of designating numbers, really do the different work of encoding cardinality quantifiers; quantifiers, which appear to quantify over numbers, really encode higher-order quantification over cardinality quantifiers; and numerical predicates, which behave syntactically like any other predicate of first level, really do the semantic job of expressions of higher levels. It would be interesting to compare Hodes' approach with David Bostock's quantificational account of the natural numbers (cf. Bostock 1974), but I must leave this for another occasion.

[33] For a critical discussion of Hodes' argument see Hale 1987, 220 ff.

[34] For a discussion of the role of the context principle in the *Basic Laws* see Schirn 1985, 1996.

reference at all. Thus interpreted, the principle embodies the primacy of sentence-reference over word-reference in the following sense: the reference of a subsentential expression consists exclusively in its contribution to determining the reference of any sentence in which it may occur. In the *Foundations*, the contextual thesis plays both a critical and a constructive role. Its critical function is to reject a psychological conception of word-reference in general and to avoid a physical view of number in particular, without slipping into a psychological view. The constructive purpose of the principle is likewise twofold. Firstly, it serves as a guideline for the logical analysis of numerical statements. Secondly, it is expressly used to defend the legitimacy of ascribing determinate references to abstract singular terms. More specifically, the context principle is intended to pave the way for Frege's introduction of numbers as logical and yet objective and self-subsistent objects. For Frege's strategy seems to be precisely this: If it has been established, by acceptable criteria for singular termhood, that number-words behave logically as singular terms, and if we have succeeded in fixing the sense of all relevant sentences in which a number-word "a" may occur, in particular the sense of all "recognition-sentences", then there can be no further legitimate question as to whether "a" really does have a reference.

Construed as a principle concerning sense, the contextual thesis lays down that the sense of a word consists in its contribution to the sense of any sentence in which it may occur. Just as the reference of a sentence possesses primacy over the references of subsentential components, so does the sense of a sentence enjoy primacy over the senses of the words out of which the sentence is put together. We cannot grasp the sense of a word save as that of a linguistic unit which is capable of forming a part of a sentence; but we understand the word independently of any particular sentence containing it. Plainly, it would be implausible to say that the sense (reference) of a sentence S consists in its contribution to the sense (reference) of a more complex sentence in which S may be embedded or to the sense (reference) of a sequence of sentences of which S may form a part. This observation, simple as it is, accounts for the distinguished role that the context principle under the two interpretations assigns to sentences. According to Frege (BLA 90; GGA I, 50), the sense of a truth-value name S is the thought that the conditions under which S denotes the True are fulfilled. The thought which S expresses can be grasped on its own; to grasp it is to know the truth-conditions of S. And the reference of S just consists in its being true or false. While after 1892 Frege constantly subscribed to the rationale that the senses of the parts of a sentence are parts of the thought expressed by the sentence (BLA, 90; GGA I, 51; PW, 207 f.; NS, 224), he rejected the analogue in the realm of reference: the references of sentence-constituents cannot reasonably be conceived of as constituents of the reference (i. e., truth-value) of the whole sentence (PW, 255; NS, 275).

Contrary to what Hodes claims, from the supposed fact that the context principle is intended to apply "to all cases of reference", it does not follow that Frege is "committed to the [implausible] claim that all thought-grasping is prior to all grasping of the senses of subsentential constituents and so prior to all reference; [...]" (Hodes 1984, 128). Hodes' attempt to salvage Frege from this apparently untenable position by restricting the scope of the contextual thesis to *mathematical* singular terms and propositions is a misguided manoeuvre. It defies credibility. Even in its quite general form, the context principle is perfectly compatible with Frege's doctrine, embodied in thesis (C), that we build up a thought out of constituents, which correspond, by and large, to the parts of the sentence expressing it. (C) appears only in Frege's late writings. It is expressed by means of the metaphor "building blocks" and designed to explain our ability to express "with a few sounds and combinations of sounds [...] a huge number of thoughts, and, in particular, thoughts which have not hitherto been grasped or expressed by any man" (PW, 225; NS, 243).

Thesis (C) implies that we grasp the sense of a sentence on the strength of our prior knowledge of the senses of its constituent expressions and the manner these are structurally combined to form the sentence. It is precisely because we cannot provide a satisfactory account of the sense of a word taken in isolation that our ability to understand new sentences by virtue of our mastery of both a finite number of semantically primitive expressions and a finite number of syntactical rules requires explanation. And it is because an acceptable account of the sense of a subsentential expression is in terms of its contribution to the thoughts expressed by sentences in which it figures that such an explanation is forthcoming.[35] Were it not that the sense of a word has to be explained as consisting in its contribution to the sense of a *sentence* containing it, Frege would have no compelling reason to represent it as a constituent of a *thought*.

[35] Here I am arguing along the lines of Dummett 1981, 373.

The upshot, then, is that Hodes fails to advance any cogent argument for his claim that the context principle, if taken to apply to subsentential expressions in general, is at variance with (C) and consequently untenable. Far from being flatly inconsistent with one another, thesis (C) and the context principle, construed as a principle concerning sense, work hand in hand. This is not to say that the role which Frege assigns to the context principle in the *Foundations* is unchallengeable, however.

On the face of it, thesis (C) appears to clash with Frege's likewise fundamental doctrine (D). (D) appears in several of his writings and was originally formulated in terms of the notion of judgeable content or judgment. It is frequently called *the thesis of the priority of judgeable contents over concepts*. In his 'Notes for Ludwig Darmstaedter', Frege states (D) as follows: I do not begin with concepts and put them together to form a thought or judgment; rather, I arrive at the parts of a thought by decomposing it (PW, 253; NS, 273).[36] From an epistemic point of view, (D) lays down that our grasp of the senses of the predicates of a given language L does not precede our grasp of the senses of complete sentences; rather, we arrive at or grasp the senses of the predicates of L only by decomposing an *antecedently* grasped thought into thought-components. Although there exists a certain kinship between the context principle and the priority thesis, these two doctrines ought not to be conflated or assimilated into one another.[37] I shall describe the distinguishing marks of the two principles in due course. But let us first see whether the apparent head-on conflict between (C) and (D) can be resolved.

In order to prevent misunderstandings, let me begin by emphasizing once again what the process of gap formation, by its very nature, is intended to be. It serves exclusively the formation of incomplete expressions, i. e., of function-, concept- and relation-names, along with the extraction of incomplete sense-components. Owing to their saturated nature, object-names are never formed by means of gap formation, i. e., they are never extracted from more complex expressions by

---

[36] Frege's terminology is rather loose here. Obviously, in that context he does not use the word "concept" in the sense of his theory of reference, according to which a concept is the reference of a concept-expression. *Concept* means here (*simply*) *unsaturated thought-constituent*, i. e., what in his theory of sense he calls the sense of a concept-word. For further information on the terminology employed in (D) see Schirn 1984, 195 f.

[37] Sluga 1980, 94 ff. and Currie 1982, 19 f. seem to be prone to commit this error; cf. Schirn 1983b, 1984, 1987.

omitting parts of the latter. In a nutshell: it is not the name we remove, but the name we arrive at that counts in every process of genuine gap formation. A suitable alternative label for this powerful syntactic device and its analogue in the realm of sense would be *the method of the extraction of function-names and of unsaturated sense-components*. It is this method, as relating to the extraction of concept- and relation-expressions (or of unsaturated thought-components) from sentences (or from thoughts), that the priority thesis embodies.

If Frege held (D) to be universally valid, that is, if he insisted that it expresses the priority of thoughts over the unsaturated senses of all predicates of a given language L, he would indeed have great trouble reconciling (C) with (D). For clearly the claim that (D) is intended to apply unrestrictedly is tantamount to saying that we grasp the sense of each predicate F of L only by dissecting, in a suitable way, a previously grasped thought expressed by a sentence of L in which F occurs. It seems to me, however, that the question as to the scope of (D) within a *natural* language would be very hard to answer. Matters stand differently for Frege's formal language. If we focus our attention on the syntactic and semantic construction of the system of the *Basic Laws*, we see that Frege does start out from eight *logically simple* functions, concepts and relations, out of which he hopes to unfold the whole richness of functions and objects with which arithmetic deals, as though from a seed (cf. TF, 181; GGA II, 149). It is plain that the names of these primitive functions are not constructed by means of the operation of gap formation, and, consequently, that their simple senses are not arrived at by decomposing a complex sense. Rather, the primitive names are introduced by means of what Frege calls elucidations, and they are, of course, well-formed by stipulation. The first composite expressions of his formal language, namely certain function-value names (truth-value names and course-of-values names), are built up by inserting a primitive one-place function-name of first level (or of second level) into the argument-place of a primitive function-name of second level (or of third level). And the first complex monadic predicates of first level are not extracted from complex object-names by omitting one or more occurrences of some constituent object-name; rather, they are obtained by inserting a proper name (compounded out of two primitive function-names) into one of the empty places of a primitive first-level relational expression.

There is thus considerable evidence that in the formalism of the *Basic Laws* (D) holds only in a restricted sense. It applies neither to the primitive concept- and relation-names and their simple senses, nor to the first complex concept-expressions of first level and their complex senses.[38] (D) applies, in effect, to all predicates which in the last stage of their entire constructional history are obtained by means of gap formation. (Needless to say, the constructional chain of a predicate may involve an iterated application of some gap formation rule.) (D) applies, in particular, to all *simple* names of complex concepts.[39] The result, then, is that in the system of the *Basic Laws* (C) can be brought into line with (D).[40] In Frege's view, a sentence and the corresponding thought are always built up out of sentence- and thought-constituents respectively. In his formal language, however, the unsaturated building-blocks, other than the primitive elements of the system, are not previously given. Rather, the unsaturated building-blocks have to be constructed by means of two interlocking ways of forming a name, by means of insertion and gap formation that is (cf. BLA, GGA I, § 30).

The foregoing exposition reveals in what precise sense (D) differs from the context principle. To summarize: The context principle is meant to apply to subsentential expressions of all kinds, though it is abstract singular terms that Frege focuses on when he comes to apply the principle. The priority thesis, by way of contrast, is concerned only with incomplete expressions. Unlike the context principle, (D) tells us

---

[38] In 'Boole's Logical Calculus and the Concept-script' (PW, 17; NS 19), Frege stresses that the properties and relations which are not further analyzable must have their own simple designations. The ideas of these properties and relations arise simultaneously with the first judgement in which they are ascribed to things.

[39] An example would be the formation of "$\xi = \xi$". Dummett's claim (1981, 294, 314, 539 f.) that Frege did not intend the method of the extraction of function-names to apply to simple predicates and function-names is untenable. This misconstrual is perhaps due to his failure to distinguish between *simple* and *logically simple* function-names with respect to the system of the *Basic Laws*. According to Frege's criterion of simplicity (TF, 171; GGA II, 79; BLA, 129 f.; GGA II, 255), a given function-name is simple if it is not put together out of signs each having a reference (sense) on its own and so contributing to determining the reference (sense) of the whole. In his 1973 book, 24 f., Dummett errs in displaying this criterion as a sufficient condition for a given expression to count as logically simple (cf. the critical remarks in Schirn 1983c, 212 f.). For Frege, a simple expression is *logically* simple if it expresses a simple sense. Although he does not explicitly lay down under what conditions the sense of an expression should be recognized as simple, the simplicity of the sense is certainly to imply that the sense is not further analyzable. Every function-name of the formalism of the *Basic Laws* which has been introduced by means of a definition is simple, but it is not logically simple. It denotes a logically complex function and expresses a complex sense, since the definition has conferred upon it the reference and the sense of a composite function-name. Admittedly, it remains to be explained how Frege could account for the complexity of the sense of a simple predicate like "$\xi = \xi$".

[40] Cf. Schirn 1992a, 1992b. I believe that (C) likewise fits together with (A), but I must refrain from discussing this issue, too.

something about the manner in which predicates are constructed and their senses arrived at. And in contrast to the context principle, (D) provides a plausible ground for Frege's doctrine that concept- and relation-words (along with both their senses and their references qua concepts and relations) are incomplete or in need of supplementation. If the context principle did supply such a ground, it would only apply to predicates. While the context principle, construed as a thesis concerning reference, incorporates the primacy of sentence-reference over word-reference in the sense explained earlier, the priority thesis fails to embody any such primacy in the realm of reference. Finally, the context principle, considered as a thesis regarding sense, includes the primacy of sentence-sense over word-sense in the following way: the sense of a subsentential expression consists wholly in its contribution to the sense of any sentence in which it may figure. Thesis (D), by contrast, displays the priority of sentence-sense over the sense of all those predicates which we extract from sentences by means of gap formation.

2.7 A dogma?

I want to conclude this part with three comments.

Firstly: We have seen that Frege's syntactic arguments in Sections 56–7 of the *Foundations* for regarding numbers as objects do not carry conviction. This does not mean, however, that Frege had no true reason for conceiving of numbers as objects. In his book *Frege: Philosophy of Mathematics*, Dummett shows that the radical adjectival strategy – it refuses to take numerical terms at face-value – can be pursued for a considerable distance in *Grundlagen* by converting the relevant definitions and proofs into their adjectival analogues (1991b, 131 f.). The analogue of the fundamental theorem that every finite number is succeeded by a finite number cannot be proved, however. Dummett succeeds in revealing the true reason for Frege's need to construe numbers as objects. Russell's ramified theory of types required that numbers, conceived of as classes of classes, be distinguished from individuals. He therefore felt obliged to introduce an Axiom of Infinity, stating that there were infinitely many individuals, in order to guarantee that there were infinitely many cardinal numbers. Frege, by contrast, recognized numbers as belonging to the domain of the individual variables. It is only for this reason that he could "spin the infinite sequence of natural numbers out of nothing, as it were" (Dummett, 1991b, 133).

Secondly: It is not evident from Frege's writings to what extent his conception of numbers as objects depends upon their identification with extensions of concepts. It seems that in order to establish the objectual status of numbers it is sufficient, in the light of his own principles, to provide an analysis of the logical form of numerical terms and statements. *Prima facie*, there is no compelling reason why he should have thought that the objecthood of extensions of concepts is scientifically better grounded than the objecthood of numbers. According to Frege's syntactic criteria, both numerical terms and class terms are to be characterized as genuine singular terms. It is by no means beyond doubt, however, that in the *Foundations* Frege believed he had secured the assumed logical nature of cardinal numbers by defining them as extensions of concepts. As we shall see in part 3, it is first and foremost the transition from an equinumerosity holding between first-level concepts to an identity between cardinal numbers which he regards as purely logical. Frege does so on the grounds that equinumerosity or one-one correspondence can be reduced to purely logical relationships, namely to the term "relation". While "concept" and "relation" are said to be expressions which intrinsically belong to pure logic, Frege refrains from making a similar explicit claim about the term "extension of a concept". And we know, of course, that after 1891 he maintained emphatically that the concept is logically prior to its extension (cf. e. g., CP, 228; KS, 209 f.).

Thirdly: It is, of course, Benacerraf who in his celebrated paper 'What Numbers Could Not Be' mounted the first really forceful attack on Frege's thesis that numbers are objects. A few remarks on what I take to be the quintessence of Benacerraf's ontological challenge to the position of the number-theoretic platonist must suffice here. The arithmetical platonist characteristically identifies numbers with sets. Thus, each individual number would be an individual set. However, Benacerraf argues that numbers cannot be sets at all, because there is no good reason to claim that any particular number is any particular set. Obviously, there is no unique set-theoretic introduction of, e. g., the natural or the real numbers. Not only does the identification of numbers with sets appear to be undermined by the arbitrariness of our actual choices, but also by the arbitrariness of our taking sets to be the basic objects of mathematics. It is possible, for instance, to regard ordinal numbers as basic and to define sets in terms of them. By way of extending the conclusion that numbers cannot be sets to objects in general, Benacerraf concludes that numbers are not objects at all. In stating the properties of numbers, we merely characterize an *abstract structure*, and "the distinction lies in the fact that the 'elements' of the structure have no properties other than those relating them to other 'elements' of the same structure" (Benacerraf 1965, 291). According to this view, then, to *be* the number 2, for instance, is no more and no less than to be preceded by 1, and possibly 0, and to be followed by 3, 4, and so forth. *Any* object *can play the role of* 2, that is, any object can be the second element in some progression or ω-sequence. This is highly reminiscent of Dedekind's analysis of number, which is presented as a characterization of the *structure* of the natural numbers. In short, according to Benacerraf, arithmetic is not a science concerned with particular objects; rather, it elaborates the abstract structure which all progressions have in common merely on the strength of being progressions.

On the one hand, Benacerraf's ontological argument does not refute the conception of numbers as objects. His view that we cannot encounter objects which are really identical with the natural numbers appears to be plausible only if we make the assumption that the proposed structural characterization provides all necessary and sufficient properties of the natural numbers, i. e., that there is no other way of characterizing the numbers than a structural one. However, Benacerraf fails to establish this assumption conclusively. The plausibility of the thesis that the structural properties enunciated in his article suffice for the characterization of the numbers is obviously due to the fact that hitherto no further features of the numbers could be found. Benacerraf's ontological argument shows merely that it is impossible to identify the numbers as determinate objects by appeal to the conditions which could be specified up to now. *On the other hand*, all serious objections raised to Benacerraf's ontological argument suffer from one decisive defect: they do not suggest what could be said about the nature of the numbers beyond their structural characterization. Unless someone has succeeded in refuting Benacerraf's argument, the conception of numbers as objects remains a dogma bequeathed to us by Frege.[41]

[41] A detailed discussion of this problem can be found in my planned paper 'Mathematical Platonism: Challenge and Defence'.

3. *The logicist definition of number and the referential indeterminacy of numerical singular terms*

3.1 First attempt: Definitions of the natural numbers

Guided by his key sentence on ascriptions of number PNP, Frege suggests the following heuristic definitions of 0, 1 and n + 1, i. e., the transition from one number to the next:

(I) $N^0_x F(x) := \forall x \neg F(x)$;
(II) $N^1_x F(x) := \neg\forall x \neg F(x) \wedge \forall x \forall y (F(x) \wedge F(y) \supset x = y)$;
(III) $N^{n+1}_x F(x) := \exists x (F(x) \wedge N^n_y (F(y) \wedge y \neq x))$.

Despite their initial appeal, Frege feels obliged to jettison these recursive definitions as unsatisfactory. He points out that the most likely definition to arouse misgivings will be (III). For, strictly speaking, the sense of the term "$N^n_x G(x)$" is just as unknown to us as that of the term "$N^{n+1}_x F(x)$". To be sure, we can explain by means of (III) and (II) the sense of "$N^{1+1}_x F(x)$" and then, by using this expression, the sense of "$N^{1+1+1}_x F(x)$", and so on. Yet, we can never decide by appeal to (I)–(III) whether Julius Caesar is a number. (I) and (II) fail to define 0 and 1 separately, i. e., they determine only the sense of the second-level predicates "$N^0_x \varphi(x)$" and "$N^1_x \varphi(x)$" (of the statements "$N^0_x F(x)$" and "$N^1_x F(x)$" respectively); "but it is not permissible to pick out in these 0 and 1 as self-subsistent objects that can be recognized as the same again" (FA, GLA, 67). As a result, we cannot prove a numerical identity and, hence, are not entitled to use the numerical term "*the* number which belongs to the concept F".[42]

There is much to find fault with Frege's way of presenting the purported shortcomings of the proposed definitions. I shall confine myself to a few comments. To begin with, the objections raised to (I)–(III) stem mainly from Frege's preconceived view, still in need of conclusive justification, that numbers must be regarded as objects. Yet, contrary to what he attempts to make the innocent reader believe, it could never have been his sincere intention to define the natural numbers qua objects by means of (I)–(III). This applies even more palpably when we rephrase the *definienda* as "There are 0 Fs", etc. Suppose that in examining (I)–(III) Frege had not intended to establish, to some degree, the

---

[42] A shorter account of Frege's logicist definition of number and the referential indeterminacy of numerical terms is given in Schirn 1988; cf. also Schirn 1993.

objectual nature of the cardinal numbers. Then both the objection that with the help of (I)–(III) we cannot prove that if $N_x^a F(x) \wedge N_x^b F(x)$, then necessarily a = b (and, hence, that we are not justified in using the term "$N_x F(x)$"), and the complaint that we are not entitled, in "$N_x^0 \varphi(x)$" and "$N_x^1 \varphi(x)$", to pick out 0 and 1 as self-subsistent objects would patently have missed their mark.[43]

In his illuminating exposition of Sections 55–57 of the *Foundations*, Dummett observes that Frege raises a cogent objection only to (III).[44] Unlike Dummett (1991b, 102), I doubt, however, that Frege believed he had convicted (I) and (II) of being *logically* faulty as well. Frege takes pains to demonstrate merely that (I) and (II) fail to serve his aim of defining 0 and 1 as objects. He is, of course, well aware that the proposed definitions (I) and (II) correctly define the second-level predicates "$N_x^0 \varphi(x)$" and "$N_x^1 \varphi(x)$", that they lay down in a precise manner conditions under which the ascriptions of number "$N_x^0 F(x)$" and "$N_x^1 F(x)$" are true. Seen from this angle, (I) and (II) are, indeed, logically unassailable for Frege.

In my view, Frege's diagnosis of his abortive attempt to define 0, 1 and n + 1 by means of (I)–(III) is designed to lead to the following result: correct definitions of the first numbers qua objects in the natural

[43] Recall that in the *Foundations* Frege appears to endorse a twofold role of a numeral. Firstly, it functions both syntactically and semantically as a genuine singular term in sentences such as "7 + 5 = 12", "3 and 5 are odd", "3 and 5 are the only factors of 15 other than 1" (cf. PW, 86; NS, 95). Secondly, it is used syncategorematically in ascriptions of number. We have seen, however, that ascriptions of number are said to be replaceable by numerical equations. When in his 'Notes for Ludwig Darmstaedter' Frege asks: "Could the numerals help to form signs for these second-level concepts [i. e., for $N_x^0 \varphi(x)$, $N_x^1 \varphi(x)$, ..., $N_x^n \varphi(x)$] and yet not be signs in their own right?" (PW, 257; NS, 277), he must, of course, be aware that as constituents figuring in numerical predicates "$N_x^0 \varphi(x)$", etc. numerals are not signs in their own right. This was already clear in the *Foundations*. The question is, rather, whether numerals and other numerical expressions ought to be recognized as proper names of objects concerning their occurrence in sentences like those I quoted above. Frege's position in his 'Notes' is that the second-level concepts $N_x^0 \varphi(x)$, $N_x^1 \varphi(x)$, etc. do not, after all, constitute the numbers of arithmetic; "we do not have objects, but concepts. How can we get from these concepts to the numbers of arithmetic in a way that cannot be faulted?" (PW, 257; NS, 277). The ensuing question "Or are there simply no numbers in arithmetic?" I take to be merely rhetorical. In the course of dealing with Russell's paradox, Frege discusses the possibility of regarding class-names and numerals as pseudo proper names, i. e., as parts of signs which had a reference only as wholes (BLA, 129 f.; GGA II, 255). He rejects this route as impassable. Let me add that I find his reference to Vol. I, § 29 of the *Basic Laws* puzzling. It by no means follows from his criteria of referentiality that proper names and function-names are merely syncategorematic parts of a sign.

[44] He shows (1991b, 103 ff.) how the defect contained in (III) can be remedied.

sequence of numbers presuppose a prior unobjectionable definition of the term "$N_xF(x)$". If we cast a glance at the definitions of the natural numbers

(IV) $0 := N_x(x \neq x)$

and

(V) $1 := N_x(x = 0)$

(and

(VI) $n + 1 := N_x(x = 0 \vee \ldots \vee x = n)$),

which Frege later sets up (FA, GLA, §§ 74, 77), we may wonder why he did not put forward these definitions in Section 55. For in (IV) and (V), the numerals "0" and "1" do not occur as undetachable parts of second-level predicates, but are defined separately. Hence the main objection which Frege raises to (I) and (II) would be met (cf. FA, GLA, § 106). On reflection, it becomes clear, however, that at the stage of Section 55 he would have had to make the questionable assumption that the sense of "$N_xF(x)$" is already known to us.[45] Put succinctly, my interpretation of Frege's strategy in Sections 55 and 56, then, is this. By starting with the definitions (I)–(III), inconclusive as this manoeuvre must appear to the critical reader, Frege pursues a twofold goal. Firstly, he wishes to gain a vantage-point for establishing his conception of numbers as objects. Secondly, he aims to suggest the need for defining the term "$N_xF(x)$", if his logicist programme is to be carried out on a sound basis. For only by framing a definition of "$N_xF(x)$" which is methodologically prior to the definitions of the natural numbers does Frege think we can attain a general means to get hold of any determinate number.

3.2 Second attempt: The contextual definition of "$N_xF(x)$"

According to Frege, it is a cardinal prerequisite for any methodologically sound introduction of abstract objects to state an adequate

[45] In the light of what Frege says in Section 57 (recall the heading to it!), it is natural to ask why in his first attempt to define 0, 1 and n + 1 he does not replace the *definienda* "$N_x^0F(x)$", "$N_x^1F(x)$" and "$N_x^{n+1}F(x)$" by "$N_xF(x) = 0$", "$N_xF(x) = 1$" and "$N_xF(x) = n + 1$". If he had done so, his principal objection raised to (I) and (II) would equally have applied to the alternative definitions (I') "$N_xF(x) = 0 := \forall x \neg F(x)$" and (II') "$N_xF(x) = 1 := \neg\forall x \neg F(x) \wedge \forall x \forall y (F(x) \wedge F(y) \supset x = y)$". Furthermore, Frege could have underlined the need to define "$N_xF(x)$", in the first place, before giving definitions of "0", "1" and "n + 1" in terms of "$N_xF(x)$". (I') and (II') are clearly contextual definitions; as such they infringe Frege's later principle of the simplicity of the *definiendum*.

criterion of identity for them, i. e., to provide a general means to conceive them, to recognize them again as the same, and to distinguish them from any other objects. The second likewise heuristic attempt to define number is intended to satisfy this precondition. For Frege, this amounts to fixing the sense (the truth-conditions) of a numerical equation "$N_xF(x) = N_xG(x)$" without using a number-word in the *definiens*. In order to license such a contextual definition of "$N_xF(x)$", he expressly invokes his context principle. Thus, he proposes the following definition ("E" is to abbreviate "equinumerous"):

(VII) $N_xF(x) = N_xG(x) := E_x(F(x),G(x))$.

Equinumerosity is later defined in terms of one-one correspondence. It is noteworthy that (VII) figures only as a *partial* contextual definition of "$N_xF(x)$", to be supplemented by explanations of all other relevant statements S about numbers. Like (VII) itself, those additional contextual definitions of "$N_xF(x)$" would be subject to the condition of substitutivity *salva veritate*: $E_x(F(x),G(x)) \supset S[N_xF(x)] \equiv S[N_xG(x)]$. As a matter of fact, Frege repudiated the legitimacy of contextual and piecemeal definitions after having developed a systematic theory of definition in the *Basic Laws*. Contextual definitions violate his principle of the simplicity of the *definiendum*, whereas piecemeal definitions offend against his principle of completeness.

Frege discusses the difficulties involved in the transition from an equivalence relation to an identity by taking parallelism and identity of directions as his paradigm for illustration. In what ensues, I shall transfer the main points of his discussion to the case of equinumerosity and numerical identity. Certainly, in doing this we ought not to lose sight of the essential differences which do exist between the two cases, in spite of their basic affinity.

Frege points out that (VII) raises three logical doubts. The first two can be met. The third doubt is that the proposed criterion of identity embodied in (VII) fails to cover all conceivable cases. It enables us to determine the truth-value of only those equations in which the expressions flanking the identity-sign are both of the form "$N_xG(x)$". Yet, like the tentative definitions (I)–(III), (VII) is powerless to decide whether, say, Julius Caesar is the number of planets. We may call this *"the Julius Caesar problem"*. Strictly speaking, (VII) does not even provide us with a means to determine the truth-value of a numerical equation such as "9 = the number of planets". For it is evident that the definitions of the natural numbers (IV)–(VI) presuppose a prior unobjectionable definition of "$N_xF(x)$". If we already had a definition of the concept *n is a number* (N(n)) satisfying Frege's principle of completeness, i. e., the requirement of sharp delimitation, so that it is determined for every object whether it falls under that concept or not, we could stipulate: if q is not a number, then "$N_xF(x) = q$" is false; if q is a number and if it is given to us appropriately, then (VII) will settle the truth-value of "$N_xF(x) = q$". At this stage of the inquiry, however, the indeterminacy arising from (VII) cannot be removed by setting up the definition

(VIII) $N(n) := \exists\varphi(N_x\varphi(x) = n)$,

since the numerical operator "$N_x\varphi(x)$", forming a part of the *definiens*, has not yet acquired a determinate meaning. In other words, in order to apply (VIII), we would first have to know in each case whether "$n = N_xF(x)$" is true or false.

Before making his final attempt to define the term "$N_xF(x)$", Frege considers a possible solution of the problem that by appeal to definition (VII) we cannot determine the truth-value of a sentence "$N_xF(x) = q$". One might be tempted to stipulate that the object q is a cardinal number if it is introduced by means of definition (VII). Frege dismisses the proposal on the grounds that we would be treating the way in which q is introduced as a property of q, which it is not. "The definition of an object as such really says nothing about it; rather it lays down the meaning of a symbol. [...] If one were to choose this way out, one would be presupposing that an object could be given in one way only; for otherwise it would not follow from the fact that q was not introduced by means of our definition that it could not be so introduced" (FA, GLA, 78). Frege adds that the import of any equation would then be that what is given to us in the same way should be recognized as the same. But this is so unfruitful that there is little to be gained by stating it. In Section 10 of the *Basic Laws*, where Frege attempts to solve the problem that his informal stipulation corresponding to Axiom V does not determine completely the references of course-of-values terms, he invokes what is basically the same argument as in Section 67 of the *Foundations*. In the second footnote to Section 10, he refuses to allow the tentative stipulation $\grave{\varepsilon}(\Delta = \varepsilon) = \Delta$ (i. e., the identification of an object Δ with its own unit class) to hold only for objects not given as courses-of-values by insisting that the way in which an object is given must not be regarded as an immutable property of it.

At any rate, Frege rejects (VII) only on the grounds that his third objection has proved sound. Of course, to deprive (VII) of its status as a definition of the term "$N_xF(x)$" does not amount to dispensing with the equivalence (T) "$N_xF(x) = N_xG(x) \equiv E_x(F(x),G(x))$" altogether.[46] On the contrary, as a provable theorem, (T) is to play a crucial role in the intended logical construction of number theory (cf. FA, GLA, § 73). Furthermore, in Section 104 of the *Foundations*, Frege declares the determination of the sense of a recognition-judgment as a guiding principle for defining fractions, irrational numbers, and complex numbers in purely logical terms. A glance at the structure of Axiom V of the *Basic Laws* – the exact formal analogue of (T)! – shows clearly that the transformation of an equivalence relation into an identity was and remained for Frege the means *par excellence* for introducing logical objects. And as may be expected, Axiom V fares no better than (T) with respect to fixing uniquely the references of abstract singular terms of a certain kind.

The numerical operator "$N_x\varphi(x)$", unlike the expression "the direction of x", is a *second*-level function-name; correspondingly, equinumerosity is, in contrast to parallelism, an equivalence relation between (first-level) *concepts*. It is precisely for this reason that (VII), unlike the proposed contextual definition of "the direction of line a": $D(a) = D(b) := a \parallel b$ (cf. FA, GLA, § 65), gives rise to another difficulty that seems to have gone unnoticed by Frege. While the contextual introduction of the direction operator does not require the existence of any more of the new objects (directions) than there were of the old ones (lines), the introduction of the numerical operator requires the existence of $n + 1$ new objects (numbers), given n objects of the original kind (cf. Dummett 1991b, 138). In fact, the introduction of "$N_x\varphi(x)$" relies crucially on the assumption that the domain of the first-order variables is infinite. While sentences involving the notion of direction could always be translated into sentences not involving it, we cannot do the corresponding sort of thing for sentences involving the notion of number. Definition (VII) provides no way of eliminating the numerical operator in all cases in which it is applied to a predicate itself containing one or more occurrences of "$N_x\varphi(x)$". In general, (VII) fails to supply any means of eliminating the operator "$N_x\varphi(x)$" from an expression of the form "$N_xF(x) = x$", where "x" is either a free or a bound variable. The other horn of the dilemma is this. To impose certain restrictions on the

46 What I refer to as the equivalence (T) is now often called "Hume's Principle". See Dummett's rejection of this label in his forthcoming article 'Neo-Fregeans: in Bad Company?'.

formation of numerical terms would constitute a serious threat to Frege's proof of the infinity of the natural sequence of numbers.

3.3 The explicit definition of "$N_xF(x)$"

Facing the predicament resulting from his third objection to (VII), Frege suggests a way out of it by eventually defining the number which belongs to the concept F as the extension of the (second-level) concept *equinumerous with the concept F*, that is, as an equivalence class of the relation of equinumerosity. (Henceforth, I use "λ" as a class-forming operator.)

(IX) $N_xF(x) := \lambda\varphi(E_x(\varphi(x),F(x)))$.

This explicit definition is obviously designed to conform to (T). "$N_xF(x)$" is defined precisely in terms of that equivalence relation which (T) displays as the criterion of identity for cardinal numbers. So, the definition of equinumerosity in terms of one-one correspondence, which Frege eventually puts forward in Section 72, is methodologically prior to the definition of the numerical operator.[47] This only confirms what is already manifest from a cursory glance at his attempted contextual definition (VII); here, too, the order of explanation is such that equinumerosity is taken to be conceptually prior to the cardinality function $N_x\varphi(x)$.

"The number of Fs" is an expression of long-established use. But just how faithful is (IX) to our understanding of that expression? As far as our ordinary, unreflective linguistic practice is concerned, the class of all concepts equinumerous with a concept F appears to be less familiar than the number belonging to the concept F. However true this may be, I do not think it is justified to interpret Frege as intending definition (IX) as an explanation of our intuitive, pre-theoretical understanding of

---

[47] In his Introduction to the *Basic Laws*, Frege confirms this expressly: "I have reduced the cardinal number [Anzahl] to the relation of equinumerosity, and reduced the latter to many-one correspondence" (BLA, 31; GGA I, 3). It should be noted that the context principle is certainly compatible with framing explicit definitions. When Frege rejects the proposed contextual definition of "$N_xF(x)$" in favour of an explicit one he does not, at the same time, dismiss the context principle as useless. To assume that he does would be short-sighted. In the *Foundations*, the context principle serves, in effect, as a useful guideline for arriving at the right form for an explicit definition of "$N_xF(x)$".

the numerical operator.[48] Certainly, not everybody would agree that (IX) does capture the sense we ordinarily attach to the expression "the number of Fs", whatever such sense may be. And many if not most of Frege's fellow mathematicians would probably have doubted that (IX) only renders explicit what is involved in our pre-analytical grasp of that expression. So much, at least, is clear: when Frege sets about criticizing the doctrines held by contemporary mathematicians he complains that not even a minimal degree of agreement exists about the nature of the most fundamental concepts of arithmetic. And he goes on to show that this is most striking in the case of the concept of number.[49]

To my mind, it is plausible to see Frege as intending definition (IX) as a precise characterization of the sense of a familiar simple expression, which usually is not clearly grasped, "but whose outlines are confused as if we saw it through a mist" (PW, 211; NS, 228). Understanding (IX) in this way is similar to regarding it as an *explication* in Carnap's sense.[50] Frege's proposed explicit definition of "$N_xF(x)$" does, however, rest on the questionable assumption that we intuitively know what the

---

[48] This seems to be Resnik's opinion in his 1980 book, 181. Frege concedes that the correctness of (IX) will, at first perhaps, not be so clear. He asks: "Do we not think of the extension of a concept as something different?" His answer is: "What we think of it becomes clear from the basic statements that can be made about them. They are the following: 1. that they are equal, 2. that the one is wider than the other" (FA, GLA, 80). Concerning equations between extensions of concepts and equations between numbers there is complete agreement. As to the second class of statements, we certainly do not say that one number includes another, in the sense in which the extension of one concept includes that of another. However, it is impossible for a case to occur that $\lambda\varphi(E_x(\varphi(x),F(x)))$ would include $\lambda\varphi(E_x(\varphi(x),G(x)))$. Frege admits that the class $\lambda\varphi(E_x(\varphi(x),F(x)))$ might be more or less inclusive than the extension of some other concept, which could then not be a number according to his definition. Furthermore, "it is not usual to call a number more or less inclusive than the extension of a concept" (FA, GLA, 81). Nevertheless, Frege feels free to adopt such a way of speaking should the occasion arise.

[49] Cf. PW, 215 ff.; NS, 232 ff.; see also PW, 72 ff., NS, 81 ff.; cf. further FA, GLA, part II.

[50] An explication of a concept serves to render more precise a predicate of natural language, whose meaning is not sharp or unambiguous, so that it can be used in an exact theory. The process of semantic specification and refinement is dependent on part of the original meaning of the predicate standing in need of explication. Thus, such a semantic specification cannot be regarded as an arbitrary stipulation of meaning, in contrast to what Frege calls a constructive definition. In my 1989 paper, I argue that in the light of Frege's distinction between constructive and analytic definitions drawn (with some reservations) in 'Logic in Mathematics', (IX) proves to be neither a constructive nor an analytic definition. In the *Basic Laws*, by way of contrast, the definition of the numerical operator "$N\xi$" is clearly a constructive one. See Schirn 1989, 73–77 in support of both claims. As to the status of definition (IX), see also Dummett, 'Frege and the Paradox of Analysis', in Dummett 1991a, 17–52.

extension of a concept is. Undoubtedly, at the time when Frege wrote the *Foundations*, he could not rely on a commonly accepted view of the nature of extensions of concepts, let alone of the nature of numbers. It is quite true that the prevailing view among contemporary mathematicians and logicians assumed the extension of a (first-level) concept to consist of objects falling under the concept. But it seems more than likely that Frege did not want to commit himself to this view. For if, in the context of his logicism, numbers are to be identified with extensions of concepts, then the latter must be of a purely logical character. The extension of a first-level concept, however, conceived of as a logical object, cannot consist of physical objects; yet this is precisely what the adherent of the predominant conception of extensions has to concede, if the objects falling under the concept are of a physical nature. Moreover, the proponent of this conception would not be justified in recognizing such a thing as the extension of an empty concept. It seems quite clear that already during the period of the *Foundations* Frege does not regard agglomerations of physical objects or, more generally, wholes made up of parts as something belonging to logic.

The upshot, so far, is evident. The rigorous standards that Frege applies to his own enquiry into the foundations of arithmetic require the introduction of new logical objects in a methodologically satisfactory manner. Thus, his assumption concerning extensions of concepts overshadows, at a crucial point, his reductionist enterprise as outlined in the *Foundations*.

On the one hand, Frege introduces extensions of concepts in order the solve the Julius Caesar problem for cardinal numbers. He stresses, moreover, the key role of extensions of concepts for the envisaged definitions of fractions, irrational numbers, and complex numbers. On the face of it, it could even seem that Frege hopes to establish the assumed logical character of the natural numbers by identifying them with equivalence classes of a relation which he considers to be purely logical and of vital importance for the intended logical construction of cardinal arithmetic. On the other hand, towards the end of his book he confesses that he attaches no decisive importance to bringing in extensions of concepts at all. This uncertain attitude towards the role of extensions is also manifest in his expressing the belief that in definition (IX) the words "extension of the concept" could be replaced with "concept" (FA, GLA, § 68). Frege's line of argument does not carry conviction here. In fact, the substitution he considers regarding (IX) leads to insuperable difficulties. For the sake of simplicity, I refer to this substitution by means of the shorthand expression "Sub". Sub yields an alternative definition of "$N_xF(x)$":

(X) $N_xF(x) :=$ the concept $E_x(\varphi(x),F(x))$.

Frege concedes that Sub may give rise to two objections: 1. that it stands in contradiction to his earlier assertion that the individual numbers are objects; 2. that concepts can have the same extension without coinciding. He believes that both objections can be met, but offers no clue whatsoever as to how this might be achieved. Frege's attitude towards solving the puzzle arising from Sub appears to be one of sheer laziness or embarrassment, as is also manifest from what he says in a preliminary draft of his article 'On Concept and Object'. There he observes that Sub was only an incidental remark on which he had based nothing, in order not to have to grapple with the misgivings to which it might have given rise. And in a footnote he adds that the question whether one should accept Sub is one of expediency (cf. PW, 106; NS, 116). Yet this is not so. Whatever Frege might have meant precisely by "expediency" in this context, the question is not merely of marginal interest. In what follows, I want to show that he would have been well-advised to disentangle his ideas concerning Sub before embarking upon any comment on it.[51]

In a passage of 'On Concept and Object' (1892), Frege denies, without further explanation, that by considering Sub he had intended to identify concept and extension of concept. He adds: "Note carefully that here [that is, in definition (IX)] the word 'concept' is combined with the definite article" (TF, 48; KS, 172). Why Frege sidesteps the issue of whether an expression of the form "the concept F" denotes the extension of F or not remains unfathomable. At any rate, closer scrutiny reveals that he apparently could have met both objections to Sub by appealing to the identification of concept and extension, but without gain for an adequate introduction of extensions qua logical objects and only by facing intolerable consequences.

To see why this is the case, we must first distinguish between an *extensional* and an *intensional* view of concepts. Let us assume, for the sake of argument, that the extensional view identifies a concept with its extension while the intensional view identifies a concept with its content. (In the strict sense, it is, of course, the concept-word that has a content.) In the light of this distinction, the second objection men-

[51] For a more detailed account see Schirn 1983a.

tioned by Frege arises from an intensional interpretation of concepts. If the statement "$N_xF(x) = N_xG(x)$" is true, then, on the basis of definition (X), the equation "The concept $E_x(\varphi(x),F(x))$ = the concept $E_x(\varphi(x),G(x))$" must also be true. Now, if a concept is identified with its intension, then there are pairs of co-extensive concepts F and G such that F and G, and, consequently, $E_x(\varphi(x),F(x))$ and $E_x(\varphi(x),G(x))$, do not coincide. Hence, this identification yields the "paradoxical" result that one and the same number is identical with two distinct, though co-extensive, concepts. Under an extensional interpretation of concepts, however, the second objection, that concepts can have the same extension without coinciding, misses its mark. Even the first objection could apparently be met, since in view of the identification of concept and extension Sub would not clash with Frege's basic tenet that numbers are objects. Evidently, according to his syntactic criteria, Frege could maintain that the *definiens* of (X) is a proper name and, therefore, of the same logical type as the *definiendum*. But this is just the minimal condition that (X) must satisfy in the first place in order to be acceptable on formal grounds. Yet the identification of a concept with its extension is incompatible with Frege's principle that concepts and objects are fundamentally different. To use his well-known metaphor: concepts are unsaturated whereas their extensions are saturated. Moreover, under an extensional interpretation of concepts it would be pointless to consider Sub at all. For the purpose of Sub could be only to avoid the dubious identification of numbers with extensions of concepts.[52]

Let us summarize: Under both an intensional and extensional interpretation of concepts, Sub turns out to be illicit. It could be useful in the *Foundations* only if (a) extensions of concepts were excluded altogether from Frege's further investigation; (b) no other "problematic" objects such as Frege's mysterious "objects of a quite special kind" (TF,

[52] After his introduction of courses-of-values in the *Basic Laws*, Frege endorsed an extensional view of (first-level) concepts which deviates from the ordinary one in set theory. Two concepts F(x) and G(x) with the same extension never coincide in the sense of identity, but stand in the second-level relation of mutual subordination or co-extensiveness which is the analogue of the first-level relation of identity: $\forall x(F(x) \equiv G(x))$. See 'Comments on Sense and Reference' (PW, 118–125; NS, 128–136) where Frege aligns himself basically with the extensionalist logicians. In any (extensional) declarative sentence, we can substitute *salva veritate* a concept-word for another, if the same extension corresponds to both. Frege observes that this circumstance seems to speak very much in favour of extensionalist as against intensionalist logicians. It implies that in relation to logical laws and inferences concepts differ only in so far as their extensions are different.

50; KS, 174) or similar inventions would replace them; (c) the identification of $N_xF(x)$ with the concept $E_x(\varphi(x),F(x))$ were justified; and (d) definition (X) proved fruitful in the sense that, with its help, the laws of cardinal arithmetic, in particular (T), could be derived in a purely logical way. It appears evident, however, that Frege could not have met all these requirements. Besides these objections, further arguments could be advanced against the alleged plausibility of replacing (IX) with (X).

It is now time to spell out the reasons why I venture to surmise that in the *Foundations* Frege did not construe expressions of the form "the concept F" – as he actually did later in his essay 'On Concept and Object' – as authentic proper names designating objects of a quite special kind. These objects are supposed to do the job of representing first-level concepts.[53] One reason for my presumption is to be seen in what I termed requirement (b) concerning definition (X). We have observed that in the *Foundations* Frege falls short of introducing extensions of concepts as logical objects in a way which would meet his own canons of methodological rigour. Certainly, it would strike us as even less convincing if he were to appeal to an intuitive acquaintance with those entirely unknown objects of a quite special kind and were to identify the numbers with them.[54] A second reason is that the use of the term "the number belonging to the concept F" would be problematic, because in the *Foundations* the bearer of a number is a concept and not an object.

Furthermore, Frege would be compelled to interpret the sentence "The concept F is equinumerous with the concept G" as a statement

[53] For an extensive discussion of Frege's introduction of "concept-representing" objects of a quite special kind see Schirn 1983a and 1990.

[54] Frege nowhere specifies the status of an expression constructed with the words "the concept" or "the function" followed by a concept-expression or a function-expression of the *second* level. However, no sound argument presents itself to justify the possible claim that, while the term "the concept *prime number*" designates an object, the terms "the concept $-\overset{\mathfrak{a}}{\smile}-\varphi(\mathfrak{a})$" and "the concept $E_x(\varphi(x),F(x))$" do not. Frege presumably favoured the view that a second-level concept, in contrast to a first-level concept, can never be represented by an *object*. At any rate, in the logical system of the *Basic Laws* only first-level functions (concepts, relations) may be represented by their courses-of-values (extensions); and second-level functions can be represented only by first-level functions. Asked about the status of an expression like "the concept $-\overset{\mathfrak{a}}{\smile}-\varphi(\mathfrak{a})$", Frege probably could have provided only the following answer (at least in the context of 'On Concept and Object'): this expression indeed designates an object, but that object does not fulfil the role of representing a concept. Thus, Frege would hardly have been able to avoid recognizing another class of mysterious objects which do not perform a "concept-representing" function, and yet are supposed to have no other meaningful role.

about those mysterious objects which both related terms allegedly denote. Since in the *Foundations* he construes equinumerosity as a second-level relation, such an interpretation would lead to a doubtful result. It is true that in the *Basic Laws* Frege is able to treat equinumerosity as a first-level relation, but as such it holds only between extensions of concepts. Hence, if in the context of the *Basic Laws* Frege intended to justify the claim that the statement "The concept F is equinumerous with the concept G" is about objects, he would first have to demonstrate conclusively that the two expressions "the concept F" and "the extension of the concept F" refer to the same object. A plausible argument allowing for this identification is not in sight, however. For one could legitimately raise the following objection: If the term "the concept F" were to designate the extension of F, then in the term "the extension of the concept F" the part "the concept F" likewise would have to refer to the extension of F such that the more complex expression would refer to the extension of an extension of a concept. This result would be unacceptable for Frege. For, in his view, the expression "the extension of –" figures as a function-name not of first level, but of second or third level, as the case may be.[55]

Elsewhere,[56] I have argued at length that the postulation of objects of a quite special kind is an aberration. It fosters the pernicious illusion that an object can be conjured into existence simply by invoking an idiosyncrasy of natural language, namely our use of the definite article whenever we intend to refer to a concept. Frege's analysis of our use of an expression "the concept F" gives rise to an unresolvable conflict between what we mean with our words and what we actually say with them. It appears evident, however, that some, if not most, of the unpalatable consequences which the advocate of such an expression as a

[55] In the expression "the extension of the concept $x = x$", the part "the extension of" is supposed to be a second-level function-name, while in the expression "the extension of the concept $E_x(\varphi(x),F(x))$" the same constituent is supposed to be a third-level function-name. However, in the logical system of the *Basic Laws* Frege assigns courses-of-values (extensions) only to first-level functions, being convinced that this suffices for laying the logical foundations of arithmetic. It is noteworthy that the abovementioned "type-error" concerning the logical classification of the expression "the extension of –" would not simply disappear if the statement "the concept F $\neq$ the extension of the concept F" were held to be true by Frege, as I, in fact, think it is held. In that case, the term on the right-hand side of the inequality would likewise designate the extension of an *object*.

[56] Schirn 1990.

genuinely referring singular term has to face had gone unnoticed by Frege.[57]

3.4 Indeterminacy and the equivalence (T)

Does definition (IX) lead out of the impasse which Frege encounters through (VII)? Surely, if he were to hold that the only mode of referring to a number consists in using a numerical term of the form "$N_xF(x)$", the third objection raised to (VII) would have lacked any force. Now, (IX) does introduce a new way of identifying numbers and, in addition, creates the possibility of defining both the concept *n is a number* and the natural numbers in a formally correct manner. Note that here I use the words "formally correct" only in the weak sense that the numerical operator, forming a part of the *definiens* of (IV), (V), (VI) and (VIII), has been defined previously. Each natural number, except 0, is defined as the number belonging to the concept under which just its predecessors fall. The obvious drawback of these definitions is the introduction of occurrences of "$N_x\varphi(x)$" resisting contextual elimination. And recall that definition (IX) is vulnerable to a serious objection. As I shall argue in a moment, Frege does not succeed in resolving the referential indeterminacy of "$N_xF(x)$" by defining the number belonging to F as an equivalence class. Hence, the referential indeterminacy of the numerical operator inevitably infects the expressions "N(n)" and "0", "1", etc. which he defines in terms of "$N_x\varphi(x)$".

(T), taken in conjunction with definition (IX) and the definitions of 0, 1, etc., provides a means of determining the truth-value of any identity-statement of the following six types:

---

[57] On the one hand, I agree with Hodes 1984, 129 f. that Frege's views on object and function, important as they are, ought not to be disparaged as the least plausible of his doctrines. On the other hand, I do not share his conviction that these views are among the most plausible doctrines bequeathed to us by Frege. In particular, it seems indisputable to me that Frege went astray when he introduced his objects of a quite special kind. In his late period, he became aware of this aberration and consequently unmasked expressions of the form "the concept F" as sham proper names (cf. PW, 249 f., 270; NS, 269, 289). However, he tended to assess the source of the paradoxes of set theory in a one-sided and oversimplified manner when he insisted that it was the use of the definite article in expressions of the form "*the* extension of the concept F" which gave rise to them (cf. PW, 269; NS, 289). Hodes correctly points out that the so-called "paradox" of the concept *horse* is *not* a paradox. Nevertheless, Frege's claim that the concept *horse* is not a concept does contain a logical error, probably originating from a mixture of the formal and the informal modes of speaking; see Schirn 1990, 40.

(1) $N_xF(x) = N_xG(x)$
(2) $N_xF(x) = \lambda\varphi(E_x(\varphi(x),G(x)))$
(3) $\lambda\varphi(E_x(\varphi(x),F(x))) = \lambda\varphi(E_x(\varphi(x),G(x)))$
(4) $N_xF(x) = n$
(5) $n = \lambda\varphi(E_x(\varphi(x),F(x)))$
(6) $n = m$

An equation of type (4), (5), or (6) is capable of being reduced to one of type (1), (2), or (3), and the truth-conditions of the latter three are determined through the right-hand branch of the biconditional (T), or, spelled out more fully, through the one-one correlation of the objects falling under F with those falling under G: $\exists R(\forall x(F(x) \supset \exists y(R(x,y) \wedge G(y))) \wedge \forall y(G(y) \supset \exists x(R(x,y) \wedge F(x))) \wedge \forall x\forall y\forall z((R(x,y) \wedge R(x,z) \supset y = z) \wedge (R(x,z) \wedge R(y,z) \supset x = y)))$.

Thus, by virtue of (IX), Frege succeeds in extending the range of applicability of the originally proposed criterion of numerical identity. Nonetheless, the criterion so extended lacks the required unrestricted generality. Take, for example, the two equations "$N_x(x \neq x) = \lambda x(x \neq x)$" and "$N_x(x \neq x)$ = Julius Caesar".[58] The first cannot be transformed into an equation of type (1), since the term "$\lambda x(x \neq x)$", unlike the term "$\lambda\varphi(E_x(\varphi(x),x \neq x))$", designates the extension of a *first*-level concept. Hence, (T) is powerless to decide whether "$N_x(x \neq x) = \lambda x(x \neq x)$" is true or false. Although Frege does not comment on such a case, he seems to assume that (IX) serves to resolve the threatening referential indeterminacy of the term "$N_xF(x)$" once and for all. If we really do know what the extension of a concept in general is, that is, if we do have a full grasp of the notion of extension qua sortal concept[59] – and Frege takes that for granted, albeit without any justification – then we ought to be able to distinguish the extension of a first-level concept from the extension of a second-level concept, and the latter from Julius

[58] Benacerraf (1965, 285 ff.) regards a statement such as "Julius Caesar = the number of planets" as *unsemantical*. According to him, an identity-statement of the form "a = b" has a sense only in contexts where it is clear that both a and b are of some kind or category. "[...] if for two predicates *F* and *G* there is no third predicate *C* which subsumes both and which has associated with it some uniform conditions for identifying two putative elements as the same (or different) *C*'s, the identity-satements crossing the *F* and *G* boundary will not make sense" (287).

[59] A full grasp of a sortal concept F involves two things: having an idea of the distinction between objects which fall under F and those which do not; and a grasp of the identity-conditions for any object which falls under F.

Caesar. Consequently, we are justified in assigning the truth-value *false* to both equations. It cannot be ruled out that this line of thought reflects Frege's tacit assessment of the situation. It is, indeed, hard to see what else he might have had in mind. In the light of his line of argument concerning (VII), it seems that he might have considered the following alternative strategy: Once the sortal concept of cardinal number has been defined without fallacy, we can solve the indeterminacy problem on the basis of (T) and definition (VIII) $N(n) := \exists\varphi(N_x\varphi(x) = n)$.[60]

Be this as it may, both strategies are equally unsatisfactory. The problem is by no means solved, but only pushed back. For we do not know whether Julius Caesar is a number unless we know whether or not he is an extension. In the *Foundations*, Frege lays down identity-conditions only for various kinds of equivalence classes, of equinumerosity and, for the sake of illustration, of parallelism and geometrical similarity. And we have seen that his assumption regarding extensions of concepts lacks foundation.

Nevertheless, this need not be the last word about Frege's introduction of extensions of concepts in the *Foundations*. Suppose that Frege succeeded in devising a stipulation that effectively resolved the third doubt arising from his proposed contextual definition of "$N_xF(x)$". It is plain that in this case he could have adumbrated his logicist programme without needing to define explicitly the number of F as the extension of a concept. For he claims to have reduced the relation of one-one correspondence to purely logical terms, and it is this very reduction with which he intends to secure the required logical nature of definition (VII). The individual numbers, including the number *infinite* ∞_1 (that is, the number belonging to the concept *finite number* or, in Cantor's terminology, the smallest transfinite cardinal number $\aleph_0$), are defined without recourse to extensions of concepts. Moreover, Frege uses his explicit definition of "$N_xF(x)$" only for the purpose of outlining his proof of the equivalence (T) (cf. FA, GLA, § 73 and GGA I, 70 ff.). In one of his excellent articles dealing with Frege's *Foundations*, George Boolos has shown why this proof fails (1987b, 137 ff.) Having derived (T), Frege never appeals to definition (IX) again, nor does he make any other use of extensions of con-

---

[60] Julius Caesar is not a cardinal number because there exists no concept φ such that the number belonging to φ is Julius Caesar.

cepts.[61] Instead he deduces familiar laws of pure number theory, in particular, the fundamental proposition that every finite number is succeeded by a finite number, from (T) together with his other definitions. It might seem therefore that Frege could have carried out his foundational project simply by making use of the originally proposed contextual definition (VII) of "$N_xF(x)$" and of the system of second-order logic set forth in his *Begriffsschrift* and implicit in the *Foundations*, hence, without invoking extensions of concepts at all. In saying this, I assume, of course, that he succeeded in removing the third decisive objection raised to (VII) by means of an appropriate additional stipulation. Unfortunately, any attempt to salvage (T) as a definition of "$N_xF(x)$" is bound to fail. As I stressed earlier, (VII) involves the intrinsic obstacle of contextually ineliminable occurrences of the numerical operator.

Following Wright and Boolos, a far more promising alternative suggests itself. Treat (T) as an axiom, adjoin it as such to standard axiomatic second-order logic, construe the numerical operator as a primitive expression governed by (T), and derive the basic theorems of pure cardinal arithmetic more or less in the fashion indicated by Frege in the *Foundations*. Yet, before the Fregean can shout for joy, the first of three essential questions ought to be answered negatively, while the second and third should be given positive answers. (1) Does formula (T), owing to its formal affinity to the notoriously inconsistent Axiom V of the *Basic Laws*, fall prey to contradiction?[62] (2) Can (T) be acknowledged as a (primitive) truth of logic? (3) Can we solve the so-called Julius Caesar problem? Let us tackle these issues one by one.

To (1): Fortunately, the answer to the first question is negative. (T) *is* consistent. Elaborating on remarks made by Hodes 1984 and Burgess

---

[61] If I have checked correctly, it is not until Section 83 of the *Foundations* that Frege uses the term "extension of concept" again. In this Section, he completes his sketch of the proof that every finite number has a successor. (The concept *finite number* is defined in Section 83, at the very end of Frege's outline of his logicist programme.) One essential step is to prove that a is the number of the concept "belonging to the natural series of numbers ending with a, but not equal to a". And to this purpose, it is necessary to prove that this concept has the same *extension* as the concept "belonging to the natural series of numbers ending with d". So what Frege has to prove is just the co-extensiveness of two *first*-level concepts.

[62] It is worth noting that the first-order fragment of the logical system of the *Basic Laws* is consistent, as first demonstrated by Terence Parsons in his 1987 paper; see also Dummett 1991b, 219. Without second-order quantification, the introduction of courses-of-values would be an idle wheel, however, because membership would be indefinable for Frege.

1984, Boolos in fact proved the consistency of (T) by constructing a model M whose domain is the set $\{0, 1, 2, \ldots, \aleph_0\}$ and which also satisfies all principles of axiomatic second-order logic.[63]

To (2): A thorough discussion of the question as to whether Frege would be entitled, by his own standards, to construe (T) as a fundamental truth of logic would be a major task in its own right. In any case, we would first have to look closely at his own criterion of analytic or logical truth. I take it that he would regard the two concepts as coextensive, though after the period of the *Foundations* he never uses the word "analytic" again.[64] In addition, we would need to elucidate the relationships between Frege's views of analyticity, self-evidence, epistemic triviality and sense- or thought-identity. Finally, we would have to take into account his conception of axioms in general and of logical axioms in particular. Only on the basis of those considerations could we appropriately decide whether the Fregean would be justified in accepting, in the light of his own principles, formula (T) as a primitive truth of logic.

Nonetheless, a few remarks on this complex issue are in order before I turn to Frege's attitude towards the status of (T) and Boolos' answer to our question.

Frege holds that both halves of (T) have the same judgeable content, i. e., express the same thought. He could therefore maintain that (T) is both self-evident and analytically true by appealing to this very fact. By the same token, he would have to allow for the possibility of converting (T) into a sentence of the form "$S \equiv S$" without altering the thought. And such a sentence would, indeed, be true by virtue of its logical structure.

---

[63] Cf. Boolos 1987a, 7. In the same article, Boolos presents a formal system FA ("Frege's arithmetic") which comprises the whole content of Frege's reductionist programme as outlined in the *Foundations* and for which a simple consistency proof can be given. The underlying logic of FA is standard axiomatic second-order logic. Boolos shows in detail that the definitions and theorems of Sections 68–83 of the *Foundations* can be formulated, respectively proved in FA, in the manner indicated by Frege (cf. 1987a, 10–14 and the appendix in Boolos 1990, 275 f.). Certainly, if analysis is consistent, so is FA. Boolos (1987a, 10; 1990, 269) demonstrates that the converse also holds. The equiconsistency of analysis and FA can be proved in Primitive Recursive Arithmetic. Furthermore, Boolos stresses the fact that in FA much further mathematics can be carried out than Frege had actually executed within his logicist programme.

[64] In the *Foundations*, Frege defines an analytic truth in terms of the notion of provability. No provision is made for those truths which are incapable of proof. Clearly, the fundamental laws of logic should also count as analytic truths.

In dealing with our question, Boolos tends to put self-evidence and analyticity on a par (1990, 272; 1987a, 19). It is worth underlining in this context that for Frege the self-evidence of a truth cannot serve as a general criterion of analyticity. According to Frege, there are, on the one hand, sentences lacking self-evidence which nevertheless have to be recognized as analytically true, such as the equation "135664 + 37863 = 173527" or what in his *Begriffsschrift* he refers to as formula (133). This formula states that the ancestral of a many-one relation is a simple ordering when restricted to the objects to which a given object is ancestrally related. On the other hand, Frege acknowledges the existence of self-evident, but non-analytic truths, such as the axioms of Euclidean geometry. They are held to be synthetic a priori. The fact that the truth of a sentence S (or of the thought it expresses) is capable of being recognized by an immediate insight does not imply, in Frege's view, that S is trivially true. Thus, he regards geometrical and logical axioms as self-evident truths, and at least the former are seen not as trivial truths but as containing real knowledge. And there is much to be said in favour of the assumption that for Frege logical axioms, too, possess epistemic value (cf. CP, 274, 277; KS, 263, 265). The converse implication does hold: epistemic triviality implies self-evidence. The sentence "Oxford = Oxford", for instance, is trivially true and, hence, self-evident; it is, of course, also analytically true. The same applies to all identity-statements of the form "a = b" in which "a" and "b" not only refer to the same object, but also express the same sense. Such statements can be transformed into statements of the form "a = a" without changing the sense. And the latter are true – at least in a logically perfect language where every well-formed singular term has exactly one reference – on the strength of their logical structure alone. For a true statement "a = b" to be analytic in Frege's sense, sameness of sense of "a" and "b" is a sufficient condition, but it is not a necessary one.

Returning to the equivalence (T), we face a difficulty. If it is to be regarded as a logical axiom, it must be self-evident from Frege's point of view. It seems that the requirement of self-evidence, when applied to (T), can be met only if "$N_xF(x) = N_xG(x)$" and "$E_x(F(x),G(x))$" express the same thought. Yet the sameness of content of these two sentences, which Frege stipulates when he attempts to *define* "$N_xF(x)$" by means of (T), ought not to imply that (T) qua axiom emerges as a trivial truth. For the epistemic triviality of (T) would be at odds with what Frege considers to be an essential feature of an axiom as such: it contains real knowledge.[65]

Frege does not go to great lengths when he comes to explain the allegedly logical nature of (T). He only observes that he has succeeded in reducing one-one correspondence to purely logical terms, to the term "relation" that is (FA, GLA, 84 f., 117). We are further told that a relational concept, like a simple concept, belongs to pure logic. The particular content of the relation is said to be of no concern here; what counts is its logical form. And the truth of whatever can be asserted about this form is analytic and known a priori (FA, GLA, 83). In the Introduction to the *Basic Laws*, Frege argues in the same vein: "Much the same holds for the word 'correspondence' as for the word 'set'; both are today used frequently in mathematics [...] If I am right in thinking that arithmetic is a branch of pure logic, then a purely logical expression must be selected for 'correspondence'. I choose 'relation' for this purpose. Concept and relation are the foundation-stones upon which I erect my structure" (BLA, 31 f.; GGA I, 3).

[65] As I have already observed, Frege holds that both halves of the biconditional (T) have the same judgeable content. Not so in the *Basic Laws* as regards Axiom V. Here he neither claims nor denies explicitly that the two sides of Axiom V express the same sense. In Section 3, he stipulates only that a universally quantified equation betweeen function-values is to have the same reference as the corresponding equation of courses-of-values. According to Dummett (1991a, 295 ff.), there is more than a kernel of truth in the assumption that after 1891 Frege tacitly accepted the following principle: If a sentence A involves a concept that a sentence B does not involve, then A and B cannot express the same thought. In the light of this principle, which Dummett calls (K), Frege would have to retract the claim of the *Foundations* that both sides of "$D(a) = D(b) \equiv a \parallel b$" and of "$N_xF(x) = N_xG(x) \equiv E_x(F(x),G(x))$" have the same judgeable content. Contrary to what Dummett suggests, it is easy to provide textual evidence that even during his mature period after 1891 Frege did not wholeheartedly subscribe to principle (K). One clear example is the pair of sentences (I) "There is at least one square root of 4" and (II) "The concept *square root of 4* is realized". In 'On Concept and Object' (CP, 188; KS, 173), Frege contends that (I) and (II) express the same thought. Furthermore, there is ample reason to suppose that he would have regarded (I) and (III) "The extension of the concept *square root of 4* is not empty" as another pair of synonymous sentences. Here, too, the second sentence of the pair involves a "concept" (the function *the extension of*) which the first sentence does not involve. Nonetheless, in Schirn 1997 it is argued at length that in all likelihood Frege tacitly changed his mind as to the transition from an equivalence statement to an identity-statement. In particular, he probably did not think that both halves of Axioms V express the same sense. Let me add that as a consequence of his distinction between the *sense* and the *reference* of an expression, Frege divided the judgeable content into *thought* and *truth-value* (CP, 187; KS, 172; BLA, 6 f.; GGA I, X; PMC, 63; WB, 96). But in retrospect, he is inclined to conceive the judgeable content primarily as what he then calls the thought (WB, 120).

Let me add one more brushstroke to the picture we may gain of Frege's views on logical truth. When in his essay 'On the Foundations of Geometry' (Second Series, 1906) Frege sets about outlining a possible way of proving the independence of a genuine axiom from other genuine axioms, he makes a number of interesting remarks on the nature of logic which are, in a certain sense, linked to his approach in the *Foundations*. There he stresses the need to supply a clear characterization of what counts as a logical inference and what is proper to logic, if we are to carry out valid independence proofs. He points out, moreover, that for this purpose it would be necessary to formulate, in a precise manner, a basic law which one might call an "emanation of the formal nature of logical laws" (CP, 337; KS, 321). At the same time, Frege emphasizes that logic is, despite first appearances, not purely formal. Just as concepts like *point, line, plane* and relations like *lies on, between, congruent* belong intrinsically to geometry, so logic, too, has its own concepts and relations such as negation, identity, subsumption, subordination of concepts for which it allows no replacement. For Frege, this is an unmistakable mark that the relation of logic towards what is proper to it is not at all formal. He concedes, however, that here we find ourselves still in unexplored territory.

I think that Boolos gives the right answer to the question whether the equivalence (T) can be accepted as a truth of logic. It "should count as a logical truth only if it is supposed that numbers do not do to concepts or sets the corresponding sort of thing, namely, fall under, or be elements, of them. [...] But Frege's proof from Hume's principle that every number has a successor [...] cannot succeed unless it is supposed that numbers are objects" (1990, 272). As we have seen, it is precisely for this reason that Frege has to establish his conception of numbers as objects, not for any of the reasons he adduces in Sections 56 and 57 of the *Foundations*.

How does Frege show that $0 \neq 1$? We know that he defines 0 as $N_x(x \neq x)$ and 1 as $N_x(x = 0)$. Since $x \neq x$ is empty and the object 0 falls under $x = 0$, the two concepts $x \neq x$ and $x = 0$ are not equinumerous, and hence the numbers belonging to them, 0 and 1, are, by (T), distinct. Boolos aptly reminds us that this argument works only on the assumption that 0 is an object that falls under $x = 0$. We arrive at 2 in like manner. We form the concept $x = 0 \vee x = 1$ and assign 2 to it as its number. Since the new concept is equinumerous neither with $x \neq x$ nor with $x = 0$, we can conclude by (T) that 2 is distinct from both 0 and 1.

In short, the equivalence (T) is not a logical truth because it cannot be made true in any finite domain. Its existential quantification is false in every finite domain. When n objects belong to the domain, there have to be n + 1 different cardinalities. I agree with Boolos (1990, 273) in another important respect. Someone advocating a conception of logical truth that characterizes (T) as a logical truth will face the difficult task of explaining why (T) is a truth of logic even though the formally analogous Axiom V is not. Obviously, it would be unsatisfactory to maintain that (T), unlike Axiom V, is a logical truth because it is consistent.[66]

Suppose that at the time when Frege wrote the *Foundations* he had convinced himself that (T) cannot, after all, be defended as a truth of logic. Might he have sacrificed his reductionist programme, especially his conception of numbers as *logical* objects, in favour of a foundation of arithmetic which, though akin to his own, would no longer deserve to be called *logicism*? I suspect that the answer is "no".[67]

To (3): It may strike us as appealing that elementary number theory and, indeed, much further mathematics can be derived in a system of axiomatic second-order logic to which (T) is adjoined as the sole non-logical axiom. Certainly, this revised foundational approach is not invulnerable to the emergence of the obstinate Julius Caesar problem. Yet, despite the fact that we have dispensed with extensions of concepts, which were introduced to determine uniquely the truth-value of any sentence of the form "$N_xF(x) = q$", it could seem that the Julius Caesar problem now loosens its grip for the Fregean. What are the prospects of removing it?

(a) At first glance, a straightforward solution for the Fregean might be this. Since cardinal numbers are not identified with equivalence

---

[66] Boolos even goes a step further in assessing "Hume's principle". In his 1990 paper, he claims "that not only do we have no reason for regarding the principle as a truth of logic, it is doubtful whether it is a truth at all" (274; cf. 1987a, 15 f.). For it is a consequence of the principle not only that there is a number belonging to the concept $x \neq x$, but also a number belonging to the concept $x = x$, a greatest number. Thus, Cantor's paradox threatens. Historically, it is noteworthy that Russell was led to his paradox in his attempt to reconcile Cantor's proof that there can be no greatest cardinal number with the plausible supposition that the class of all objects has necessarily the greatest possible number of members.

[67] Recent attempts to vindicate logicism (cf. Bostock 1974 and 1979, Wright 1983) strike me as unconvincing. They are retrograde steps. What threatens logicism is not the lurking pitfalls of the set-theoretic paradoxes, but rather the commitment of arithmetic to the existence of infinitely many objects. This is a widely shared view and I believe that it is well-sustained.

classes nor with sets of any other type, (T), taken in conjunction with definition (VIII), fixes the truth-conditions of every identity-statement of the form "$N_xF(x) = q$", where "q" is any singular term which does not have the syntactic form of "$N_xG(x)$". This proposal appears to tally with what Frege says in Section 66 of the *Foundations*, assuming that definition (VIII) satisfies his principle of completeness.[68] Hopes of overcoming the difficulty are dashed, however. If we assess (T) qua axiom in the light of what Frege says at the beginning of Section 10 of the *Basic Laws* about his informal stipulation corresponding to Axiom V, we see that (T) effects only a partial determination of the reference of a term "$N_xF(x)$". Just at the point where (T) forsakes us, namely concerning the determination of the truth-conditions of equations of the form "$N_xF(x) = q$", (VIII) would have to fill the gap. Yet whatever indeterminacy affects the numerical operator will be transferred to the predicate "N(n)".

It is plain enough that (T) does not settle whether, e. g., the number of proved theorems in the *Basic Laws* coincides with the last number which Julius Caesar had in mind before he died. But even in the case of numerical equations of the form "$N_xF(x) = N_xG(x)$", (T) does not, for every instance, determine its truth-value. Take the equation (p) "$N_x(x = x) = N_xFN(x)$". "FN(x)" is to abbreviate the predicate "finite number".[69] As Boolos shows (1987a, 16), (p) is an undecidable sentence in the formal system FA ("Frege's arithmetic"); it is true in some models of FA, and false in others. Boolos rightly conjectures that Frege would have probably considered (p) to be false. Frege's remarks on Cantor in Section 86 of the *Foundations* suggest this. There Frege welcomes the fact that Cantor's investigations have paved the way for the introduction of higher transfinite cardinal numbers. About twenty years later, in the course of outlining his theory of real numbers (GGA II, § 164), Frege argues for the falsity of "$N_xFN(x) = N_xCFN(x)$" ("CFN(x)" is to abbreviate the predicate "class of finite numbers") without mentioning Cantor's proof that for any set M, $\mathfrak{P}M$ is always of

[68] Cf. TF, 159 ff.; GGA II, 69 ff.; CP, 300 ff.; KS, 291 f. See the solution of the Julius Caesar problem that has been suggested by Crispin Wright 1983, 116 f.

[69] Frege defines this predicate in terms of the number 0 and the successor relation (cf. FA, GLA, § 83). In the *Basic Laws*, he does not introduce a special sign for the concept of finite or natural number. In the *Foundations* (§ 84), Frege defines ∞_1 as $N_xFN(x)$. In the *Basic Laws*, he uses the symbol "∞" to refer to the number endless; see his definition of ∞ in GGA I, § 122.

a power greater than M itself. "If we call the extension of a concept which is subordinate to the concept *finite number*, a class of finite numbers, then to the concept *class of finite numbers* there belongs an infinite number which is bigger than endless; i. e., the concept *finite number* can be mapped into the concept *class of finite numbers*, but the latter cannot be mapped into the former" (GGA II, 161).

(b) A possible second option would amount to a stipulation similar to the one which Frege makes in Section 10 of the *Basic Laws* in order to be able to determine uniquely the references of course-of-values terms. In this Section, he proposes to identify the two truth-values the True and the False with their own unit classes. He does so by appeal to what has come to be known as his "permutation argument". Now, imagine that Frege had aimed at laying the foundations of number theory without identifying cardinal numbers with extensions of concepts. Suppose that he had intended to carry out his plan within a fragment of the system of the *Basic Laws*, more specifically, within a system of axiomatic second-order logic with identity, but without both the course-of-values operator "$\grave{\varepsilon}\varphi(\varepsilon)$" and the description operator "$\backslash\xi$", and, hence, without Axioms V and VI. However, the equivalence (T) has been adjoined as an axiom to the Axioms I–IV. I call this system BLS. Let us assume further that the exposition of the Begriffsschrift[70] has roughly reached the stage of Section 25 of the *Basic Laws*. Section 3 of the *Basic Laws* where Frege introduces his courses-of-values, Section 9 where he explains his notation for courses-of-values, Section 10 where he tries to solve the problem that Axiom V does not fully determine the reference of a course-of-values name, and Section 11 which serves to elucidate the function-name "$\backslash\xi$" are to be ignored.[71] The primitive functions $—\xi$, $\top\xi$, $\xi = \zeta$, $\sqsubset^{\xi}_{\zeta}$, $—\overset{\mathfrak{a}}{\smile}—\varphi(\mathfrak{a})$, and $—\overset{\mathfrak{f}}{\smile}—\mu_{\beta}(\mathfrak{f}(\beta))$, whose values for suitable arguments are exclusively the True and the False, have been explained in the familiar fashion. The numerical operator "$N_x\varphi(x)$" has been introduced as a primitive second-level function-name governed by (T) just as in the *Basic Laws* "$\grave{\varepsilon}\varphi(\varepsilon)$" figures as a primitive expression governed by Axiom V. The relation of equinumerosity between first-level concepts has been defined in terms of one-one correlation. We

---

[70] In the *Basic Laws*, the exposition of the Begriffsschrift comprises Sections 1–52 of the first Volume.

[71] Of course, one would have to ignore other bits of text, such as the last three lines of Section 18 and the last five lines of Section 20. But such minor details need not concern us here.

may further assume that the concept of cardinal number has been defined in the same way as in the *Foundations.*

Now, the question is whether, in BLS and with respect to cardinal numbers, Frege might have argued in a manner similar to his line of argument in Section 10 of the *Basic Laws* concerning courses-of-values: (T) does not fix uniquely the reference of a numerical term "$N_xF(x)$". In particular, (T) is powerless to decide whether or not either truth-value is a cardinal number. We are free to stipulate, by invoking a special "permutation argument" and without contradicting (T), that the True shall be identical with a number belonging to an arbitrary first-level concept F, and the False with a number belonging to any other first-level concept G which is not equinumerous with F. The answer to this question may be left to the reader.

Acknowledgements

I am grateful for opportunities I have had to read earlier (and shorter) versions of this paper (some in German or Spanish or Portuguese) at various universities: Oxford, Cambridge, St. Andrews, Boston (Colloquium for the Philosophy of Science), City University of New York (Graduate Center), University of Pennsylvania, University of California (Berkeley, Logic Colloquium), University of California (Irvine), University of Southern California, Chicago, Notre Dame, University of Minnesota (Minneapolis), Michigan State University, Universidad Nacional Autónoma de México (Instituto de Investigaciones Filosóficas), Buenos Aires (Universidad Nacional), Sociedad Argentina de Análisis Filosófico (Buenos Aires), Pontificia Universidad Católica de Chile (Santiago), Universidad de Puerto Rico (Río Piedras), Barcelona (Departamento de Lógica, Historia y Filosofía de la Ciencia), Universidade de Santiago de Compostela (Departamento de Lóxica e Filosofía da Ciencia), Campinas (UNICAMP, Centro de Lógica, Epistemologia e Historia da Ciência), Universidade de São Paulo, Pontifícia Universidade Católica de São Paulo, Pontifícia Universidade Católica de Rio de Janeiro, Universidade de Brasília, Universidade Federal de Pernambuco (Recife, Departamento de Informática), Universidade Estadual Paulista (Rio Claro, Departamento de Matemática), Salzburg, Graz, Bonn, Bayreuth, Jena, Munich, Kaiserslautern. Part of the material has also been discussed at the university of Padua and at the conference ANPOF in Aguas de Lindóia (Brazil). I am grateful to a number of friends and colleagues for support, advice and helpful comments and I want to mention them all here: Patricia Blanchette, John Addison, Robert Bublak, Charles Chihara, Michael Cremer, Max Cresswell, Michael Detlefsen, Hartry Field, Pierdaniele Giaretta, Bob Hale, Richard Heck, Jaakko Hintikka, Ignacio Jané, Jerry Katz, Richard Mendelsohn, Charles Parsons, Francisco Rodríguez Consuegra, Marco Ruffino, Stephen Schiffer, Howard Stein, Bill Tait, Michael Wrigley. Special thanks are due both to George Boolos for commenting on my paper in the Boston Colloquium and to Akihiro Kanamori for chairing the session, also to Ivette Fred (who patiently attended various versions of my lecture), Bob Cohen, Warren Goldfarb, Dirk Greimann, Brian Loar, Raúl Orayen, Terry Parsons, Thomas Ricketts, Guillermo E. Rosado Haddock, Peter Simons, Hans Sluga, Timothy Smiley, Roberto Torretti, Luis Villegas-Forero, Peter Woodruff and Crispin Wright. I owe a very special debt to Daniel Isaacson and Michael Resnik for reading an earlier version of this paper; their comments helped me to improve it. Last but not least, I would like to thank the Deutsche Forschungsgemeinschaft and the Deutscher Akademischer Austauschdienst for travel support and Uwe Lück for typing both the original version and the final draft.

References

I use the following abbreviations for references to Frege's works:

BG: 'Über Begriff und Gegenstand', in KS, 167–178, trans. in TF, 42–55 and CP, 182–194.

BLA: *The Basic Laws of Arithmetic. Exposition of the System*, trans. and ed. M. Furth, University of California Press, Berkeley and Los Angeles, 1964.

BS: *Begriffsschrift. Eine der arithmetischen nachgebildete Formelsprache des reinen Denkens*, L. Nebert, Halle a. S., reprinted in G. Frege, *Begriffsschrift und andere Aufsätze*, ed. I. Angelelli, Darmstadt and Hildesheim, 1964.

CN: *Conceptual Notation and Related Articles*, trans. and ed. T. W. Bynum, Oxford University Press, Oxford, 1972.

CP: *Collected Papers on Mathematics, Logic, and Philosophy*, ed. B. McGuinness, trans. M. Black et al., Basil Blackwell, Oxford, 1984.

FA: *The Foundations of Arithmetic. A logico-mathematical enquiry into the concept of number*, trans. J. L. Austin, Basil Blackwell, Oxford, 1950, 1953.

GGA: *Grundgesetze der Arithmetik. Begriffsschriftlich abgeleitet*, Vol. I, H. Pohle, Jena, 1893, Vol. II, H. Pohle, Jena, 1903, reprinted Darmstadt and Hildesheim, 1962.

GLA: *Die Grundlagen der Arithmetik. Eine logisch mathematische Untersuchung über den Begriff der Zahl*, W. Koebner, Breslau, 1884, reprinted Darmstadt and Hildesheim, 1961.
KS: *Kleine Schriften*, ed. I. Angelelli, G. Olms, Hildesheim, 1967.
NS: *Nachgelassene Schriften*, eds. H. Hermes, F. Kambartel and F. Kaulbach, Felix Meiner, Hamburg, 1969.
PMC: *Philosophical and Mathematical Correspondence*, ed. B. McGuinness, trans. H. Kaal, Basil Blackwell, Oxford, 1980.
PW: *Posthumous Writings*, trans. P. Long and R. White, Basil Blackwell, Oxford, 1979.
TF: *Translations from the Philosophical Writings of Gottlob Frege*, ed. and trans. P. Geach and M. Black, Basil Blackwell, Oxford, 1952, 1960.
WB: *Wissenschaftlicher Briefwechsel*, eds. G. Gabriel, H. Hermes, F. Kambartel, C. Thiel and A. Veraart, Felix Meiner, Hamburg, 1976.

I refer by author and year of publication to the following works:

Benacerraf, P.: 1965, 'What Numbers Could Not Be', *The Philosophical Review* **74**, 47–73; reprinted in Benacerraf and Putnam 1983, 272–294.
Benacerraf, P.: 1973, 'Mathematical Truth', *The Journal of Philosophy* **70**, 661–679; reprinted in Benacerraf and Putnam 1983, 403–420.
Benacerraf, P.: 1981, 'Frege: The Last Logicist', *Midwest Studies in Philosophy* **VI**, eds. P. French, T. Uehling and H. Wettstein, University of Minnesota Press, Minneapolis, 17–35.
Benacerraf, P. and Putnam, H. (eds.): 1983, *Philosophy of Mathematics: Selected Readings*, Cambridge University Press, Cambridge.
Boolos, G.: 1987a, 'The Consistency of Frege's Foundations of Arithmetic', in J. J. Thomson (ed.), *On Being and Saying. Essays for Richard Cartwright*, The MIT Press, Cambridge, Massachusetts, 3–20.
Boolos, G.: 1987b, 'Saving Frege from Contradiction', *Proceedings of the Aristotelian Society* 1986–87, 137–151.
Boolos, G.: 1990, 'The Standard of Equality of Numbers', in Boolos (ed.), *Meaning and Method: Essays in Honor of Hilary Putnam*, Cambridge University Press, Cambridge, 261–277.
Bostock, D.: 1974, *Logic and Arithmetic: Natural Numbers*, Oxford University Press, Oxford.
Bostock, D.: 1979, *Logic and Arithmetic: Rational and Irrational Numbers*, Oxford University Press, Oxford.
Burgess, J. P.: 1984: Review of C. Wright 1983, *The Philosophical Review* **93**, 638–640.
Currie, G.: 1982, *Frege: An Introduction to his Philosophy*, The Harvester Press, Brighton.
Dummett, M.: 1973, *Frege: Philosophy of Language*, Duckworth, London.
Dummett, M.: 1981, *The Interpretation of Frege's Philosophy*, Duckworth, London.
Dummett, M.: 1991a, *Frege and Other Philosophers*, Oxford University Press, Oxford.
Dummett, M.: 1991b, *Frege: Philosophy of Mathematics*, Duckworth, London.

Dummett, M.: (1997), 'Neo-Fregeans: in Bad Company?', forthcoming in *Philosophy of Mathematics Today,* ed. M. Schirn, Oxford University Pess, Oxford.
Gödel, K.: 1947, 'What is Cantor's Continuum Problem?', *American Mathematical Monthly* **54**, 515–525; revised and expanded version in Benacerraf and Putnam 1983, 470–485.
Haaparanta, L. and Hintikka, J. (eds.): 1986, *Frege Synthesized. Essays on the Philosophical and Foundational Work of Gottlob Frege,* D. Reidel, Dordrecht, Boston.
Hale, B.: 1987, *Abstract Objects,* Basil Blackwell, Oxford.
Hodes, H.: 1984, 'Logicism and the Ontological Commitments of Arithmetic', *The Journal of Philosophy* **81**, 123–149.
Hodes, H. 1990: 'Where do the Natural Numbers Come From?', *Synthese* **84**, 347–407.
Parsons, T.: 1987, 'On the Consistency of the First-Order Portion of Frege's Logical System', *Notre Dame Journal of Formal Logic* **28**, 161–168.
Putnam, H.: 1967, 'Mathematics Without Foundations', *The Journal of Philosophy* **64**, 5–22; reprinted in Putnam, *Mathematics, Matter and Method. Philosophical Papers,* Vol. 1, Cambridge University Press, Cambridge, 43–59.
Resnik, M. D. 1980, *Frege and the Philosophy of Mathematics,* Cornell University Press, Ithaca.
Schirn, M.: 1983a, 'Begriff und Begriffsumfang. Zu Freges Anzahldefinition in den *Grundlagen der Arithmetik', History and Philosophy of Logic* **4**, 117–143.
Schirn, M.: 1983b, 'Juicio, concepto y curso de valores', *Diálogos* **42**, 187–216.
Schirn, M.: 1983c, Review of Dummett 1973, second edition, 1981, *Archiv für Geschichte der Philosophie* **65**, 210–217.
Schirn, M.: 1983d, Review of Dummett 1981, *Erkenntnis* **20**, 243–251.
Schirn, M.: 1984, 'Sluga über Freges These der Priorität von Urteilen gegenüber Begriffen', *Archiv für Geschichte der Philosophie* **66**, 194–215.
Schirn, M.: 1985, 'Semantische Vollständigkeit, Wertverlaufsnamen und Freges Kontextprinzip', *Grazer Philosophische Studien* **23**, 79–104.
Schirn, M.: 1987, 'Logik, Mathematik und Erkenntnistheorie', *Archiv für Geschichte der Philosophie* **69**, 92–109.
Schirn, M.: 1988, Review of G. Frege, *Die Grundlagen der Arithmetik. Eine logisch mathematische Untersuchung über den Begriff der Zahl,* centenary edition, with supplementary text critically edited by Christian Thiel, Felix Meiner, Hamburg 1986, *The Journal of Symbolic Logic* **83**, 993–999.
Schirn, M.: 1989, 'Frege on the Purpose and Fruitfulness of Definitions', *Logique et Analyse* **125–126**, 61–80.
Schirn, M.: 1990, 'Frege's Objects of a Quite Special Kind', *Erkenntnis* **32**, 27–60.
Schirn, M.: 1992a, 'Gottlob Frege', in *Philosophy of Language. An International Handbook of Contemporary Research,* eds. M. Dascal, D. Gerhardus, K. Lorenz, G. Meggle, Walter de Gruyter, Berlin, New York, 467–494.
Schirn, M.: 1992b, 'El método de la descomposición de pensamientos en Frege', *Análisis Filosófico* **12**, 31–41.

Schirn, M.: 1993, 'Contexto proposicional e a questão do significado de termos singulares abstratos', *Manuscrito* **16**, 19–46.
Schirn, M.: 1996, 'O principio do contexto nas *Grundgesetze* de Frege', forthcoming in *Theoria* **27**.
Schirn, M.: 1997, *Gottlob Frege: Grundlagen der Logik und Mathematik*, forthcoming, Walter de Gruyter, Berlin, New York.
Sluga, H.: 1980, *Gottlob Frege*, Routledge & Kegan Paul, London.
Wright, C.: 1983, *Frege's Conception of Numbers as Objects*, Aberdeen University Press, Aberdeen.

Nominalism and the Contingency of Abstract Objects*

Bob Hale and Crispin Wright

Traditional mathematical nominalism[1] regards a mathematical theory as acceptable only if it can be so interpreted as to involve no purported quantification over or singular reference to abstract objects. To whatever extent a theory fails this test, it is viewed as literally unintelligible. Hartry Field's[2] nominalism is atraditional in this respect. Field disbelieves in abstract objects, of course, and accepts that it is a constraint upon defensible mathematical thought that it avoid commitment to them. But he does not hold it to be a condition on the intelligibility of a mathematical theory that it allow of a non-Platonist *semantics*. On the contrary, he is prepared to grant that a Platonist semantics for, e. g., number theory is descriptively correct: that number-theoretic statements do indeed purport singular reference within and quantification over a specific domain of abstract objects. Since, in his view, there are no such objects, Field has to hold that such statements' truth conditions are systematically unrealized. And so, of course, he does.

How can anyone who holds this view – that math is massively false – hope to save the discipline without semantic reconstruction? By accepting a distinction, roughly parallel to that advocated by Bas van Fraassen[3] in the philosophy of science, between *endorsing* a theory and *accepting it as true*. For van Fraassen, one need not, in endorsing a physical theory, be committed to more than its empirical adequacy – its correctness as far as what is observable is concerned. And for Field, one need not, in endorsing a mathematical theory, be committed to more

---

Reprinted with the kind permission of the editors from *The Journal of Philosophy* 89 (1992), 11–35.

* Thanks to the participants at colloquia at Michigan, Oxford, and the 1991 Frege conference at Munich, organized by Matthias Schirn, at which versions of this material were presented. Special thanks to Hartry Field for much interesting discussion.

[1] See, e. g., W. v. Quine and Nelson Goodman's classic nominalist manifesto, "Steps towards a Constructive Nominalism," *Journal of Symbolic Logic* (1947): 105–22.

[2] As originally expounded in his *Science without Numbers* (New York: Blackwell, 1980).

[3] *The Scientific Image* (New York: Oxford, 1980).

than its *conservativeness*: where a theory is, roughly, conservative with respect to a discourse just in case any inferences among statements of the discourse which it mediates could at least in principle validly be constructed – though perhaps at great cost in length and notational simplicity – without appeal to that theory.[4] For Field, a mathematical theory can be acceptable, even if its ingredient statements have Platonistic truth conditions, provided it is in this way conservative with respect to *nominalistic* discourse: discourse all of whose ingredient statements have truth conditions that make ontological demands only within the domain of the acceptably concrete. Field's program accordingly has the objective of showing that as much as possible of classical mathematics can be saved as acceptable by this standard.

A question arises, however, about the interpretation of the idea of conservativeness. No mathematical theory that employs a material conditional can be conservative unless consistent. But what understanding is one who disbelieves in abstract objects to have of the notion of consistency? Orthodox proof-theoretic and semantic accounts each involve quantification over nominalistically forbidden entities – *sequences* of sentences in the case of proof-theoretic accounts, and *models* in the case of semantic ones. Field ought, therefore, to forgo both of these approaches. And this is precisely what he does, construing consistency and, associatedly, the notion of consequence utilized in the concept of conservativeness, in *primitively modal* terms. Roughly: a statement is a consequence of others if and only if it is not *possible*, in a primitively modal sense, for it to be false while they are all true, and is inconsistent with others if and only if its negation is a consequence of them.

These proposals have a simple corollary. A consistent theory will be one whose axioms are *possibly collectively true*, where this possibility is to be understood in Field's primitive sense.[5] And a believer in the conservativeness of a particular mathematical theory is committed, on Field's account, to endorsing this possibility.

---

[4] In Field's own formulation: a mathematical theory S is *conservative* if, for any nominalistic assertion A and any body of such assertions N, A is not a consequence of $N + S$ unless A is a consequence of N alone. *Realism, Mathematics and Modality* (New York: Blackwell, 1989), p. 125.

[5] Reflect that a set of sentences – in particular, a mathematical theory – is consistent provided that each of its ingredients, S, is consistent with the remainder, $\{S_1, \ldots, S_n\}$. So if $\{S, S_1, \ldots, S_n\}$ is a consistent theory, then, following through on the foregoing construals, not-S will not be a consequence of $\{S_1, \ldots, S_n\}$, and hence it will be possible for $S_1, \ldots, S_n$ and not-not-S, = S, to be simultaneously true.

The objection that is to occupy us arises at this point from the consideration that Field's primitive operator of possibility is *object-linguistic*. It has to be, of course. The program would lose any advantage over more traditional versions of nominalism were it to construe claims about consistency metalinguistically. For if consistency is so construed, a nominalist who believes in the widespread conservativeness, and hence consistency, of mathematics is committed to a belief in nothing other than the feasibility of the traditional nominalist program – of providing concrete (or otherwise nominalistically acceptable) *models* of mathematical theories. And that is just what Field wanted to distance himself from. But when the operator is object-linguistic, Field's belief in the conservativeness of, e. g., number theory would appear to commit him to the view that the Peano axioms, say, might have been true as standardly interpreted, and hence that the non-existence of numbers is a mere *contingency*. As a nominalist, he disbelieves in the existence of abstract objects. But number theory, he holds, is conservative, so consistent, so – if the reasoning just offered is sound – what its axioms collectively say might have been true. And if that had been so, there would have been numbers.

This may seem odd. But why is it an objection – why should the contingency of numbers and other abstract mathematical objects be awkward for Field? For at least two reasons. First, granted that standard mathematical theories are conservative, and that their deployment in the formulation of, e. g., physical theories can be avoided in the manner Field envisages, there is a question about the possibility of anyone having *grounds* for nominalism in the first place. If whether numbers exist is contingent, the question is presumably beyond a priori resolution. An appraisal will therefore demand empirical evidence. Field agrees with the present authors, however, in rejecting any idea of a *direct*, intuitional or "perceptual" epistemology of abstract objects.[6] And the very conservativeness of number theory, along with its dispensability in formulating scientific theories, would have the effect

---

[6] Along lines, for example, famously suggested – or so it is often supposed – by Gödel (cf. "What is Cantor's Continuum Problem?" in Paul Benacerraf & Hilary Putnam, eds., *Philosophy of Mathematics: Selected Readings* (New York: Cambridge, 1983, 2nd ed.), pp. 483–4); or perhaps as developed by Penelope Maddy, "Perception and Mathematical Intuition," *Philosophical Review, LXXXIX* (1980): 163–96. It would take us too far afield to rehearse our objections to these views here. For some of them, see Wright, "Why Numbers Can Believably Be," *Revue Internationale de Philosophie*, CLXVII (1988): 425–73, esp. pp. 464–5; and Hale, *Abstract Objects* (New York: Blackwell, 1987).

that there can be no nominalistically statable, *indirect* evidence for or against the existence of numbers. Whether or not the alleged contingency is realized would seem, therefore, to be in principle beyond evidence of any sort, and consequently utterly imponderable. So why is Field a nominalist? Why does he presume to a view on the matter?

In his recent book, reacting to this thought, Field writes as follows: "if the dispensability programme *can* be carried out" – that is, if it can be shown not merely that mathematics is conservative but that the hypothesis of mathematical objects is dispensable without loss in explanations, in descriptions of our observations, in accounts of metalogic, and so on:

> that gives us reason to not literally believe mathematics but only to adopt a fictionalist attitude towards it. Admittedly, we can't have *direct evidence* against mathematical entities. We also can't have direct evidence against the hypothesis that there are little green people living inside electrons and that are in principle undiscoverable by human beings; but it seems to me undue epistemological caution to maintain agnosticism rather than flat out disbelief about such an idle hypothesis. I think that platonism has seemed a plausible position because it has been assumed that the existence of mathematical entities is not an idle hypothesis. But if ... the hypothesis is dispensable without loss ... it is natural to go beyond agnosticism and assert that mathematical entities do not exist (*Realism, Mathematics and Modality*, pp. 44–5).

One observation about this response would be that the general form of transition it makes, from recognition of the necessary absence of all evidence germane to a hypothesis to a "natural" belief in its negation, is arguably completely unwarranted unless there is independent reason to think that there *would* be germane evidence if the hypothesis were true. That is precisely not the situation in the relevant case. But matters are worse than that. For idle hypotheses, in the sense Field means, have idle negations, too. So "flat out disbelief" – belief in the negation – of what is idle cannot in general be justified merely because it is idle. Agnosticism does not here reflect "undue epistemological caution"; it is demanded by mere consistency.[7]

There is a second and deeper reason, however, why the contingency of abstract mathematical objects might be thought an objectionable outcome. This is the reflection which the contingency in question

[7] Or if there is a principle that sanctions disbelief in some such cases, it must involve reference to something more than idleness – and Field does not identify, still less support any such principle.

would seem to be, metaphysically, absolutely *surd* – that there is no prospect of any account of what the alleged contingency could be contingent on. It is deeply rooted in our ordinary conception of contingency that there ought in general to be explanations of *why* things which might have been the case are not the case, and of *why* things are the case which might not have been the case. But in the present kind of instance – the existence of the integers, or real numbers, for example – there is not even a glimmering of how such an explanation might proceed. This seems a strong sign that the notion of contingency would be *misapplied* in the present context, and hence that there is philosophical error in any assumptions that commit Field to so applying it.

Field has responded tersely to this criticism, charging that the idea that he suffers any such commitment depends on an equivocation. He writes

> Hale and Wright both note that I regard mathematics, and the existence of mathematical entities, as consistent ... and that I take consistency as a primitive modal notion, a sort of possibility. They then argue that since I regard it as false that there are mathematical entities, I must hold the existence of such entities to be 'contingently false'; and they both proceed to interpret 'contingently' in some non-logical sense (i. e. some sense other than 'neither logically true nor logically contradictory'), to make the position seem absurd. ... It should be noted that if [their] objection were good, it would apply equally well against any platonist who is not a logicist: that is, any platonist who agreed that mathematics goes beyond mere logic, and hence that the denial of mathematics is logically consistent and hence 'contingent'. But of course it is not good, for as I've said it turns on an equivocation on the meaning of 'possible' (*Realism, Mathematics and Modality*, pp. 43–5).[8]

Field's response, then, is that the sense of 'possible' for which he acknowledges commitment to the possible truth of the Peano axioms is not one that subserves the objection, which is consequently defused.

[8] The claim that "if [our] objection [i. e., the contingent falsehood objection] were good, it would apply equally against any platonist who is not a logicist" tacitly assumes, in effect, that no relevant notion of necessity (that is, no "absolute" notion) is available, other than the austere notion Field describes. Obviously enough, if it is *not* – save in that austere sense – a contingent matter whether numbers exist or not, or whether or not number-theoretic statements are true (i. e., if they are only *weakly* contingent), then the existence of numbers, and the truth of number-theoretic statements, is, if not impossible (conceptually) necessary. There is thus clear space for a version of Platonism, falling well short of logicism as Field conceives it, that is not prey to the contingent falsehood objection we have urged against Field's own position.

But how exactly? There are two – and, so far as we can see, only two – possible lines of thought that would be relevant.

(A) It might be maintained (i) that Field's preferred "austere" notion of logical contingency is the only one available; and (ii) that the objection needs a richer but dubious notion of contingency.

(B) It might be granted that there are available other, less austere but still respectable notions of contingency, but then maintained that while the nonexistence of numbers and other mathematical entities is contingent in the original austere sense, it is not a contingency – a fortiori not a metaphysically surd contingency – in the richer sort of sense demanded by the objection.

The considerations to follow will depend on no assumption about which, if either, of these lines might be Field's own preference.[9] Our interest is rather in the moves available in principle to the nominalism Field has advocated, irrespective of other aspects of its author's actual views. We shall argue that both strategies, (A) and (B), are fraught with difficulty. Since Field's claim – that the original objection is defused by reflecting that it is only to the "logical" contingency of, e. g., numbers that his belief in the conservativeness of number theory commits him – requires that one or the other strategy be defensible, it should be rejected as ill-considered.

The program is as follows. The next section will argue that claims A(i) and A(ii) are highly implausible at best. Section II will briefly explore certain problems with strategy B. This will lead naturally into a discussion, in section III, of certain internal tensions of Field's views about the semantics of mathematical theories with the sorts of consideration by which he seemingly regards Platonism as discredited.

Even if the original objection is not to be dismissed as relying on an equivocation, the nominalist still, of course, has the option of biting the bullet: of allowing that the existence of numbers and other abstract objects is indeed contingent in a richer sense than Field's austerely logical one and of attempting to maintain that the resulting position is unexceptionable. Section IV will attempt to deepen and refine the original objection in various respects. In particular, we shall begin to explore

---

[9] Actually, Field has made it quite clear in conversations and correspondence that he would want nothing to do with the former.

whether the nominalist has any prospect of a satisfactory head-on response to it via play with the idea of *brute contingency*.

I. Strategy A

Suppose that there were no other relevant notion of contingency than Field's austerely logical notion – according to which, approximately, a statement is contingent just in case it is neither (first-order) logically true nor (first-order) logically contradictory.[10] How might someone proceed to argue – as per A(ii) – that the contingency objection would then be stifled?

One idea might be that, if we introduce 'it is logically contingent that' as a sentential operator, simply and solely as a definitional abbreviation for 'it is neither first-order logically true nor first-order logically contradictory that', we have – so far, anyway – done nothing to provide a sense for statements of the form 'the fact that ... is contingent upon the fact(s) that ...', and so we cannot so much as intelligibly formulate the question, "What is the (allegedly) contingent fact (that there are no numbers) contingent *upon*?" Nothing in the way in which we would have introduced the contingency operator would prepare for the idea that when it is (merely) contingent that the things are thus and so, there is – still less has to be – something else upon which their being thus and so *depends*.

This, however, has no force. Asking what the putative contingency, that there are no numbers, is contingent on is a convenient and natural way to raise the kind of question that fuels the objection. But we can just as well ask *why* there are no numbers,[11] given that it is, in Field's

[10] In fact, this oversimplifies Field's notion. Field's explanation (*Realism, Mathematics and Modality*, pp. 32–8) of "LTrue" (read 'it is logically true that'), in terms of the laws governing it, makes it clear that many statements involving this operator would rank as (austerely) logically true, and that various statements involving second-order quantification would be among them. It is thus strictly incorrect to suggest that a statement is logically contingent for Field if it is neither (first-order) logically true not (first-order) logically contradictory. The fact remains that (austere) logical truth, contingency, and impossibility are, for Field, essentially a matter of logical form – the point is merely that this is not exhaustively characterizable in first-order terms. It should be clear that none of our arguments makes essential play with this simplifying assumption adopted here – in particular, there is no question but that the examples of (austere) logical contingency discussed in this section would rank as such, even given Field's own slightly more generous characterization of the notion.

[11] As indeed, in one of the passages to which Field is responding, Wright in effect does. The passage, which Field actually quotes (cf. *Realism, Mathematics and Modality*,

austere sense, contingent that there are none. That is a perfectly well-formed and legitimate request for an explanation, for which no further preparation, supplementary to Field's account of contingency, is needed. According to Field, it is logically contingent but false that there are numbers. Well, there are many statements that are logically contingent but false, and it is typically perfectly proper to ask *why* these statements are false. So a special consideration would be needed – and none is so far in sight – to justify rejecting the question as improper in the present case.

Admittedly, there is a gap between its being "typically perfectly proper" to ask why an austerely contingent matter is as it is and there being, on the other, an *internal conceptual connection* ensuring the availability of at least an appropriate *category* of explanation for each austerely contingent matter. We make no claim to have closed this gap, and it is thus theoretically open to a supporter of Field to attempt to exploit it – more about that later. The fact remains that asking for an explanation of why an austerely logical contingency is unrealized is, at least in a wide class of cases, perfectly legitimate. If there are other austere contingencies where the demand for explanation is out of order, there ought to be some stateable distinction between the two sorts of cases. And argument will be needed that the existence of numbers belongs in the latter class. There is no progress otherwise.

It is in any case very implausible that, in line with A(i), a restriction to austerely logical contingency could defensibly be maintained. Any statement is austerely logically contingent whose most explicit first-order formalization is neither a theorem of first-order logic nor refutable therein.[12] So, for instance, both

(i) There are vixens weighing in excess of eighty pounds.

and

(ii) There are male vixens.

rank as austerely logical contingencies. More generally, austerely logical contingency will embrace many statements whose truth it is attractive

---

p. 43), runs: "Field has no prospect of an account of what the alleged contingency is contingent *on*. The world does not, in Field's view, but might have contained numbers. But there is no explanation of *why* it contains no numbers; and if it had contained numbers, there would have been no explanation of that either. There are no conditions favorable for the emergence of numbers, and no conditions which prevent their emergence" (Wright, *op. cit.*, p. 465).

[12] But see note 9.

to regard as, in old-fashioned terminology, *conceptually* impossible but a first-order demonstration of whose impossibility, if one could be given at all, would require supplementary content-explicating postulates of a nonlogical sort. Now, in each of the kinds of example typified by (i) and (ii), respectively, the request for an explanation of *why* the statement in question is false is apparently perfectly intelligible, and it is easy to say how at least the beginnings of an answer should proceed. Given that first-order logic does not preclude the truth of such a statement, the question is, what does? And the answers are, roughly, the typical genetic make-up and feeding habits of vixens, and the nature of the concept of a vixen, respectively.

So much is merely common sense. But the question now is: How should a proponent of A(i) respond to the demand for an explanation of why there are no male vixens? How to respond without betraying an understanding of a broader notion of contingency than the austerely logical variety – a notion of contingency that, if numbers are so contingent, will subserve the original objection, even if, contrary to what was just suggested, austere logical contingency will not?

Well, reflect on the possibilities. In effect, the question – "Why are there no male vixens?" – sets a trilemma:

(i) There is the option of rejecting the demand for an explanation as illegitimate.

(ii) There is the option of attempting an explanation similar to that of why there are no vixens weighing in excess of eighty pounds.

(iii) There is the option of attempting an explanation that preserves a distinction between the two kinds of example.

Option (i) is surely hopeless. It simply is not credible to suppose that there is nothing at all to be said about why there are no male vixens.[13] But it would be equally incredible to suppose that a proper explanation might assimilate the two examples, attributing the non-occurrence of male vixens to *causal* factors, specifically, to the genetic make-up and typical environmental conditions of foxes. Someone who thought that a satisfactory explanation might proceed along those lines would owe a story about what kind of variation in those factors might secure the emergence of male vixens. But there is no such story to tell.

---

[13] A parent who so responded to her child's question would surely be failing to explain something that a responsible parent ought to want to explain.

The fact is that only option (iii) is a credible response. More specifically, however indigestible the thought and unpalatable its terminology may be to philosophers weaned on the thin nutrient of W. V. Quine's "Two Dogmas of Empiricism,"[14] the only credible response is to acknowledge that it is not in the nature of the concepts *male* and *vixen* to allow coinstantiation. There may or may not be more to say in response to the request for explanation in this particular case – but that much must be at least the basis of any satisfactory response.

As soon as it is granted, however, that there is work to be done by notions of conceptual possibility and impossibility which apply differentially within the class of austerely logical contingencies, houseroom has been given to just the kind of richer notion of contingency that it was essential to strategy A to repudiate. Say that a statement is *strongly contingent*, just in case it is both austerely logically contingent and not conceptually impossible: no impediment to its truth is generated either by first-order logic or by anything internal to the character of the nonlogical concepts that it involves. If this notion is granted and, with it, the resources for a distinction, among austerely logical contingencies, between those like "There are vixens weighing in excess of eighty pounds" which are also strongly contingent and those like "There are male vixens" which are not, the objection immediately resurges in the form of a dilemma, namely, to which camp should "There are natural numbers" be assigned?

If to the latter – if the existence of numbers is regarded not as a strong contingency but as a conceptual impossibility – then that is to abandon strategy A in favor of strategy B. But if to the former – if the existence of numbers is conceived as a strong contingency – that immediately invites the assimilation of "There are natural numbers" to "There are vixens weighing in excess of eighty pounds" – a statement whose falsehood allows of a broadly *causal* explanation. And the idea that such an explanation might be possible of why there are no numbers is not merely in tension with our ordinary conception of abstract objects as acausal but is demonstrably inconsistent with the belief that number theory is conservative with respect to nominalistic discourse.[15]

---

[14] In *From a Logical Point of View* (Cambridge: Harvard, 1980, rev. ed.).

[15] *Proof*. Suppose that a causal explanation of the nonexistence of numbers could be given. Then assuming that such an explanation will be of the familiar deductive-nomological pattern, there will be some nominalistically formulated laws (briefly, L) and nominalistic statement of initial or standing conditions (briefly, A) from whose conjunction the statement that there are no numbers could be derived. We may

Even if the presumption of *causal* explicability can be defeated, there is no way round the point that to conceive of the existence of numbers as a strong contingency is a commitment to the view that nothing in a correct understanding of number theory or its underlying logic poses any obstacle to its truth. The idea would have to be, rather, that *it is merely a feature of nature* that, if it is, the world is numberless. And to try to take that conception of the matter seriously is simply to point one's chin at the original question: *Why* has nature turned out that way – what kind of consideration might in principle explain the nonexistence of numbers once they are acknowledged to be a strongly contingently nonexisting class of entities? And how can it be legitimate to think of the matter as, so to speak, merely natural, unless that question has some kind of constructive answer?

II. Strategy B

The question can be avoided by reverting to strategy B. This, recall, involves granting that there is, in addition to any austerely logical sense of contingency like Field's, a richer notion; and then claiming that while the existence or nonexistence of numbers is a contingent matter in the former sense, it is *not* so in the richer sense. The question – "Why are there no numbers?" – would thus admit of a direct answer: the nominalist could simply adduce whatever considerations show that numbers are impossible in the richer sense.

The problem, obviously, is to locate a suitable richer notion – one that allows successful argument that numbers and other abstracta are indeed richly impossible. The simplest line, naturally, would be to attempt to maintain that the existence of numbers is conceptually impossible in the relatively straightforward manner of male vixens, or squares with unequal sides. On this view, all existentially committed classical arithmetical theorems would conceal definitionally elicitable inconsistency. One difficulty with this is that it is crucial to Field's project that arithmetic be formally consistent (else it cannot be conservative). And presumably its formal consistency should be sufficiently

assume that the conjunctive premise is consistent, and that neither conjunct alone entails this conclusion. Let N be elementary number theory. Then, since N entails that there are numbers, not-A is a nominalistic statement entailed by $N + L$ but not by L alone, whence – bearing in mind Field's definition (cf. note 4 above) – N is not conservative.

robust not to be disrupted merely by expansion of its statements in the light of correct definitions of certain of the concepts involved. But there seems, in any event, no prospect of making the necessary case, i. e., of a successful argument that while there is no overt formal inconsistency in, e. g., the (normally regarded as true) statement that $117 > 5$, it can nevertheless, with the help of acceptable definitions of the specifically arithmetical terms involved, be made to give way to explicitly contradictory consequences, in the manner in which the statement that there are squares with unequal sides gives way to an explicit contradiction on replacement of the geometrical term 'square' by its normal definiens.

If there is to be a conceptual impossibility here, it has to be of a subtler sort. Actually, it is hard to see how the alleged impossibility could be held to derive from anything but the very abstractness of numbers, so that the existence of any kinds of abstract object would have to be impossible. Of course, taken in one way, this would merely inherit the difficulties just glossed. For it seems equally implausible to suppose that the very concept of *abstract object* is self-contradictory after the fashion of *male vixen* or *four-sided triangle*. But perhaps it could be argued that the concept, though not self-contradictory, is nevertheless defective in such a way as to guarantee that it can have no instances.

How? Field's writings endorse *passim* a number of familiar anti-Platonist lines.[16] In particular, he contends – in company with many others, of course – that the acausality and lack of spatiotemporal position of Platonistically conceived mathematical entities make for serious, if not insuperable, difficulties over knowledge of and reference to them. He also endorses the well-known arguments against Platonism given by Paul Benacerraf[17] and by Hilary Putnam.[18] Unless bolstered with some substantial further premise, however, none of these arguments will deliver the type of conclusion that strategy B requires, namely, that there can be no such objects as numbers, sets, etc. For instance, the proper conclusion of a cogent argument based upon some presumed causal constraint on knowledge would be merely that, *if* there are such objects as numbers, we can have no knowledge concerning them. Likewise, the upshot of the arguments – whether based upon some proposed causal theory of reference, or upon the kinds of considerations adduced by

[16] *Realism, Mathematics and Modality*, pp. 68–70, also pp. 23 and 230 ff.
[17] "What Numbers Could Not Be," *Philosophical Review*, LXXIV (1965): 47–73.
[18] "Models and Reality," *Journal of Symbolic Logic*, XLV (1980): 464–82.

Benacerraf or Putnam – purporting to establish that (determinate) reference to numbers is impossible, is just *that*, i. e., that, *if* there are such things as numbers, there is no way we can make (determinate) reference to them. If these arguments are to yield the further conclusion that there simply cannot be any such objects at all, then further premises are needed, perhaps directly to the effect that no objects can exist of which knowledge is in principle impossible or to which there cannot, even in principle, be identifying reference, or perhaps yielding the desired result some other way. In any case, it is clear that proving the impossibility of numbers, or abstract objects generally, is a very tall order which no extant line of philosophical argument meets. If nominalists had such an argument, we would have heard about it.

But if the case were somehow made, there would be a further issue, worth briefly airing. Field, remember, has no quarrel with Platonism as an account of our actual understanding of mathematical language. And this literalism is an attractive and essential element in his kind of nominalism, distinguishing it from nominalist programs of the more traditional reconstructive (and failed) sort. What, in his view, is wrong with Platonism is not that it misrepresents our actual understanding of mathematical statements, or credits them with a metaphysically impossible content, but merely that it tells ontological untruths. But suppose we had the impossibility proof that strategy B demands. What repercussions ought such a discovery, once it were generally known, to have on the semantics of mathematical statements? Could it be appropriate to continue to take the literalist view?

It may seem that exactly this option – combining a literal construal with denying the possible existence of mathematical entities – is what is secured by going *fictionalist.* No doubt it would be absurd to represent a community as *sincerely* making mathematical claims referring to entities that they regarded as impossible. But a face-value construal of their mathematical language need not involve doing so, provided we can avoid interpreting the commitments that their mathematical assertions express as *beliefs in the truth* of those sentences. And that is just what mathematical fictionalism enables us to do. Fictionalism opens a gap between (i) acknowledging that, if the belief prevails in a community that there are, necessarily, no abstract objects, it cannot be part of the best semantic account of the community's mathematical assertions to represent their *makers* as purporting reference to and quantification over such objects, and (ii) abandoning the face-value construal of the terms and quantifiers that their *mathematical language* contains. The gap is opened precisely because fictionalism makes a distinction between endorsing a mathematical statement and accepting it as true.

There is a residual concern, however. Only fictionalism, it seems, can reconcile three crucial ingredients in Field's philosophy of mathematics: (i) a face-value construal of the semantics of number-theoretic language, (ii) the – in his view – justified disbelief that any actual objects answer to the terms and quantifiers discerned therein, and (iii) our endorsement of the axioms and theorems of number theory. But the envisaged case, remember, is one in which we are supposed to be rightly persuaded not merely that there are not but that there *could not* be numbers – the fiction has to be one of acknowledged *impossibilia*. There will therefore be issues about the *intelligibility* of mathematical language, construed at face-value, which fictionalism as such does nothing to address.

The interactions between intelligibility and the perception of various forms of conceptual impossibility are of course subtle. One might believe in the incoherence of the idea of time travel, for instance, yet still, to all intents and purposes, understand and enjoy a fiction in which such technology plays a central part. By contrast, many of us are persuaded of deep conceptual singularities in the idea of the *Cartesian ego* as a temporal continuant – persuaded, e. g., that there is no coherent notion of what it would be for such an entity to endure through a given period of time. Would there nevertheless be no difficulty about devising an intelligible fiction in which such entities were to be the central – repeatedly referred to and quantified over – characters? Again, assuming – plausibly – that there are conceptual obstacles in the way of interpreting an imagined community as genuinely operating with a deviant arithmetic, there is a real question concerning the intelligibility and coherence of a fiction in which the protagonists practice an arithmetic in which 7 + 5 = 13, etc.[19]

Obviously, in the absence of any clear indication of the kind of conceptual impossibility that a successful prosecution of strategy B might disclose, or the kind of considerations that might be involved, there has to be an element of shadow boxing about this. Nevertheless, there are, unquestionably, kinds of conceptual incoherence that would jeopardize the intelligibility even of fictional reference and quantification. So there

[19] Of course, we can readily make sense of a fiction in which the protagonists are depicted as making, more or less systematically, what would, by their own lights, be reckoned arithmetical mistakes; but that is beside the point.

can be no a priori assurance that an impossibility "of a more subtle sort" which pursuit of strategy B might disclose will not have such an effect – no assurance that the strategy could be prosecuted without casualty to the semantical literalism that distinguishes Field's position from nominalism of more traditional kinds.

III. Literalism and the Standard anti-Platonist arguments

That there is cause to doubt the coherence of literalism with other aspects of Field's views may be brought out by recalling some of the considerations to which Field actually appeals as discrediting Platonism – in particular, the allegedly irremediable indeterminancy disclosed by Benacerraf's argument and the idea that their acausality and lack of spatial position preclude any satisfactory account of how reference to numbers is possible. For these considerations, if cogent at all, seem by themselves quite sufficient to make trouble for a face-value construal of, e. g., number theory, even in the context of an overall mathematical fictionalism. If refer*ence* to numbers is ruled out on conceptual grounds – whether or not the further step is taken of concluding that the existence of the refer*ents* themselves is impossible – it is immediately cast into doubt whether the *face-value* construal adds up to a *coherent* semantic story, let alone a preferable one. Nominalism-cum-literalism has it that arithmetical statements, as ordinarily understood, have determinate though unrealized (or only vacuously realized) Platonistic truth conditions. But in order for any statement to have determinate truth conditions, there must be something correct, ergo coherent, to say about the reference of its singular terms, the satisfaction conditions of its predicates, the range of its quantifiers, and so on – and no *coherent* account of these matters, in the nature of the case, can stand in violation of general conditions on the very possession of such semantical characteristics. A face-value construal of mathematical statements – even if offered in a fictionalist spirit – is bound to offend in this respect if it involves attempting to assign reference to the relevant terms in a way that clashes with deep intrinsic features of the reference relation itself; and just that, according to proponents of the Benacerraf and acausality arguments, is what it must involve if it represents those terms as purporting reference to *abstract* objects.

To put the matter another way: one simply has not given a semantics for a discourse if the details of the proposed account controvert essential features of the semantical notions that provide its very framework. But, according to the arguments in question, to represent, e. g., the numerals as having reference, if to anything, then to abstract objects, is precisely so to controvert essential features of the relation of singular reference. Somebody who accepts such arguments should therefore hold that, rather than deliver unrealized (perhaps necessarily unrealizable) truth conditions for the sentences of arithmetic, the face-value construal gives implicit offense against the semantic relations on which it needs to rely in order to say anything determinate about what the truth conditions of arithmetical sentences are – and thus fails to deliver anything at all.

If this is right, then any anti-Platonist driven by the familiar arguments cited approvingly by Field is left, so far as we can see, with just three options. The first is *traditional nominalism*, according to which the appearance of singular reference and quantification presented by mathematical discourse may be genuine, provided the objects referred to and quantified over are concrete. The second is a *reductive* program to disclose that appearance as spurious, and eliminate it. The third is to accommodate the appearance in the only other way that seems open – to go *formalist* and reject altogether the idea that mathematical statements are bona fide contentful statements having genuine truth conditions. On the latter view, such statements possess a determinate syntax, which may be exploited in formal derivations, but no semantics – a fortiori, not a Platonistic one.

What is not apparently an option, for anyone driven by the familiar arguments, is Field's own view, or any view that shares the idea that mathematics, while a fiction, is a *Platonistic* fiction.[20]

IV. Brute Contingencies

Let us see where we have arrived. The literalist can avoid Platonism only at the cost of denying that mathematical theories are true. The question is then unavoidable: Do such theories fail of truth – specifically, do mathematical singular terms fail of reference – necessarily or merely contingently? The preceding suggests that the former option, in

---

[20] Doubts along the lines of the last few paragraphs are aired by Hale in "Nominalism," in A. D. Irvine, ed., *Physicalism in Mathematics* (Cambridge: Kluwer, 1990), pp. 136–7.

effect that pure mathematics is a fiction of impossibilia, is not the way for a supporter of Field to go. It remains to reconsider the alternative, that mathematical and abstract objects generally exist or fail to exist by grace of strong though inexplicable contingency – precisely the position targeted by the main objection.

The objection could be neutralized if it could be shown that the demand for explanation is here illegitimate; that there are, in any case, certain strongly contingent matters – matters to which the case of mathematical objects may properly be compared – of which, however they are resolved, there is no conceivable explanation. "No conceivable explanation" because not every contingency need, of course, be explicable *in fact*. For instance, if only subsumption under ulterior physical laws can explain the obtaining of any particular physical law, then it will be vain to seek an explanation of any *fundamental* physical law, if such there be. Still, there will be *conceptual* space – as it happens unoccupied – for an explanation of such a law; (for the law might not have been fundamental). By contrast, someone who wants to dismiss the question – "Why are there no numbers?" – as inviting an explanation of a matter that is actually a brute contingency, will need to meet the point that there seems to be no corresponding conceptual space – we seem to have no conception of how an answer might run in principle.

Pursuing this direction will involve a two-stage operation: first, a search for other relevant examples of brute contingency – contingencies where we conceive not merely that no explanation need be possible of the direction in which it is resolved but, more, that there simply is no *category* of explanation that might in principle serve; and second, an argument that the existence of numbers, and of abstract objects generally, belongs with this group. What are the prospects?

In correspondence, Field has remarked:

> On the question of why there are no numbers, my view is that this is like why is there matter, or why are there no immaterial minds, or why is there no God: you simply can't expect there to be an answer to such basic existence and non-existence questions ('basic' in the sense of: applied to a whole 'ontological category').[21]

This is a very natural thought, and it is therefore important to see that it comes to not very much. For unless the claim is to be simply dogmatic, it needs the backing of an *account* of "basic ontological category" which both discloses why explanations of "basic" matters of existence

[21] Letter of September 8, 1990.

and nonexistence are indeed not to be expected and also ensures that numbers, and abstract objects generally, turn out to be a basic ontological category in the relevant sense. But it is not clear how such an account might run. Indeed, it is obscure what intuitive conception of a basic category covers each of the cited examples. Notice that certain quite attractive proposals will not subserve Field's thought in any case. Persons, and living organisms, will both be basic on many accounts, including – though its author expresses some uncertainty – the largest-class-of-objects-sharing-criteria-of-identity idea of Michael Dummett.[22] But presumably it is not absurd to seek explanations both of why there are any persons and why there is life.

There are, however, other cases which may be held to exhibit brute contingency, and so to provide the model Field needs. Consider the ageless conundrum – "Why is there something, rather than nothing?" An unanswerable question if physical existence is at issue and a satisfactory explanation why a physical state of affairs obtains has to advert to a causally prior situation in which it does not obtain. For a physically empty world would presumably have no causal progeny. Or consider the "great contingency" – the global state of affairs that verifies all actually true contingent propositions.[23] Since the obtaining of each of its ingredients is a matter of contingency, the exact character of the global state is no doubt itself a contingency. But there can be no explaining its exact character – at least not if contingencies can be explained only by reference to further contingencies. For there are, precisely, no *further* contingencies to appeal to; everything further is a matter of necessity.

Do such examples offer any comfort to a supporter of Field? To stress: it will not do just to observe that a certain sort of brute contingency is illustrated by examples of this sort, and is therefore at the call of the nominalist according to convenience. The crux will be whether "There are no numbers," or "There are no abstract objects," comes in the same rather special sort of category. And argument is required. The original objection was that the concept of contingency has no proper application to the cases in question, on the ground that contingencies invite the question – "Why?" – and that we have, in the cases in question, no concept of the form a satisfactory answer might assume. The onus is therefore on the nominalist to show that this is a bad

---

[22] *Frege: Philosophy of Language* (London: Duckworth, 1973), pp. 75–6.

[23] Here we are indebted to Peter Carruthers and Gideon Rosen for suggestions and helpful discussion.

reason – that the correct account of our lack of any such concept is not that the idea of contingency is being stretched beyond the limits of its intelligible application, but that – as in the case of the two mooted examples – explanation is here pre-empted in perfectly intelligible ways, quite compatibly with the claim of contingency.

Reflect on what would be demanded by a serious analogy with those two examples. There the essential work is done by certain restrictive principles: roughly,

> The explanation of the obtaining of a (physical) state of affairs must advert to a causally prior state of affairs in which it does not obtain.

and

> The explanation of a contingency must proceed in terms of further contingencies.

The key point is that such restrictions pre-empt explanation as soon as something sufficiently general is put up as the explanandum, and that this – rather than any misappropriation of the notion of contingency – is at the root of our inability to see how a satisfactory explanation of the contingencies in question might proceed. Accordingly, to establish a strict parallel, the nominalist needs to call attention to a restriction on explanations of mathematical existence which has a corresponding effect: our inability to explain the existence, or nonexistence, of numbers, or abstract objects generally, should be traced to the fact that the demand to do so implicitly transgresses bounds within which a satisfactory explanation would have to work.

The concept of mathematical explanation is a difficult one. But we can at least take it that explaining pure mathematical facts is always a matter of giving proofs, even if not all proofs are explanatory. That might suggest a suitable line for the nominalist. Mathematical proofs, it might be argued, accomplish demonstrations, and therefore explanations, only *within a framework* supplied by the basic assumptions, including assumptions of existence, made by the relevant mathematical theories. Mathematical why-questions, the suggestion continues, are *internal* questions: grant the framework and ontology presupposed by the Peano axioms, and we can explain the infinity of the primes; grant the framework and ontology of Zermelo-Fraenkel set theory and we can – or so it is sometimes thought – explain the truth of the Peano axioms. But just as in the ascent to "Why is there any physical substance at all?" or "Why is the totality of contingencies composed just the way it is?" escalating generality of this sort must sooner or later clash with the possibility of explanation. It has done so by the time we ask why there are (no) mathematical objects.

Although this may seem to pass muster, on closer inspection the analogy limps at the crucial point. What exactly is the restrictive principle in play? Presumably, something like,

> The explanation of any matter of mathematical existence or nonexistence can only be accomplished within a framework in which the existence of a range of mathematical objects is assumed.

But that principle overlooks an asymmetry in the character of mathematical explanations of existence and nonexistence, respectively. No doubt it is only if the existence of the natural numbers is granted that someone can set about trying to explain why the ratio of primes $\leq n$ becomes smaller and smaller as n increases. But what about the explanation offered by Euclid's proof of the *non*existence of a greatest prime? Here it is inessential to the success of the explanation to presuppose that there really are numbers. For the proof succeeds in any case: if the natural numbers exist, then, by Euclid's reasoning, it is in their nature that there is no finite limit to the occurrence of primes; and if they do not, then there are no prime numbers and, a fortiori, no greatest among them.

The point is perfectly general. Mathematical explanations of nonexistence proceed, without existential presupposition, by showing that the corresponding existence claim clashes with essential aspects of our *concept* of the type of mathematical objects concerned, that the concept's being fully and properly exemplified would be inconsistent with the truth of the existence claim. There is no range of objects demanded by all such explanations, whose existence or nonexistence is thereby pushed beyond explanation. The candidate restrictive principle is false.

No firm conclusion can be drawn, of course, on the basis of this one failed comparison. But is there a better one? The heart of the objection is the claim that, while there is no simple connection between contingency and the possibility of explanation, there is nevertheless a *presumptive* connection: explicability is the default status, and the exceptions have to be *explicable* exceptions – something has to be done to show *why* ordinary explanation is precluded. The implicit principle, P, is that, for any putative (strong) contingency C, it should be possible *either*

(a) to indicate at least a category of explanation that an explanation of C might in principle instantiate – to gesture at the lines along which an explanation of *C* might run, even if as a matter of fact no correct explanation of that kind can be given; *or*

(b) to explain why there cannot be any such category – why any explanation of *C* would be required, by virtue of certain special features of the case, to meet unsatisfiable conditions.

And the backing for the principle is merely the natural idea that since, when it is a (strong) contingency that *C*, things might have been otherwise, it ought to be perfectly appropriate to ask *why* they are not otherwise, i. e., why *C*. Since this question is appropriate only if a suitable category of explanation exists, the onus has to be on someone who accepts that no such category exists but who is determined that *C* is nevertheless a (strong) contingency, to show why, in this special case, all space for explanation is closed off, i. e., to satisfy the second disjunct of the principle. This option can be nicely accomplished in the case of the two examples reviewed. But we have so far no clear suggestion how it might be accomplished for the putative contingency that there are no numbers.[24]

All this, however, is undeniably inconclusive. Can we conjure a wider class of examples that may help to place the putative contingency of mathematical objects in a more luminous perspective? We already have

(1) The "great contingency";
(2) "Why is there anything material?" – the Conundrum;
(3) Any fundamental scientific law, *L*;
(4) The existence of God – or, presumably, of any Being whose concept, like that of the Christian God, requires that its existence depend on nothing else.

Other examples do occur. Suppose physics suggests, for instance, that the universe is approximately *n* years old. Then the question

(5) Why did the universe begin approximately *n* years ago?

will admit of no causal explanatory answer since it presupposes that, prior to that period, there was no physical process – hence nothing to determine that process should start then rather than earlier. Or consider

[24] It is important to remember, moreover, that, since we are at this point concerned with a theorist who regards the strong contingency of numbers as the correct account of their modal status, hence as something imposed by an ordinary understanding of number – it has to be possible to draw on *agreed aspects* of that understanding to discharge the obligations of course (b).

an experiment in which under radioactive decomposition a particle escapes through one slit in a screen rather than another, and suppose that – so we conceive – activity at this micromaterial level is subject only to statistical – indeterministic – law. Then there need be no answer to the question

(6) Why did the particle pass through the top slit?

No doubt such examples put paid to any lingering attraction in the simple thought that, for any bona fide contingency, there has to be something *on* which it is contingent. But collectively they merely reinforce the presumptive connection asserted by the principle, *P*. For in each case we can either indicate quite definite lines along which explanation might have been possible – *L* might have been subsumable under ulterior physical law, for instance, or the laws governing the relevant level of micromaterial activity might have been deterministic – or else, as in cases (1), (2), (4), and (5), we can point to clear conceptual considerations which, drawing just on our ordinary understanding of the example, preclude explanation in principle.

But there *are* more awkward customers. Consider any contingency, *C*, for which, as in cases (3) and (6), we possess a general conception of how an explanation might in principle proceed but where no correct explanation of that sort can actually be given. Then the circumstance – call it *C** – that there is no explanation of *C* will itself be a contingency, and it will be liable to be brute. For how might an explanation of *C** go? Not a priori, presumably, since *C** is a contingency – there might have been an explanation of *C*. But what sort of *empirical* explanation might be given of the contingent absence of all empirical explanation of a given (as it happens fundamental) law, or of the (as it happens) indeterministic character of some range of micromaterial events? Thus, for any *C* of the type illustrated by (3) and (6), the corresponding contingency of the form of *C** already poses a problem to the taxonomy implicit in *P* – it will be, prima facie, a case where it is clear neither what a suitable category of explanation might be nor that there can be, a priori, no such category. Thus we can add to the list, for any fundamental law *L*,

(7) There is no explanation of *L*'s obtaining.

and likewise, in respect of example (6),

(8) There is no explanation of the particle's passing through the top slit.

More generally, for any of a wide range of physical phenomena which we can seemingly readily conceive might have been otherwise and yet of which our best current conception is that they have no physical preconditions – the whole phenomenon of gravitational attraction, for instance, or space's being non-Euclidean, or *n*-dimensional; in all such cases, the claim that there is no explanation of the phenomenon will seemingly be itself a contingency about which we are clear neither along what general lines an explanation might be given nor about what if any conceptual considerations preclude explanation in principle.

Well, are not these reflections enough to dislodge principle *P*? Do we not have to admit now that the intuition of contingency can perfectly justifiably be robust in circumstances where we are unclear whether either limb of principle *P* can be satisfactorily made out? Of course, these, as it were, second-order contingencies are of a special sort, and bear no immediate analogy to the putative contingency that there are no abstract objects. Still, the principle that drives the objection is not, it appears, exceptionless in any case. So what cause for skepticism if the nominalist postulates a further class of exceptions?

In an earlier discussion Wright wrote:

> ... the idea of a contingent state of affairs that is contingent on nothing, admits of no evidence even in principle, and is beyond anything we could justifiably regard as cognition, is simply not a credible piece of metaphysics. ...[25]

Focusing, as we have been, on the question of explanation belongs with the thought that bona fide contingencies are presumptively contingent on something, and that the exceptions should be few and clearly explicable as such. But we should not lose sight of the second component in the quoted complaint – the utter epistemic isolation of the existence or nonexistence of mathematical objects, when construed as a (strong) contingency. Moreover, it is important to be clear that the offense given thereby need not depend on any implicit verificationism – any general aversion to the idea of evidence-transcendent fact. The reason there can be no cognition of, nor evidence for or against, the existence of mathematical objects, conceived as a strong contingency, is because *nothing whatever* – in particular, no uncontroversial contingency – varies as a function of their existence or nonexistence, save

[25] Wright, "Why Numbers Can Believably Be," p. 465; cf. also Hale, *Abstract Objects*, pp. 111–2.

precisely that.[26] It is merely one consequence of this that their existence or nonexistence exerts no differential influence on what is cognitively possible. The conception that mathematical objects exist or not as a matter of strong contingency thus clashes not merely with the presumption that contingencies should be contingent on something but, equally, with the converse principle that bona fide contingencies are presumptively *things on which other bona fide contingencies are contingent.*

There is, of course, a vast range of ordinary unproblematic contingencies that respect both these conditions. That is the normal case. But what is striking is that *all* the examples we have considered, even those which prima facie make trouble for *P*, respect the converse principle. There are manifold uncontroversial contingencies that would not be so but for the detail of the "great contingency," the existence of matter, the obtaining of a particular scientific law, the universe's being *n* years old, space's being *n*-dimensional, the particle's going through the top slit (it leaves traces), God's existing (if he does), *L*'s being fundamental (for if *L* were not fundamental, other superior laws would obtain and would be manifest in differential behavior of matter), and there being no explanation why the particle went through the top slit (for if there were an explanation, the relevant class of phenomena would not be indeterministic and all sorts of things would be different).

The suggestion, then, is that both the following presumptive principles constrain the ascription of strong contingency:

(1) That a putative contingency will be contingent on something – there will be circumstances on which its obtaining depends (and whose citation can thus contribute to its explanation).

(2) That a putative contingency will have things contingent on it – there will be circumstances whose obtaining depends on it (would be otherwise if it did not obtain.)

Brute contingencies are counterexamples to the first. Call a contingency *barren*, by contrast, if it is a counterexample to the second. Then none of the brute contingencies reviewed in our list of eight seems to be barren. Uncontroversial examples of barren contingencies are, indeed, hard to muster. The occurrence of a mental state or episode, viewed in the special way characteristic of epiphenomenalism,

[26] If this were not so, then mathematical theories would not be conservative with respect to nominalistically stateable consequences.

would be one possible type of example. But even that would not also, of course, be brute. What is striking about "There are (no) numbers," viewed as a strong contingency, is that, at least with respect to all other uncontroversially contingent matters, the contingency it putatively expresses is absolutely *insular* – is both barren *and* brute. We conjecture that there is no other example of a contingency – at least, no case that would pass without controversy as a contingency – that has both those features.

It would be satisfying to disclose some definite metaphysical incoherence in the notion that there can be absolutely insular contingencies. At the very least, it is at odds with the idea that the realm of contingency forms a single integrated system – a tree-like structure in which the connecting links are dependencies and in which every node is linked to others. There is a potential place in such a structure for initial nodes – nodes ancestrally dependent on no others, which thus correspond to brute contingencies – and for terminating branches, whose terminal nodes thus correspond to barren contingencies. What there is no place for is a node that is both initial and terminal; for by forgoing any relations of dependency, such a node forfeits any case to be regarded as part of the structure at all.

What should we conclude? The picture of the realm of contingency as comprising such a unified web seems to go deep enough in our ordinary thinking to ensure that it will be very difficult to vindicate the conception of mathematical objects as strongly contingent merely by description of anything we already find intuitive. But it is hard to avoid feeling that there is a deeper incongruity in any play with absolutely insular contingency, yet to be brought out. At the very least, a supporter of Field who goes in this direction has to swallow some very lumpy metaphysics for the sake of his ontological conscience.

The main purpose of this paper has been to raise the question whether someone who takes a nominalist-cum-literalist position has the resources coherently to classify the modal status of the proposition that there are numbers, or abstract mathematical objects generally. By contrast, the form of deflationary Platonism – Fregean Platonism, or "Platonism for cheap," as Field styles it – whose claims we have each defended elsewhere, does, we believe, provide a coherent and coherently argued view: it is that the existence of numbers is a necessity, whose truth may be recognized a priori. But the further defense of that contention is for another occasion: in particular, we ought to stress that the argument of this paper is not, and is not meant to be, a positive argument for this, or any other, version of Platonism.[27] On the contrary, it relies upon no assumption that could not be endorsed by an adherent of classical reductive nominalism, formalism, or – so far as we can see – virtually any other position that stands opposed to the unorthodox variety of nominalism against which it is directed.

[27] Experience has taught us that there is a tendency toward misunderstanding of this point. So it may help to stress what sort of view comes within the range of our argument. Not any form of anti-Platonism but only views which specifically allow four things:

(i) That the sentences of number theory, analysis, etc., have sufficiently determinate truth conditions to give sense to the question whether they are satisfied or not;

(ii) that those truth conditions are correctly reflected by the overt syntax of the sentences in question;

(iii) that those truth conditions are systematically unrealized; and

(iv) that they are so as a matter either of conceptual necessity or strong contingency.

– only an anti-Platonism that grants *all* these is in trouble if the argument of the paper fully succeeds. Our discussion, if cogent, ought therefore to leave most anti-Platonists undismayed, just because the effect of the most widely influential anti-Platonist arguments – in particular, those deriving from causal accounts of singular reference and Benacerrafian worries about the inscrutability of reference of mathematical singular terms – is (or ought to be) to call one or more of these presuppositions into question. Benacerraf himself, for instance, rejected clause (ii); and section III argued in effect that one who holds that reference to abstract objects is necessarily precluded by their causal inertness cannot coherently endorse both (i) and (ii).

Definition by Induction in Frege's *Grundgesetze der Arithmetik*

RICHARD G. HECK, JR.

0. Opening

Dedekind's *Was Sind und Was Sollen die Zahlen?* has long been celebrated for the proofs of two important theorems. The first of these is the proof of the validity of the definition, by induction, of a function defined on the natural numbers, that is, of the recursion theorem for ω. The second is the proof that all "simply infinite systems" – that is, structures which satisfy the Dedekind-Peano axioms – are isomorphic.[1] Dedekind's proofs of these theorems are carried out set-theoretically, the set theory in question being unformalized. It is well-known that the proofs of these theorems can be carried out in standard set-theories, such as ZF, and Dedekind is usually credited with having been the first to give a set-theoretic proof of these results.

It is almost unknown, however, that, in his *Grundgesetze der Arithmetik*,[2] Frege gives formal proofs of both of these results. Of course, the formal system of the *Grundgesetze* is inconsistent, Russell's Paradox being derivable, in second-order logic, from Frege's infamous Axiom V. For present purposes, Axiom V may be taken to be:

$$(\dot{\varepsilon}.F\varepsilon = \dot{\varepsilon}.G\varepsilon) \equiv \forall x(Fx \equiv Gx)$$

Axiom V, so formulated, governs terms which purport to stand for the extensions of concepts,[3] and such terms occur throughout the *Grund-*

[1] Richard Dedekind, "The Nature and Meaning of Numbers", in *Essays on the Theory of Numbers*, tr. by W. W. Beman (New York: Dover, 1963), *Was Sind?* was first published in 1887; the second edition, in 1893. The former is Theorem 126; the latter, Theorem 132.

[2] Gottlob Frege, *Grundgesetze der Arithmetik* (Hildesheim: Georg Olms, 1966). References are in the text, marked "*Gg*" with volume and section numbers.

[3] Note that the *first-order* theory which contains Axiom V as its sole non-logical axiom is actually consistent. See Michael Dummett, *Frege: Philosophy of Mathematics* (Cambridge MA: Harvard University Press, 1992), p. 219; Terence Parsons, "On the Consistency of the First-Order Portion of Frege's Logical System", *Notre Dame Journal of Formal Logic* 28 (1987), pp. 161–8.

gesetze. As I have shown elsewhere, however, Frege makes reference to value-ranges, for the most part, only for convenience, and such reference may be eliminated, uniformly, from the great majority of his proofs.[4]

Frege makes essential use of Axiom V only in his proof of what has come to be known as *Hume's Principle*. Hume's Principle may be formalized in second-order logic as:

$$Nx{:}Fx = Nx{:}Gx \text{ iff } (\exists R)[\forall x\forall y\forall z(Rxy \,\&\, Rxz \rightarrow y=z) \,\&\, \forall x\forall y\forall z(Rxz \,\&\, Ryz \rightarrow x=y) \,\&\, \forall x(Fx \rightarrow \exists y(Rxy \,\&\, Gy)) \,\&\, \forall x(Gx \rightarrow \exists y(Ryx \,\&\, Fy))]$$

The second-order theory whose sole "non-logical" axiom is Hume's Principle we may call *Fregean Arithmetic*: Fregean Arithmetic is equiconsistent with second-order arithmetic and is thus almost certainly consistent.[5] Frege's proofs of the axioms of arithmetic, in the *Grundgesetze*, can thus be reconstructed as proofs in Fregean Arithmetic: Indeed, it can be argued that Frege knew full well that the axioms of arithmetic are derivable, in second-order logic, from Hume's Principle. That is to say: The main theorem of the *Grundgesetze*, which George Boolos has rightly urged us to call *Frege's Theorem*, is that Hume's Principle implies the axioms of second-order arithmetic.

That Frege offered proofs of the axioms of arithmetic in the *Grundgesetze* is well-known, even if the fact that he proved Frege's Theorem has not been. However, only one-third of Part II of the *Grundgesetze*, entitled "Proofs of the Basic Laws of Number", is concerned with the proofs of these axioms. The remainder of Part II contains proofs of a number of additional results, among them Theorem 167, that there is an infinite cardinal (namely, the number of natural numbers); Theorem 359, of which the least number principle is an instance; and Theorem 469, the main theorem required for the definition of addition. Of interest to us here is Theorem 263.

To state Theorem 263 and the form of Hume's Principle which Frege employs in his proof of it, we need a number of definitions. These definitions, and the theorems themselves, I give in translation

---

[4] See my "The Development of Arithmetic in Frege's *Grundgesetze der Arithmetik*", *Journal of Symbolic Logic* 58 (1993), pp. 579–601.

[5] See George Boolos, "The Consistency of Frege's *Foundations of Arithmetic*", in *On Being and Saying: Essays in Honor of Richard Cartwright* (Cambridge MA: MIT Press, 1987), pp. 3–20.

into second-order logic. The first is that of the *converse of a relation* (*Gg* I § 39):[6]

$$\text{Conv}_{\alpha\varepsilon}(R\alpha\varepsilon)(x,y) \equiv\text{df } Ryx$$

Thus, x stands in the converse of the relation Rξη to y if y stands in the relation Rξη to x. The second definition is the familiar one of the *functionality* of a relation (*Gg* I § 37):

$$\text{Func}_{\alpha\varepsilon}(R\alpha\varepsilon) \equiv\text{df } \forall x\forall y\forall z(Rxy \,\&\, Rxz \rightarrow y=z)$$

The third is that of a relation's *mapping* one concept into another (*Gg* I § 38):

$$\text{Map}_{\alpha\varepsilon xy}(R\alpha\varepsilon)(Fx,Gy) \equiv\text{df } \text{Func}_{\alpha\varepsilon}(R\alpha\varepsilon) \,\&\, \forall x[Fx \rightarrow \exists y(Rxy \,\&\, Gy)]$$

Thus, a relation Rξη maps a concept Fξ into a concept Gξ just in case Rξη is functional and every F stands in it to some (and therefore exactly one) G.[7] Using these definitions, Frege concisely formulates Hume's Principle as:

$$Nx{:}Fx = Nx{:}Gx \text{ iff } (\exists R)[\text{Map}(R)(F,G) \,\&\, \text{Map}(\text{Conv } R)(G,F)]$$

This formulation is equivalent to the more familiar formulation given above, but it has certain technical advantages over it.[8]

Additionally, we need Frege's famous definition of the *strong ancestral* of a relation (*Gg* I § 45):

$$\mathcal{F}_{\alpha\varepsilon}(Q\alpha\varepsilon)(a,b) \equiv\text{df } \forall F[\forall x(Qax \rightarrow Fx) \,\&\, \forall x\forall y(Fx \,\&\, Qxy \rightarrow Fy) \rightarrow Fb]$$

That is, *b follows after a in the Q-series* if, and only if, b falls under every concept (i) under which all objects to which a stands in the Q-relation fall and (ii) which is *hereditary in the Q-series*, i. e., under which every object to which an F stands in the Q-relation falls. Frege defines the *weak ancestral* as follows (*Gg* I § 46):

$$\mathcal{F}^{=}_{\alpha\varepsilon}(Q\alpha\varepsilon)(a,b) \equiv\text{df } \mathcal{F}_{\alpha\varepsilon}(Q\alpha\varepsilon)(a,b) \vee a=b$$

6 I insert the bound variables in the definitions but will drop them when it causes no confusion.

7 Note that the definition does *not* say that Rξη is functional and is *onto* the Gs.

8 See "The Development of Arithmetic" for discussion of these. See also *Gg* I § 66.

Thus, *b belongs to* (or *is a member of*) *the Q-series beginning with a* if, and only if, either b follows after a in the Q-series or b is identical with a.

We now turn to the definitions of more particularly arithmetical notions. The number zero is defined by Frege as the number of objects which are not self-identical (*Gg* I § 41; see *Gl* § 74):[9]

$$0 = Nx{:}x \neq x$$

The relation of predecession is defined as (*Gg* I § 43; see *Gl* § 74):

$$Pred(m,n) \equiv df\ \exists F \exists y[n = Nx{:}Fx\ \&\ Fy\ \&\ m = Nx{:}(Fx\ \&\ x \neq y)]$$

That is, *m immediately precedes n in the number-series* if, and only if, there is some concept Fξ, whose number is n, and an object y falling under Fξ, such that m is the number of Fs other than y. The concept of a *finite* or *natural number* may then be defined as:[10]

$$\mathbb{N}x \equiv df\ \mathcal{F}^{=}(Pred)(0,x)$$

So a number is a natural number, is finite, if, and only if, it belongs to the Pred-series (the number-series) beginning with zero. (Famously, induction is a near immediate consequence of this definition.) And, finally, the first transfinite number, which Frege calls "Endlos", may be defined as (*Gg* I § 122):

$$\infty = df\ Nx{:}\ \mathcal{F}^{=}(Pred)(0,x)$$

Thus, Endlos is the number of natural numbers.

Theorem 263 may then be formulated as:

$$\exists Q[Func(Q)\ \&\ \neg\exists x.\mathcal{F}(Q)(x,x)\ \&\ \forall x(Gx \rightarrow \exists y.Qxy)\ \&$$
$$\exists x \forall y(Gy \equiv \mathcal{F}^{=}(Q)(x,y))] \rightarrow Nx{:}Gx = \infty$$

Suppose that there is a relation Qξη which satisfies the following conditions: First, it is functional; second, no object follows after itself in the Q-series; thirdly, each G stands in the Q-relation to some object; and, finally, the Gs are the members of the Q-series beginning with some object. Then, says Theorem 263, the number of Gs is Endlos.

---

[9] Gottlob Frege, *The Foundations of Arithmetic*, tr. by J. L. Austin (Evanston IL: Northwestern University Press, 1980). References are in the text, marked by "*Gl*" and a section number.

[10] Frege does not have any special symbol for this concept, though he reads "$\mathcal{F}^{=}$(Pred)(0,ξ)" as "ξ is a finite number [*endliche Anzahl*]". See *Gg* I § 108.

1. Frege's Proof of Theorem 263

It is worth quoting Frege's initial explanation of Theorem 263 in full:

> We now prove ... that Endlos is the number which belongs to a concept, if the objects falling under this concept may be ordered in a series, which begins with a particular object and continues without end, without coming back on itself and without branching.

By an "unbranching series", Frege means one whose determining relation is functional; by a series which does not "come back on itself", he means one in which no object follows after itself. By a series which "continues without end", he means one every member of which is immediately followed by some object.

> What it is essential to show is that Endlos is the number which belongs to the concept *member of such a series....* For this purpose, we use proposition (32) and have to establish that there is a relation which maps the number-series into the *Q*-series beginning with *x* and whose converse maps the former into the latter.

Proposition (32) is one direction of Hume's Principle, namely: If there is a relation which maps the Fs into the Gs and whose converse maps the Gs into the Fs, then the number of Fs is the same as the number of Gs.

> It suggests itself to associate 0 with *x*, 1 with the next member of the *Q*-series following after *x*, and so always to associate the number following next with the member of the *Q*-series following next. Each time, we combine a member of the number-series and a member of the *Q*-series into a pair, and we build a series from these pairs.

That is, the theorem is to be proven by defining, by induction, a relation[11] between the numbers and the members of the Q-series beginning with *x*: The number, *n*, which is the immediate successor of a given number, *m*, will be related to that member of Q-series, call it x_n, which follows immediately after the member of the Q-series to which *m* is related, say, x_m:

$$0 \to 1 \to 2 \to \ldots \to m \to n \to \ldots$$
$$x_0 \to x_1 \to x_2 \to \ldots \to x_m \to x_n \to \ldots$$

[11] In connection with Frege's use of the word "associate", see *Gg* I § 66.

The proof of the theorem will require a proof of the validity of such definitions. The idea is to define the relation by defining a *series of ordered pairs*: Namely, the series $<0;x_0>$, $<1;x_1>$, etc. The relation will then hold between objects x and y just in case <x;y> is a member of this series of ordered pairs; as one might put it, the members of this series will be the extension of the relation to be defined. To define this series of ordered pairs, Frege thus needs to introduce ordered pairs into his system and to define the relation in which a given member of the series stands to the next member of the series.

Unfortunately, Frege's definition of ordered pairs is, as George Boolos once remarked, "extravagant" and can not be consistently reconstructed, either in second-order logic or in set-theory. According to Frege's definition, the ordered pair (a;b) is the class to which all and only the extensions of relations in which a stands to b belong. Obviously, this is a proper class. Frege's proof can, however, be carried out if we take ordered pairs as primitive and subject to the usual ordered pair axiom:

OP: (a;b) = (c;d) ≡ [a=c & b=d]

Indeed, Frege derives OP from his definition,[12] and the fact that ordered pairs, so defined, satisfy OP is all he really uses (just as the fact that numbers satisfy Hume's Principle is all he uses).[13]

After introducing ordered pairs, Frege continues by defining the relation in which a given member of his series of pairs stands to the next:[14]

---

[12] From left-to-right, the axiom is proven as Theorem 218; from right-to-left, as Theorem 251.

[13] Frege does, it should be said, also use the following trick, which explains why he defines ordered pairs as he does. Indeed, it is really too bad that his definition does not work. For certain purposes, when making use of ordered pairs, one needs, given a relation Rξη, to define a concept Fξ such that:

F[(a;b)] ≡ Rab

Frege's definition makes this extremely easy: For Rab just in case the extension of the relation Rξη is a *member of* the ordered pair (a;b), since the ordered pair is the class of all extensions of relations in which a stands to b. That is, where ἀἐ.Rεα is the extension of Rξη:

[ἀἐ.Rεα ∈ (a;b)] ≡ Rab

(ἀἐ.Rεα is the *double value-range* of the relation Rξη. Note that Frege does *not* use ordered pairs to define the extension of a relation. See *Gg* I § 36.)

[14] I have translated Frege's definition into second-order logic, eliminating the reference to value-ranges.

> The series-forming relation is thereby determined: a pair stands in it to a second pair, if the first member of the first pair stands in the Pred-relation to the first member of the second pair, and the second member of the first pair stands in the Q-relation to the second member of the second pair. If, then, the pair (*n*;*y*) belongs to our series beginning with the pair (0;*x*), then *n* stands to *y* in the mapping relation to be exhibited.

That is to say: (m;n) will stand in the "series-forming relation" to (x;y) just in case Pred(m,n) and Qxy. Frege goes on to define this relation for the general case:

> ... For the ... relation, which, in the way given above, is, as I say, *coupled* from the *R*-relation and the *Q*-relation, I introduce a simple sign, by defining:
> $\Sigma_{\xi\eta\zeta\tau}(R\xi\eta,Q\zeta\tau)(a,b) \equiv df\ \exists x\exists y\exists z\exists w[a = (x;y)\ \&\ b = (z;w)\ \&\ Rxz\ \&\ Qyw)]$

If, then, we have a series x_0, x_1, etc., where Qx_mx_{m+1}, we have that Σ(Pred,Q)[(0;x_0), (1;x_1)], since Pred(0,1) and Qx_0x_1; similarly, Σ(Pred,Q) [(1;x_1), (2;x_2)]. Note, however, that we also have Σ(Pred,Q)[(0;x_{16}), (1;x_{17})], since Pred(0,1) and $Qx_{16}x_{17}$. To define the wanted relation, we therefore need to restrict attention to members of the *series* (0;x_0), (1;x_1), (2;x_2), etc. As always, Frege employs the ancestral for this purpose:

> Accordingly,
> $\mathscr{F}^{=}[\Sigma(\text{Pred},Q)][(0;x), (\xi;\eta)]$
> indicates our mapping relation ...

The relation in question is thus that in which ξ stands to η just in case the ordered pair (ξ;η) belongs to the Σ(Pred,Q)-series beginning with (0;x). It can indeed be proven that, under the hypotheses of Theorem 263, this relation maps the natural numbers into the members of the Q-series beginning with x and that its converse maps the latter into the former. Thus, in terms of ordered pairs and his definition of the coupling of two relations, Frege is able *explicitly to define* a relation which correlates the Gs one-to-one with the natural numbers.

2. *Frege's Use of Ordered Pairs*

As said, Frege's proof of Theorem 263 can be carried out in Fregean Arithmetic, if we add the ordered pair axiom to it. More interestingly, however, the proof can also be carried out in Fregean Arithmetic itself.

Frege introduces ordered pairs for two reasons. First, he uses them to give his definition of the coupling of two relations. The use of ordered pairs is obviously inessential to this definition, which can instead be given as:[15]

$$(R\ \pi\ Q)(a,b;c,d) \equiv\text{df } Rac\ \&\ Qbd$$

Secondly, as we saw, the relation which is to correlate the natural numbers one-one with the Gs is defined by Frege as:

$$\mathscr{F}^{=}[\Sigma(\text{Pred}, Q)][(0;x), (\xi;\eta)]$$

Thus, Frege uses ordered pairs in order to be able to use the ancestral – which is the ancestral of a *two-place* relation – to define this new relation. Given our definition of $(R\pi Q)(\xi,\eta;\zeta,\tau)$, it is a *four-place* relation, so we can not apply Frege's definition of the ancestral.

As mentioned, Frege essentially uses nothing about ordered pairs in his proofs other than that they satisfy the ordered pair axiom: Indeed, much of his effort is devoted to *eliminating* reference to ordered pairs from certain of his theorems. A particularly nice example is an instance of induction for series determined by the couplings of relations. The definition of the strong ancestral yields the following, Frege's Theorem 123:

$$\mathscr{F}(Q)(a,b)\ \&\ \forall x\forall y(Fx\ \&\ Qxy \rightarrow Fy)\ \&\ \forall x(Qax \rightarrow Fx) \rightarrow Fb$$

Taking $Q\xi\eta$ to be $\Sigma(R,Q)(\xi,\eta)$; a to be (a;b); b to be (c;d); we have:

$$\mathscr{F}[\Sigma(R,Q)][(a;b),(c;d)]\ \&\ \forall x\forall y(Fx\ \&\ \Sigma(R,Q)(x,y) \rightarrow Fy]\ \&$$
$$\forall x[\Sigma(R,Q)[(a;b),x] \rightarrow Fx] \rightarrow F[(c;d)]$$

Here, the bound variables will range, for all intents and purposes, over ordered pairs, since the domain and range of $\Sigma(R,Q)(\xi,\eta)$ consist only of ordered pairs; $F\xi$, in turn, will be a concept under which ordered pairs fall. Let us define the concept $Col_{\eta\zeta}(F\eta\zeta)(\xi)$ – the collapse of $F\xi\eta$ – to be that concept under which an ordered pair (x;y) stands just in case Fxy. Formally:

$$Col_{\alpha\varepsilon}(F\alpha\varepsilon)(a) \equiv\text{df } \exists x\exists y[a = (x;y)\ \&\ Fxy]$$

We then substitute $Col(F)(\xi)$ for $F\xi$ to get (*):

[15] This is derived by Frege from his definition: From right-to-left, it is Theorem 208; from left-to-right, Theorem 223.

$$\mathscr{F}[\Sigma(R,Q)][(a;b),(c;d)] \ \& \ \forall x \forall y(Col(F)(x) \ \& \ \Sigma(R,Q)(x,y) \rightarrow Col(F)(y)] \ \& \ \forall x[\Sigma(R,Q)[(a;b),x] \rightarrow Col(F)(x)] \rightarrow Col(F)[(c;d)]$$

Consider, now, the third conjunct of (*):

$$\forall x[\Sigma(R,Q)[(a;b),x] \rightarrow Col(F)(x)]$$

What we wish to show is that this conjunct follows from:

$$\forall y \forall z[Ray \ \& \ Qbz \rightarrow Fyz]$$

We suppose that $\Sigma(R,Q)[(a;b),x]$ and must show that $Col(F)(x)$. Since $\Sigma(R,Q)[(a;b),x]$, we have, by the definition of coupling:

$$\exists y \exists z \exists u \exists v[(a;b) = (u;v) \ \& \ x = (y;z) \ \& \ Ruy \ \& \ Qvz]$$

By the ordered pair axiom, u=a and v=b, so by the laws of identity: $\exists y \exists z[x = (y;z) \ \& \ Ray \ \& \ Qbz]$. But, by hypothesis, if Ray & Qbz, then Fyz; so: $\exists y \exists z[x = (y;z) \ \& \ Fyz]$. But, therefore, $Col(F)(x)$, by definition.

Similarly, the second conjunct of (*) follows from:

$$\forall x \forall y \forall z \forall w(Fxy \ \& \ Rxz \ \& \ Qyw \rightarrow Fzw)$$

Thus, applying these two results and the definition of '$Col(F)(\xi)$' to (*), we have:

$$\mathscr{F}[\Sigma(R,Q)][(a;b),(c;d)] \ \& \ \forall x \forall y \forall z \forall w(Fxy \ \& \ Rxz \ \& \ Qyw \rightarrow Fzw) \ \& \ \forall y \forall z[Ray \ \& \ Qbz \rightarrow Fyz] \rightarrow Fcd$$

This is Frege's Theorem 231, and it is one of the forms of induction he uses in his proof of Theorem 263. Note that all reference to ordered pairs has been eliminated, except in the first conjunct, where it is needed for the application of the ancestral.

Frege is thus using ordered pairs to define a two-place relation, $\Sigma(R,Q)(\xi,\eta)$, from a four-place relation, $R\xi\eta \ \& \ Q\xi\tau$, so that he can use his definition of the ancestral and the theorems proven about it (in this case, Theorem 123) to prove results about series determined by such relations. That is to say, just as Frege's definition of ordered pairs is used, essentially, only in the proof of the ordered pair axiom, what is important about ordered pairs is that one can use the ancestral to define *series* of ordered pairs; and what, in turn, is important about series of ordered pairs is that they satisfy theorems such as Theorem 231, from which reference to ordered pairs has been almost entirely eliminated. It therefore seems natural to abandon the use of ordered pairs entirely and to formulate a definition of the ancestral, for *four-place* relations, on the model of Frege's Theorem 231.

We thus define the *strong 2-ancestral* as follows:

$$\mathcal{F}_2(R\xi,\eta;\zeta,\tau)(a,b;c,d) \equiv df\ \forall F[\forall x\forall y\forall z\forall w(Fxy\ \&\ Rxyzw \rightarrow Fzw)\ \&\ \forall y\forall z[Rabxy \rightarrow Fxy] \rightarrow Fcd$$

We similarly define the *weak* 2-ancestral as:

$$\mathcal{F}^{=}{}_2(R\xi,\eta;\zeta,\tau)(a,b;c,d) \equiv df\ \mathcal{F}_2(R)(a,b;c,d) \vee (a{=}c\ \&\ b{=}d)$$

These definitions are plainly analogous to Frege's definition of the ancestral for two-place relations.

As we shall see, Theorem 263 can be proven using this definition of the 2-ancestral and so without the use of ordered pairs. Of course, if we are to use the 2-ancestral to prove Theorem 263, we must prove the analogues of those theorems about the ancestral which Frege uses in his proof. As an example, consider Frege's Theorem 131:

$$Qad \rightarrow \mathcal{F}(Q)(a,d)$$

The analogue, for the 2-ancestral, we may call Theorem 131_2:

$$Qab;cd \rightarrow \mathcal{F}_2(Q)(a,b;c,d)$$

To prove this, we use Theorem 127_2, which is immediate from the definition of the 2-ancestral:

$$\forall F\{\forall x\forall y[Qab;xy \rightarrow Fxy]\ \&\ \forall x\forall y\forall z\forall w[Fxy\ \&\ Qxy;zw \rightarrow Fzw] \rightarrow Fcd\} \rightarrow \mathcal{F}_2(Q)(a,b;c,d)$$

For the proof of (131_2), suppose that $F\xi\eta$ is hereditary in the Q-series and that, if Qab;xy, then Fxy; we must show that Fcd. But, by hypothesis, Qab;cd; so Fcd. Done.

This proof of Theorem 131_2 simply mirrors Frege's proof of Theorem 131. Indeed, it will *always* be possible to prove analogues of Frege's theorems concerning the ancestral by following his proofs of the original theorems, making use of analogues whenever he makes use of a theorem about the ancestral. The reason for this is, of course, that the definition of the 2-ancestral is itself a precise analogue of his definition of the ancestral. To see exactly in what sense this is so, define the *n-ancestral*, on the same model, as follows:

$$\mathcal{F}_n(R\xi_1\ldots\xi_n;\eta_1\ldots\eta_n)(a_1\ldots a_n;b_1\ldots b_n) \equiv df\ \forall F\{\forall x_1\ldots\forall x_n[R(a_1\ldots a_n;x_1\ldots x_n) \rightarrow Fx_1\ldots x_n]\ \&\ \forall x_1\ldots\forall x_n\forall y_1\ldots\forall y_n[R(x_1\ldots x_n;y_1\ldots y_n)\ \&\ Fx_1\ldots x_n \rightarrow Fy_1\ldots y_n] \rightarrow Fb_1\ldots b_n\}$$

We now write '**x**' for '$x_1 \ldots x_n$' and '**x=y**' for '$x_1=y_1$ & ... & $x_n=y_n$'; the definition of the n-ancestral can then be written as:

$$\mathcal{F}_n(R\xi;\eta)(\mathbf{a};\mathbf{b}) \equiv df\ \forall F[\forall \mathbf{x}(R\mathbf{ax} \rightarrow F\mathbf{x})\ \&\ \forall \mathbf{x}\forall \mathbf{y}(R\mathbf{xy}\ \&\ F\mathbf{x} \rightarrow F\mathbf{y}) \rightarrow F\mathbf{b}]$$

And, of course, we define the weak n-ancestral as:

$$\mathcal{F}_n^{=}(R\xi;\eta)(\mathbf{a};\mathbf{b}) \equiv df\ \mathcal{F}_n(R\xi;\eta)(\mathbf{a};\mathbf{b}) \vee \mathbf{a=b}$$

To say that this schematic definition was highly reminiscent of Frege's definition of the ancestral would be an understatement. In any event, it is because of the similarity which the present notational trick reveals that Frege's proofs of theorems concerning the ancestral can be immediately transformed into (schematic) proofs of theorems concerning the n-ancestral.

Theorem 231, of course, is *not* an analogue of a theorem concerning the ordinary ancestral, since reference is made to ordered pairs in it. Nevertheless, there is a proof of Thoerem 231 which is parallel to Frege's proof. As said above, that proof involves the elimination of references to ordered pairs from an instance of Theorem 123. The proof of the analogue, in this case, shows it to be a nearly immediate consequence of Theorem 123_2:

$$\mathcal{F}_2(R\xi,\eta;\zeta,\tau)(a,b;c,d)\ \&\ \forall x\forall y\forall z\forall w(Fxy\ \&\ Rxyzw \rightarrow Fzw)\ \&\ \forall y\forall z[Rabxy \rightarrow Fxy] \rightarrow Fcd$$

If we take $R\xi,\eta;\zeta,\tau$ to be a coupling of two relations – that is, if we substitute '$(R\pi Q)(\xi,\eta;\zeta,\tau)$' for '$R\xi,\eta;\zeta,\tau$' – then we have:

$$\mathcal{F}_2[R\pi Q](a,b;c,d)\ \&\ \forall x\forall y\forall z\forall w(Fxy\ \&\ (R\pi Q)(x,y;z,w) \rightarrow Fzw)\ \&\ \forall y\forall z[(R\pi Q)(a,b;x,y) \rightarrow Fxy] \rightarrow Fcd$$

As above, we have, by the definition of coupling;

$$\mathcal{F}_2[R\pi Q](a,b;c,d)\ \&\ \forall x\forall y\forall z\forall w(Fxy\ \&\ Rxz\ \&\ Qyw \rightarrow Fzw)\ \&\ \forall y\forall z[Rax\ \&\ Qby \rightarrow Fxy] \rightarrow Fcd$$

This is the mentioned analogue of Frege's Theorem 231.

It seems to me that Frege almost certainly knew that his uses of ordered pairs were inessential to his proof. He uses a number of other forms of induction, for series determined by the couplings of relations. In each of these cases (e. g., Theorem 257), reference to ordered pairs is similarly eliminated from the theorem, except in terms in which the ancestral itself occurs. Surely, it would have been *obvious* to Frege that the use of ordered pairs was unnecessary, if not at the outset, then at least by reflection on the pattern which these theorems display, namely, the pattern used above to motivate the definition of the 2-ancestral. For this reason I shall reproduce Frege's proof of Theorem 263 using the definition of the 2-ancestral, rather than ordered pairs. One might wonder, however, why, if Frege knew he could do without ordered pairs, he made use of them. One can only speculate about such a question, but the following two part answer has some plausibility: Firstly, ordered pairs were in wide use in mathematics, no suitable definition of them had at that time (1893) been given, and the provision of such a definition contributes to Frege's claim to be able to formalize, in the formal theory of the *Grundgesetze*, all of mathematics. Secondly, using ordered pairs in this context relieves Frege, first, of having to introduce a new definition of the ancestral and, secondly, of having to prove a number of theorems which are simple analogues of ones he had already proven. More generally, the use of ordered pairs *unifies* the treatment of the 2-ancestral and the usual ancestral, as well as the various other n-ancestrals. The use of ordered pairs thus has not insignificant advantages: If one can define them, why not use them?

We, however, are not in the same position Frege was. To use ordered pairs in the proof of Theorem 263, we should have to add the ordered pair axiom to Fregean Arithmetic: So far as I know, ordered pairs are *not* definable in FA. Indeed, it is an interesting and open question whether FA plus the ordered pair axiom is even a conservative extension of FA.[16]

3. Definition by Induction

We now return to the proof of Theorem 263. I shall discuss only parts of Frege's proof here, namely, those parts which are of some con-

[16] An affirmative answer would imply the definability of addition for transfinite cardinals in FA, which is, for reasons I shall not discuss, equivalent to the following:

$$\exists F[Nx{:}Fx = Nx{:}\neg Fx \,\&\, Nx{:}Fx = Nx{:}x{=}x]$$

That is: It is equivalent to the theorem that the domain can be partitioned into two equinumerous classes each of which is equinumerous with the whole domain. The problematic case is that in which Choice fails, whence the domain can not be well-ordered.

It is, however, easy to prove this theorem in FA+OP: Let $F\xi$ be defined as $\exists x[\xi = (x;0)]$; $G\xi$ as $\exists x[\xi = (x;1)]$. Plainly, $F\xi$ and $G\xi$ are equinumerous with $\xi{=}\xi$ and so with each other. Since $\forall x(Gx \rightarrow \neg Fx)$, by the Schröder-Bernstein Theorem (which is provable in second-order logic), $\neg F\xi$ is equinumerous with $\xi{=}\xi$.

cceptual interest. Those parts of the proof which proceed primarily by brute force (i. e., by repeated applications of the definition of the ancestral) will be omitted.

Recall that Theorem 263 is:

$$\exists Q[\mathrm{Func}(Q)\ \&\ \neg\exists x.\mathscr{F}(Q)(x,x)\ \&\ \forall x(\neg\exists y.Qxy \rightarrow \neg Gx)\ \&\ \exists x\forall y(Gy \equiv \mathscr{F}^{=}(Q)(x,y))] \rightarrow Nx{:}Gx = \infty$$

The Theorem follows immediately from Frege's Theorem 262, which is:

$$\mathrm{Func}(Q)\ \&\ \neg\exists x.\mathscr{F}(Q)(x,x)\ \&\ \forall x[\neg\exists y.Qxy \rightarrow \neg\mathscr{F}^{=}(Q)(a,x)] \rightarrow Nx{:}\mathscr{F}^{=}(Q)(a,x) = \infty$$

That is: If $Q\xi\eta$ is functional, no object follows after itself in the Q-series, and every member of the Q-series beginning with a stands in the Q-relation to some object, then the number of members of the Q-series beginning with a is Endlos. – For, if the Gs are the members of the Q-series beginning with a, then certainly the number of Gs is Endlos. The Theorem then follows by existential quantification in the antecedent.

Hence, the Theorem follows from the definition of Endlos and Frege's Theorem 262, η:

$$\mathrm{Func}(Q)\ \&\ \neg\exists x.\mathscr{F}(Q)(x,x)\ \&\ \forall x(\neg\exists y.Qxy \rightarrow \neg\mathscr{F}^{=}(Q)(a,x)) \rightarrow Nx{:}\mathscr{F}^{=}(Q)(a,x) = Nx{:}\mathscr{F}^{=}(\mathrm{Pred})(0,x)$$

This Theorem is to be proven, as Frege indicates, by showing that, if the antecedent holds, then there is a relation which maps the natural numbers into the members of the Q-series beginning with a and whose converse maps the latter concept into the former. Recall that $(R\pi Q)$ is the coupling of the relations $R\xi\eta$ and $Q\xi\eta$, which is defined as:

$$(R\pi Q)(a,b;c,d) \equiv_{df} Rac\ \&\ Qbd$$

The mapping relation is then to be:

$$\mathscr{F}_2^{=}[\mathrm{Pred}\ \pi\ Q](0,a;\xi,\eta)$$

Because relations defined in this way are of such importance in this connection, Frege introduces an abbreviation, which we reproduce as (see *Gg* I § 144):

$$\mathscr{F}^{\wedge}{}_{\alpha\varepsilon\gamma\delta}[R(\alpha,\varepsilon;\gamma,\delta);b,c](x,y) \equiv_{df} \mathscr{F}_2^{=}(R)(b,c;x,y)$$

Thus, an object x stands in the $\mathscr{F}^{\wedge}(R;b,c)(\xi,\eta)$-relation to y just in case the pair of x and y follows after the pair of b and c in the

R-series. The relation which correlates the natural numbers one-to-one with the members of the Q-series beginning with a is thus to be: $\mathcal{F}^{\wedge}[(\text{Pred}\ \pi\ Q);0,a](\xi,\eta)$.

We turn, then, to the proof of Theorem 262,η, which is, again:

$$\text{Func}(Q)\ \&\ \neg\exists x.\mathcal{F}(Q)(x,x)\ \&\ \forall x[\neg\exists y.Qxy \rightarrow \neg\mathcal{F}^{=}(Q)(a,x)] \rightarrow \\ Nx{:}\mathcal{F}^{=}(Q)(a,x) = Nx{:}\mathcal{F}^{=}(\text{Pred})(0,x)$$

Frege's proof of this Theorem requires three lemmas, the first of which is Theorem 254:

$$\text{Func}(R)\ \&\ \text{Func}(Q)\ \&\ \neg\exists y(\mathcal{F}^{=}(R)(m,y)\ \&\ \mathcal{F}(R)(y,y))\ \& \\ \forall y(\neg\exists z.Qyz \rightarrow \neg\mathcal{F}^{=}(Q)(a,y)) \rightarrow \text{Map}[\mathcal{F}^{\wedge}(R\pi Q,\ m,a)](\mathcal{F}^{=}(R) \\ (m,\xi),\ \mathcal{F}^{=}(Q)(a,\xi))$$

That is: If $R\xi\eta$ and $Q\xi\eta$ are functional, no member of the R-series beginning with m follows after itself in the R-series, and the Q-series beginning with a is, as we say, *endless*, then $\mathcal{F}^{\wedge}(R\pi Q;\ m,a)(\xi,\eta)$ maps the members of the R-series beginning with m into the members of the Q-series beginning with a. The second lemma is Theorem 256:

$$\text{Func}(Q)\ \&\ \forall y(\neg\exists z.Qyz \rightarrow \mathcal{F}^{=}(Q)(a,y)) \rightarrow \text{Map}[\mathcal{F}^{\wedge}(\text{Pred}\ \pi\ Q; \\ 0,a)](\mathcal{F}^{=}(\text{Pred})(0,\xi),\ \mathcal{F}^{=}(Q)(a,\xi))$$

That is: If $Q\xi\eta$ is functional and the Q-series beginning with a is endless, then $\mathcal{F}^{\wedge}(\text{Pred}\ \pi\ Q;\ 0,a)(\xi,\eta)$ maps the natural numbers into the members of the Q-series beginning with a. The third, and last, lemma is Theorem 259:

$$\forall y\forall z\{\mathcal{F}^{\wedge}(Q\pi R;\ a,m)(y,z) \equiv \text{Conv}[\mathcal{F}^{\wedge}(R\pi Q;\ m,a)](y,z)\}$$

That is: $\mathcal{F}^{\wedge}(Q\pi R;\ a,m)(\xi,\eta)$ is the converse of $\mathcal{F}^{\wedge}(R\pi Q;\ m,a)(\xi,\eta)$.

To prove Theorem 262,η, we assume the antecedent and show that $\mathcal{F}^{\wedge}[\text{Pred}\ \pi\ Q;0,a](\xi,\eta)$ correlates the natural numbers one-to-one with the members of the Q-series beginning with a. Now, $Q\xi\eta$ is functional and the Q-series beginning with a is endless. So, $\mathcal{F}^{\wedge}(\text{Pred}\ \pi\ Q;\ 0,a)(\xi,\eta)$ maps the natural numbers into the members of the Q-series beginning with a, by Theorem 256. Moreover, if, in Theorem 254, we take $Q\xi\eta$ as $\text{Pred}(\xi,\eta)$, $R\xi\eta$ as $Q\xi\eta$, m as a, and a as 0, then we have:

$$\text{Func}(Q)\ \&\ \text{Func}(\text{Pred})\ \&\ \neg\exists y(\mathcal{F}^{=}(Q)(a,y)\ \&\ \mathcal{F}(Q)(y,y))\ \& \\ \forall y(\neg\exists z.\text{Pred}(y,z) \rightarrow \mathcal{F}^{=}(\text{Pred})(0,y)) \rightarrow \text{Map}[\mathcal{F}^{\wedge}(Q\ \pi\ \text{Pred};\ a,0)] \\ (\mathcal{F}^{=}(Q)(a,\xi),\ \mathcal{F}^{=}(\text{Pred})(0,\xi))$$

The antecedent holds, since, by hypothesis, $Q\xi\eta$ is functional and no object follows after itself in the Q-series (*a fortiori*, no member of the

Q-series beginning with a does); moreover, Pred(ξ,η) is functional (Frege's Theorem 71) and every natural number is immediately succeeded by a natural number (Frege's Theorem 156); hence, $\mathcal{F}$^(Q π Pred; a,0)(ξ,η) maps the members of the Q-series beginning with a into the natural numbers. Since, by Theorem 259, this relation is the converse of the relation $\mathcal{F}$^(Pred π Q; 0,a)(ξ,η), the relation $\mathcal{F}$^(Pred π Q; 0,a)(ξ,η) maps the natural numbers into the members of the Q-series beginning with a and its converse maps the latter into the former. Done.

Most of the interest of Frege's proof of Theorem 263 lies in these three lemmas. Recall that Theorem 256 is:

> Func(Q) & ∀y(¬∃z.Qyz → ¬$\mathcal{F}^{=}$(Q)(a,y)) → Map[$\mathcal{F}$^(Pred π Q; 0,a)]($\mathcal{F}^{=}$(Pred)(0,ξ), $\mathcal{F}^{=}$(Q)(a,ξ))

It is this theorem which justifies the definition of a function(al relation), defined on the natural numbers, by induction. For, eliminating 'Map' *via* the definition, we have:

> Func(Q) & ∀y(¬∃z.Qyz → ¬$\mathcal{F}^{=}$(Q)(a,y)) → {Func($\mathcal{F}$^(Pred π Q; 0,a)) & ∀x[$\mathcal{F}^{=}$(Pred)(0,x) → ∃y($\mathcal{F}^{=}$(Q)(a,y) & $\mathcal{F}$^((Pred π Q); 0,a)(x,y)]}

If Qξη is functional and the Q-series beginning with a is endless, then $\mathcal{F}$^(Pred π Q; 0,a)(ξ,η) *is functional and every natural number is in its domain*. Moreover, it is not difficult to see that its range consists entirely of members of the Q-series beginning with a.[17]

Theorem 256 is thus a version of what is known as the recursion theorem for ω. The usual set-theoretic statement of this theorem is:

> Let g(ξ) be a function, g: A → A; let a ∈ A. Then there is a unique function φ: ℕ → A such that φ(0) = a and φ(n+1) = g(φ(n)).

Assume the antecedent and define Qξη ≡ (η = g(ξ)). Qξη is then functional, since g(ξ) is, and the Q-series beginning with a is endless, as can easily be seen. Thus, the antecedent of Theorem 256 is satisfied, so $\mathcal{F}$^(Pred π Q; 0,a)(ξ,η) is functional and its domain contains all the natural numbers. Uniqueness is obvious.

It is not hard to see that this relation satisfies the recursion equations. We must show that $\mathcal{F}$^(Pred π Q; 0,a)(0,a), that is, φ(0) = a. And, we must show that $\mathcal{F}$^(Pred π Q; 0,a)(n+1,g(x)), if $\mathcal{F}$^(Pred π Q; 0,a)(n,x), that is, φ(n+1) = g(φ(n)). The former is obvious, since $\mathcal{F}$^(Pred

[17] See Theorem 232, to be mentioned below.

π Q; 0,a)(0,a) if $\mathcal{F}^{=}$(Pred π Q)(0,a;0,a), which follows from Theorem 140_2: $\mathcal{F}_2^{=}$(R)(y,z;y,z).[18] Suppose, then, that $\mathcal{F}$^(Pred π Q; 0,a)(n,x). Then, by definition, $\mathcal{F}_2^{=}$(Pred π Q)(0,a;n,x); furthermore, Pred(n,n+1) and Qx,g(x), by the definition of Q$\xi\eta$; hence, (Pred π Q)(n,x;n+1,g(x)), by the definition of coupling. We have, then, Theorem 137_2:

$$\mathcal{F}_2^{=}(R)(b,c;y,z)\ \&\ Ryz;uv \rightarrow \mathcal{F}_2^{=}(R)(b,c;u,v)$$

Thus, $\mathcal{F}_2^{=}$(Pred π Q)(0,a;n+1,g(x)), since $\mathcal{F}_2^{=}$(Pred π Q)(0,a;n,x) and (Pred π Q)(n,x;n+1,g(x)); hence, $\mathcal{F}$^(Pred π Q; 0,a)(n+1,g(x)), by definition.[19]

As an example, consider the recursion equations:

$$\varphi(0) = a$$
$$\varphi(Sm) = S(\varphi(m))$$

(Sξ here is of course the successor-function.) As above, we define Q$\xi\eta$ as η = Sξ; thus, Q$\xi\eta$ is just Pred(ξ,η). Hence, by the above, $\mathcal{F}$^[Pred π Pred; 0,a](ξ;η) is functional and satisfies these equations. We may write them, in more familiar form, as:

$$a + 0 = a$$
$$a + Sm = S(a + m)$$

These are, of course, the standard recursion equations for addition. Thus, $\mathcal{F}$^[Pred π Pred; 0,a](ξ;η) holds just in case η = a + ξ; so

$$\mathcal{F}\hat{\ }[(Pred\ \pi\ Pred);0,\zeta](\xi;\eta)$$

defines $\eta=\zeta+\xi$, for natural numbers.[20]

---

[18] For proofs of the various theorems concerning the 2- and n-ancestrals, see Appendix II.

[19] Note that not all functions intuitively given by induction can be defined in this way. Coupling can not be used to define functions when the recursion equations are of the form:

$$\varphi(0) = a$$
$$\varphi(n+1) = g(n,\varphi(n))$$

However, let T$\xi\eta$;$\zeta\tau$ be defined as:

$$Pred(\xi,\zeta)\ \&\ \tau = g(\xi,\eta)$$

It is not hard to show that $\mathcal{F}$^[T;0,a](ξ,η) is functional and satisfies the recursion equations.

[20] Frege was aware of this. See *Gg* II § 33: "'The sum of two numbers is determined by them'. The thought of the proposition [in question here] is most easily recognized in this expression, and it may be quoted for this purpose, although the definite article in the subject actually anticipates the assertion of determination, and although the word 'sum' here is used differently than we will later use it in connection with [real] numbers. Namely, we here call [Nx:(Fx ∨ Gx)] the sum of [Nx:Fx] and [Nx:Gx], if no object falls both under the [F-] and under the [G-]concept. Also, infinite numbers

Frege's proof of Theorem 256 can be carried out in Fregean Arithmetic. The *general* theorem licensing the definition of a function by induction is, however, Theorem 254, from which Theorem 256 easily follows. Substituting 'Pred(ξ,η)' for 'Rξη' and '0' for 'm' in (254), we have:

$$\text{Func(Pred)} \ \& \ \text{Func(Q)} \ \& \ \forall y(\neg\exists z.Qyz \rightarrow \neg\mathcal{F}^{=}(Q)(x,y)) \ \& \ \neg\exists y(\mathcal{F}^{=}(\text{Pred})(0,y) \ \& \ \mathcal{F}(\text{Pred})(y,y)) \rightarrow \text{Map}(\mathcal{F}^{\wedge}(\text{Pred} \ \pi \ Q; 0,x))(\mathcal{F}^{=}(\text{Pred})(0,\xi), \mathcal{F}^{=}(Q)(x,\xi))$$

But Pred(ξ,η) is provably functional, and no natural number follows after itself in the Pred-series, by Frege's Theorem 255:

$$\neg\exists y[\mathcal{F}^{=}(\text{Pred})(0,y) \ \& \ \mathcal{F}(\text{Pred})(y,y)]$$

Theorem 256 follows immediately.

The validity of the definition of a function on the natural numbers, by induction, thus follows from Theorem 254, the functionality of Pred(ξ,η), and the fact that no finite number follows after itself in the Pred-series: Note that the *endlessness* (infinity) of the series of natural numbers is *not* needed for the proof of Theorem 256. Theorem 254 itself states that, if Rξη is functional and no member of the R-series beginning with m follows after itself in the R-series, then, if Qξη is functional and the Q-series beginning with a is endless, $\mathcal{F}^{\wedge}(R\pi Q; m,a)(\xi,\eta)$ is a functional relation whose domain contains the whole of the R-series beginning with m: Thus, Theorem 254 justifies the definition of a functional relation, by induction, on *any* "unbranching" series which does not "turn back on itself"; it is for this reason that, above, I referred to it as the general theorem licensing definition by induction. What is remarkable about Theorem 254 is that it is itself a theorem of *second-order logic*: The general theory of definition by induction can be developed entirely within second-order logic, and it was Frege who first showed that it can be (*modulo* his use of ordered pairs).

It is worth proving Theorem 262,η, and so Theorem 263, from the lemmas, in a slightly different way, to see just what Frege has done here. First, we exchange 'Rξη' with 'Qξη', and 'm' with 'a', in Theorem 254, yielding:

$$\text{Func(R)} \ \& \ \text{Func(Q)} \ \& \ \neg\exists y(\mathcal{F}^{=}(Q)(a,y) \ \& \ \mathcal{F}(Q)(y,y)) \ \& \ \forall y(\neg\exists z.Ryz \rightarrow \neg\mathcal{F}^{=}(R)(m,y)) \rightarrow \text{Map}[\mathcal{F}^{\wedge}(Q \ \pi \ R; a,m)](\mathcal{F}^{=}(Q)(a,\xi), \mathcal{F}^{=}(R)(m,\xi))$$

---

hereby come into consideration. If we wanted to prove the proposition only for finite numbers, then another method would be more sensible". I presume he means this one.

By Theorem 259, then:

$$\text{Func}(R)\ \&\ \text{Func}(Q)\ \&\ \neg\exists y(\mathcal{F}^{=}(Q)(a,y)\ \&\ \mathcal{F}(Q)(y,y))\ \&\ \forall y(\neg\exists z.Ryz \rightarrow \neg\mathcal{F}^{=}(R)(m,y)) \rightarrow \text{Map}[\text{Conv}[\mathcal{F}^{\wedge}(Q\ \pi\ R;\ a,m)]](\mathcal{F}^{=}(Q)(a,\xi),\ \mathcal{F}^{=}(R)(m,\xi))$$

Putting this together with Theorem 254, we have:

$$\text{Func}(Q)\ \&\ \neg\exists y(\mathcal{F}^{=}(Q)(a,y)\ \&\ \mathcal{F}(Q)(y,y))\ \&\ \forall y(\neg\exists z.Qyz \rightarrow \neg\mathcal{F}^{=}(Q)(a,y))\ \&\ \text{Func}(R)\ \&\ \neg\exists y(\mathcal{F}^{=}(R)(m,y)\ \&\ \mathcal{F}(R)(y,y))\ \&\ \forall y(\neg\exists z.Ryz \rightarrow \neg\mathcal{F}^{=}(R)(m,y)) \rightarrow \text{Map}(\mathcal{F}^{\wedge}(R\pi Q;\ m,a))(\mathcal{F}^{=}(R)(m,\xi),\ \mathcal{F}^{=}(Q)(a,\xi))\ \&\ \text{Map}[\text{Conv}(\mathcal{F}^{\wedge}(R\pi Q;\ m,a))](\mathcal{F}^{=}(Q)(a,\xi),\ \mathcal{F}^{=}(R)(m,\xi))$$

Let us call this theorem the *Isomorphism Theorem*. We say that the P-series beginning with t is *simple*[21] if (i) $P\xi\eta$ is functional and (ii) no object which belongs to the P-series beginning with t follows after itself in the P-series; the P-series beginning with t is *simply endless* if it is simple and endless. The Isomorphism Theorem thus says that, if the R-series beginning with m and the Q-series beginning with a are simply endless, $\mathcal{F}^{\wedge}(R\pi Q;\ m,a)(\xi,\eta)$ maps the former series into the latter and its converse maps the latter series into the former. Since, as it is not particularly diflicult to see, these relations are order-preserving – where the ordering relations are the strong ancestrals of $R\xi\eta$ and $Q\xi\eta$ – this theorem implies that all simply endless series are isomorphic.[22] Hence the name.

Theorem 262,η then follows by substituting 'Pred(ξ,η)' for '$R\xi\eta$', and '0', for 'm', and noting that the Pred-series beginning with 0 is simply endless. Frege thus in effect proceeds by *first* proving that all simply endless series are isomorphic and *then* conduding that, since the series of natural numbers is a simply endless series, every simply endless series is isomorphic to it. The Isomorphism Theorem is an immediate and trivial consequence of Frege's Theorems 254 and 259, and these two Theorems are, as has been said, theorems of second-order logic. The Isomorphism Theorem is thus *itself* a theorem of second-order logic, and Frege's proof in effect shows it so to be. Facts about the natural numbers, derived from Hume's Principle, are used only in

---

[21] See *Gg* I § 158: "...a simple series (unbranching, not coming back on itself)...".

[22] The fact that $\mathcal{F}^{\wedge}(R\pi Q;\ m,a)(\xi,\eta)$ is an order-isomorphism is a relatively obvious consequence of the way it is defined. But see Appendix I for a proof, which is not given by Frege. The proof is easy, given certain theorems he does prove. It is perhaps worth mentioning that Dedekind does not explicitly prove this result either.

the derivation of Theorem 256 from Theorem 254 and of Theorem 263 from the Isomorphism Theorem.

4. *The Definition of Relations by Induction*

We now turn to the proof of the central lemma in Frege's proof of Theorem 263, Theorem 254. (The proof of Theorem 259 is not difficult and will be omitted.) Recall that Theorem 254 is:

$$\mathrm{Func}(R) \mathbin{\&} \mathrm{Func}(Q) \mathbin{\&} \neg\exists y[\mathcal{F}^{=}(R)(m,y) \mathbin{\&} \mathcal{F}(R)(y,y)] \mathbin{\&} \forall y(\neg\exists z.Qyz \rightarrow \neg\mathcal{F}^{=}(Q)(a,y)) \rightarrow \mathrm{Map}(\mathcal{F}^{\wedge}(R\pi Q;\ m,a))(\mathcal{F}^{=}(R)(m,\xi),\ \mathcal{F}^{=}(Q)(a,\xi))$$

Theorem 254 is derived from two lemmas, the first of which is Theorem 253:

$$\mathrm{Func}(R) \mathbin{\&} \mathrm{Func}(Q) \mathbin{\&} \neg\exists y[\mathcal{F}^{=}(R)(m,y) \mathbin{\&} \mathcal{F}(R)(y,y)] \rightarrow \mathrm{Func}(\mathcal{F}^{\wedge}[R\pi Q;\ m,a])$$

That is: If $R\xi\eta$ and $Q\xi\eta$ are functional and no member of the R-series beginning with m follows after itself in the R-series, then $\mathcal{F}^{\wedge}[(R\pi Q);\ m,a](\xi,\eta)$ is functional. The second needed lemma is Theorem 241:

$$\mathrm{Func}[\mathcal{F}^{\wedge}(R\pi Q;\ m,a)] \mathbin{\&} \forall x[\neg\exists y.Qxy \rightarrow \neg\mathcal{F}^{=}(Q)(a,x)] \rightarrow \mathrm{Map}(\mathcal{F}^{\wedge}[R\pi Q;\ m,a])(\mathcal{F}^{=}(R)(m,\xi),\ \mathcal{F}^{=}(Q)(a,\xi))$$

If $\mathcal{F}^{\wedge}(R\pi Q;\ m,a)(\xi,\eta)$ is functional and the Q-series beginning with a is endless, $\mathcal{F}^{\wedge}(R\pi Q;\ m,a)(\xi,\eta)$ maps the R-series beginning with m into the Q-series beginning with a. Theorem 254 is a simple consequence of these two lemmas.

Theorem 241 is an immediate consequence of the definition of 'Map' and Theorem 241,ζ:

$$\forall x[\mathcal{F}^{=}(Q)(a,x) \rightarrow \exists y.Qxy] \rightarrow \forall x\{\mathcal{F}^{=}(R)(m,x) \rightarrow \exists y[\mathcal{F}^{\wedge}(R\pi Q;\ m,a)(x,y) \mathbin{\&} \mathcal{F}^{=}(Q)(a,y)]\}$$

For the consequent states that, if x belongs to the R-series beginning with m, then x stands in the $\mathcal{F}^{\wedge}(R\pi Q;\ m,a)(\xi,\eta)$-relation to some member of the Q-series beginning with a; hence, by definition, if $\mathcal{F}^{\wedge}(R\pi Q;\ m,a)(\xi,\eta)$ is functional, it maps the R-series beginning with m into the Q-series beginning with a. Thus: If the Q-series beginning with a is endless, the *domain of* $\mathcal{F}^{\wedge}(R\pi Q;\ m,a)(\xi,\eta)$ *contains the whole of the R-series beginning with m*. Now, Frege's Theorem 232 is:

$\mathcal{F}(R\pi Q; m,a;x,y) \rightarrow [\mathcal{F}(R)(m,x) \,\&\, \mathcal{F}(Q)(a,y)]$

It follows easily from this theorem that all and only members of the R-series beginning with m are in the domain, and the range contains only members of the Q-series beginning with a. Thus, the proof of the Theorem 241,ζ amounts to a proof of the validity of the definition, not of a function, but of a *relation* by induction. The only condition on such definitions is that the Q-series beginning with a not end.[23] Frege thus derives the validity of the inductive definition of a *functional relation* from this general theorem about the validity of inductive definitions of relations by proving that, under certain conditions (namely, those mentioned in Theorem 253), the relation so defined will be functional.[24]

An example should help to explain the theorem. Let Qξη relate m to n just in case there is some prime p such that n=pm. That is:

$Qmn \equiv df\ \exists p[\mathrm{Prime}(p) \,\&\, n=pm]$

Now, it is a theorem of arithmetic that, for every natural number m≥1, there is some n such that Qmn:

$\mathcal{F}^{=}(\mathrm{Pred})(1,m) \rightarrow \exists y.Qmy$

Hence, the Q-series beginning with 1 is endless. By Theorem 241,ζ, then:

$\forall x\{\mathcal{F}^{=}(\mathrm{Pred})(0,x) \rightarrow \exists y[\mathcal{F}^{\wedge}(\mathrm{Pred}\ \pi\ Q; 0,1)(x,y) \,\&\, \mathcal{F}^{=}(Q)(1,y)]\}$

That is, $\mathcal{F}^{\wedge}\{\mathrm{Pred}(\xi,\eta)\ \pi\ \exists p[\mathrm{Prime}(p) \,\&\, \tau=p\zeta];\ 0,1\}(\alpha,\beta)$ is a relation whose domain consists of all natural numbers and whose range is wholly contained in the Q-series beginning with 1. Moreover,

$\mathcal{F}^{\wedge}[(\mathrm{Pred}(\xi,\eta)\ \pi\ \exists p(\mathrm{Prime}(p) \,\&\, \tau=p\zeta));\ 0,1](m,x)$

holds just in case then x is a product of m (not necessarily distinct) primes, as can easily be seen.[25]

---

23 For this reason, certain relations so defined are not defined inductively, in any natural sense, e. g., if the relation Rξη is dense. But let us ignore such complications.

24 Thus, by substitution:

$\forall x[\mathcal{F}^{=}(\mathrm{Pred})(0,x) \rightarrow \exists y.\mathrm{Pred}(x,y)] \rightarrow \forall x\{\mathcal{F}^{=}(R)(m,x) \rightarrow \exists y[\mathcal{F}^{\wedge}(R\pi\mathrm{Pred};\ m,0)(x,y) \,\&\, \mathcal{F}^{=}(\mathrm{Pred})(0,y)]$

Since the antecedent is a theorem of FA, we have, as a theorem of FA:

$\forall x\{\mathcal{F}^{=}(R)(m,x) \rightarrow \exists y[\mathcal{F}^{\wedge}(R\pi\mathrm{Pred};\ m,0)(x,y) \,\&\, \mathcal{F}^{=}(\mathrm{Pred})(0,y)]$

That is: For *any* relation Rξη and *any* object m, $\mathcal{F}^{\wedge}(R\pi\mathrm{Pred};m,0)(\xi,\eta)$ is a relation whose domain is exactly the R-series beginning with m and whose range is contained in the natural numbers.

25 It is worth mentioning, too, that $\mathcal{F}^{\wedge}(Q\pi\mathrm{Pred};\ 1,0)(\xi,\eta)$ – the converse of $\mathcal{F}^{\wedge}(\mathrm{Pred}\ \pi\ Q;\ 0,1)(\xi,\eta)$ – is also, by (241,η), a relation whose domain consists of all members of

The proof of Theorem 241,ζ itself is of no great interest and will be omitted.

5. *Functionality and the n-Ancestral*

We turn then to the proof of Frege's Theorem 253, for which we need two lemmas. Recall that Theorem 253 is:

$$\mathrm{Func}(R)\ \&\ \mathrm{Func}(Q)\ \&\ \neg\exists y[\mathscr{F}^{=}(R)(m,y)\ \&\ \mathscr{F}(R)(y,y)] \rightarrow \mathrm{Func}(\mathscr{F}^{\wedge}[R\pi Q;\ m,a])$$

The first of the lemmas, Theorem 252, is essentially:

$$\mathrm{Func}(R)\ \&\ \mathrm{Func}(Q) \rightarrow \mathrm{Func}(R\pi Q)$$

That is: If $R\xi\eta$ and $Q\xi\eta$ are functional, then $(R\pi Q)(\xi,\eta;\zeta,\tau)$ is functional. However, since the notion of *functionality* has so far been defined only for two-place relations, the consequent is not yet well-formed. Now, for Frege, the coupling of two relations is a relation between ordered pairs, whence it is indeed a two-place relation: To say that $(R\pi Q)[\xi,\eta]$ is functional is then to say that:

$$\forall x\forall y\forall z[(R\pi Q)(x,y)\ \&\ (R\pi Q)(x,z) \rightarrow y=z]$$

Since 'x' and 'y' here range over ordered pairs, we may, as earlier, reduce this proposition to:

$$\forall x\forall y\forall z\forall w\forall u\forall v[(R\pi Q)[(x;y),(z;w)]\ \&\ (R\pi Q)[(x;y),(u;v)] \rightarrow z=u\ \&\ w=v]$$

We therefore define:

$$\mathrm{Func}_{2,2}(R\xi,\eta;\zeta,\tau) \equiv_{df} \forall x\forall y\forall z\forall w\forall u\forall v[Rx,y;z,w\ \&\ Rx,y;u,v \rightarrow z=u\ \&\ w=v]$$

That is: A four-place relation $R\xi,\eta;\zeta,\tau$ is 2,2-functional if, and only if, whenever the pair of x and y stands in it to the pair of z and w and to

---

the Q-series beginning with 1 and whose range consists wholly of natural numbers, since every natural number has a successor. Note, too, that its range consists of *all* natural numbers and that $\mathscr{F}^{\wedge}(Q\pi\mathrm{Pred};\ 1,0)(\xi,\eta)$ is functional. Thus, $\mathscr{F}^{\wedge}(Q\pi\mathrm{Pred};\ 1,0)(\xi,\eta)$ is a counterexample to a converse of Theorem 248: $\mathscr{F}^{\wedge}(Q\pi R;\ m,n)(\xi,\eta)$ may be a function, even if $(Q\pi R)(\xi,\eta;\zeta,\tau)$ is not a 2,2-function. It is a nice question whether some general, informative theorem can be formulated concerning the conditions $Q\xi\eta$ and $R\xi\eta$ must satisfy if $\mathscr{F}^{\wedge}(R\pi Q;\ m,a)(\xi,\eta)$ is to be functional.

the pair of u and v, the pair of z and w is the pair of u and v. We may then write Theorem 252 as:

$$\text{Func}(R) \,\&\, \text{Func}(Q) \rightarrow \text{Func}_{2,2}(R\pi Q)$$

The second needed lemma is Theorem 248:

$$\text{Func}_{2,2}(R\pi Q) \,\&\, \neg\exists x[\mathcal{F}^{=}(R)(m,x) \,\&\, \mathcal{F}(R)(x,x)] \rightarrow \text{Func}(\mathcal{F}^{\wedge}[R\pi Q;\, m,a])$$

That is: If $(R\pi Q)(\xi,\eta;\zeta,\tau)$ is 2,2-functional and no member of the R-series beginning with m follows after itself in the R-series, then $\mathcal{F}^{\wedge}[(R\pi Q);\, m,a](\xi,\eta)$ is functional. Theorem 253 is an immediate consequence of these two lemmas.

The proof of Theorem 252 is straightforward.[26]

The proof of Theorem 248, in outline, is as follows. By definition, if $\mathcal{F}^{\wedge}[R\pi Q;\, m,a](\xi,\eta)$ is not functional, then we must have $\mathcal{F}^{\wedge}[R\pi Q;\, m,a](b,d)$ and $\mathcal{F}^{\wedge}[R\pi Q;\, m,a](b,c)$, for some b, c, and d, where $c \neq d$. By the definition of '$\mathcal{F}^{\wedge}$', then, $\mathcal{F}^{=}(R\pi Q)(m,a;b,c)$ and $\mathcal{F}^{=}(R\pi Q)(m,a;b,d)$. Now, during the inductive "process" of assigning values for the relation $\mathcal{F}^{\wedge}[(R\pi Q);\, m,a](\xi,\eta)$ to members of the R-series beginning with m, b must have been assigned a value *twice*. For, at each step in the process, exactly one member of the R-series beginning with m is assigned exactly one value: It is to this that the assumption that $(R\pi Q)(\xi,\eta;\zeta,\tau)$ is 2,2-functional amounts. But since the process of assigning values proceeds from one member of the R-series beginning with m to a *unique* next member of the R-series beginning with m (since $R\xi\eta$ is functional), b must occur *twice in this series*. But that will imply that b follows after itself in the series, contradicting the supposition no member of the R-series beginning with m follows after itself.

---

26 Viz.: Suppose that $R\xi\eta$ and $Q\xi\eta$ are functional; suppose that $(R\pi Q)(x,y;z,w)$ and $(R\pi Q)(x,y;u,v)$. Then, by the definition of coupling, Rxz and Rxu, and Qyw and Qyv. Since $R\xi\eta$ and $Q\xi\eta$ are functions, z=u and w=v. QED.

Frege's own proof consists in the application of the definition of 'Func' to yield:

$$\text{Func}(R) \,\&\, \text{Func}(Q) \,\&\, Rco \,\&\, Rce \,\&\, Qda \,\&\, Qdi \rightarrow o=e \,\&\, a=i$$

He then applies the ordered pair axiom (Theorem 251) to get:

$$\text{Func}(R) \,\&\, \text{Func}(Q) \,\&\, Rco \,\&\, Rce \,\&\, Qda \,\&\, Qdi \rightarrow (o;a)=(e;i)$$

This is Frege's Theorem 251,α. The remainder of the proof consists in the introduction of free variables "A", "D", and "E" for the ordered pairs, which turns out to be exceedingly tedious.

The central lemma in Frege's proof of Theorem 248 is the famous Theorem 133 from Frege's *Begriffsschrift*, here appearing as Theorem 243:[27]

$$\text{Func}(R) \mathbin{\&} \mathcal{F}^{=}(R)(m,t) \mathbin{\&} \mathcal{F}^{=}(R)(m,n) \rightarrow [\mathcal{F}(R)(n,t) \vee \mathcal{F}^{=}(R)(t,n)]$$

That is: If $R\xi\eta$ is functional and both t and n belong to the R-series beginning with m, then either n belongs to the R-series beginning with t or t follows after n in the R-series. It is this theorem which will imply that any object which "occurs twice" in the R-series beginning with m follows after itself in the R-series.

What we need is in fact not Theorem 243, but its analogue for the 2-ancestral, Theorem 243_2:

$$\text{Func}_{2,2}(R) \mathbin{\&} \mathcal{F}^{=}(R)(m,a;n,y) \mathbin{\&} \mathcal{F}^{=}(R)(m,a;s,t) \rightarrow [\mathcal{F}(R)(n,y;s,t) \vee \mathcal{F}^{=}(R)(s,t;n,y)]$$

We need two other lemmas:

$$244\text{: } \mathcal{F}(R\pi Q)(x,y;z,w) \rightarrow \mathcal{F}(R)(x,z)$$
$$246\text{: } \mathcal{F}^{=}(R\pi Q)(m,a;x,y) \rightarrow \mathcal{F}^{=}(R)(m,x)$$

The derivation of Theorem 248 from these lemmas provides a nice illustration of the power of the definition of the 2-ancestral. Suppose that $(R\pi Q)(\xi,\eta;\zeta,\tau)$ is 2,2-functional, that no member of the R-series beginning with m follows itself, and (as above) that $\mathcal{F}^{=}(R\pi Q)(m,a;x,y)$ and $\mathcal{F}^{=}(R\pi Q)(m,a;x,z)$. Note that, by Theorem 246, $\mathcal{F}^{=}(R)(m,x)$. By Theorem 243_2, one of the following three cases must hold:

(i) $\mathcal{F}(R\pi Q)(x,y;x,z)$
(ii) $\mathcal{F}(R\pi Q)(x,z;x,y)$
(iii) $x=x \mathbin{\&} y=z$

[27] Gottlob Frege, *Begriffsschrift, eine der arithmetischen nachgebildete Formelsprache des reinen Denkens*, first published in 1879, reprinted in *Begriffsschrift und andere Aufsätze*, ed. by Ignacio Angelelli (New York: Georg Olms, 1988). For an English translation, see *Frege and Gödel: Two Fundamental Texts in Mathematical Logic*, tr. and ed. by J. van Heijenoort (Cambridge MA: Harvard University Press, 1970). One might wonder if the fact that this theorem is *not used* in the proofs of the axioms of arithmetic does not indicate something about how many of the proofs of the theorems of the *Grundgesetze* Frege had in 1879. However, Wright does use Theorem 133 in his reconstruction of the proof that every natural number has a successor, and his proof closely follows Frege's suggestions in the *Grundlagen*: The proof given in the *Grundgesetze* is somewhat different. Whether Theorem 133 is needed for the *Grundlagen* version of the proof is another question, but we are unlikely, at this point, to be able to establish anything very definitive about the date of Frege's discovery of Theorem 263.

But, if either (i) or (ii), then $\mathcal{F}(R)(x,x)$, by Theorem 244, contradicting the assumption that the R-series beginning with m is simple. Hence, (iii) must hold, y=z, and $\mathcal{F}^{\wedge}(R\pi Q;m,a)(\xi,\eta)$ is functional.

The proofs of the remaining lemmas are not difficult and, except for that of Theorem 243_2 (which is in Appendix II), will be omitted.

6. *Theorem 263 in the Context of Frege's Development of Arithmetic*

We have thus completed our discussion of Frege's proofs of the Isomorphism Theorem and Theorem 263. As we have seen, Frege's proof of Theorem 263 can be reconstructed in Fregean Arithmetic, with or without the use of the ordered pair axiom. Moreover, Frege's proofs of the crucial Theorems 254 and 259, of which the Isomorphism Theorem is an immediate consequence, can be reconstructed in second-order logic, either with or without the use of the ordered pair axiom. Since Frege was perfectly aware that his uses of Axiom V in such proofs are dispensible,[28] Frege did, in effect, prove Theorem 263 in Fregean Arithmetic, augemented by the ordered pair axiom, and the Isomorphism Theorem in second-order logic, augmented by the ordered pair axiom. I have suggested, further, that Frege knew his use of ordered pairs to be dispensible.

The significance of Theorem 263, in the context of Frege's project, is far from clear, however; Frege does not say very much about it. In the Introduction, for example, Frege writes:

> It might be thought that the propositions concerning the number 'Endlos' could have been omitted; to be sure, they are not necessary for the foundation of arithmetic in its traditional compass, but their derivation is for the most part simpler than that of the corresponding propositions for finite numbers and can serve as preparation for it (*Gg* p. v).

But, surely, the Theorem is of greater significance than *that*: The *Grundgesetze* is not just a random collection of results, and it is implausible in the extreme that Frege should have included the proof of Theorem 263 in what was intended to be his *magnum opus* just because it is similar to, and simpler than, the proof of a corresponding result for finite numbers.[29]

---

28 See again "The Development of Arithmetic".

29 The corresponding result in question is Theorem 327, which states that the number of a concept Fξ is *finite* if:

$$\exists Q \exists x \exists y \{Func(Q) \,\&\, \neg\mathcal{F}(Q)(y,y) \,\&\, \forall z[Fz \equiv \mathcal{F}^{=}(Q)(x,z) \,\&\, \mathcal{F}^{=}(Q)(z,y)]\}$$

Theorem 263, together with Frege's Theorem 207, does yield a characterization of concepts whose number is Endlos. Theorem 207 is just the converse of Theorem 263:

$$Nx{:}Gx = \infty \rightarrow \exists Q\{Func(Q)\ \&\ \neg\exists x.\mathscr{F}(Q)(x,x)\ \&\ \forall x[Gx \rightarrow \exists y.Qxy]\ \&\ \exists x\forall y[Fy \equiv \mathscr{F}^{=}(Q)(x,y)]\}$$

Thus, the number of Gs is Endlos just in case the Gs can be ordered as a simply endless series beginning, say, with an object a. It follows from Frege's Theorem 243 that the Q-series beginning with a is linearly ordered by $\mathscr{F}(Q)(\xi,\eta)$; moreover, Theorem 359, which is a generalization of the least number principle, implies that the Q-series beginning with a is *well-ordered* by $\mathscr{F}(Q)(\xi,\eta)$. And so, Theorems 207, 263, 243, and 359 together imply that the number of Gs is Endlos if and only if the Gs can be ordered as an ω-sequence. Presumably, *part* of Frege's reason for proving Theorem 263 is that he intends so to characterize countably infinite concepts.

Further, Theorem 263 plays an important role in the proof of Theorem 428:

$$Nx{:}Gx = \infty\ \&\ \forall x(Fx \rightarrow Gx) \rightarrow [Nx{:}Fx = \infty \vee \mathscr{F}^{=}(Pred)(0,Nx{:}Fx)]$$

That is: If the number of Gs is Endlos and every F is a G, then, either the number of Fs is Endlos or the number of Fs is finite; i. e., every concept subordinate to a concept whose number is countably infinite is countable. This result is of some importance, but its significance is not immediately apparent.[30]

I suggest, however, that the real point of Theorem 263 is rather different. A modern reader would naturally take the theorem – or, more generally, the Isomorphism Theorem – to show that any two structures

---

Indeed, Frege defines:

$$Btw(Q\xi\eta;a,b)(x) \equiv df\ Func(Q)\ \&\ \neg\mathscr{F}(Q)(b,b)\ \&\ \mathscr{F}^{=}(Q)(a,x)\ \&\ \mathscr{F}^{=}(Q)(x,b)$$

Frege reads "Btw(Q;a,b)(x)" as "x belongs to the Q-series running from a to b", but I write it as I do because it seems more intuitive to read it as "x is between a and b in the Q-series" or just "x is Q-between a and b". Theorem 327 is then:

$$\exists Q\exists x\exists y\forall z[Fz \equiv Btw(Q;x,y)(z)] \rightarrow \mathscr{F}^{=}(Pred)(0,Nx{:}Fx)$$

The converse of Theorem 327 is Theorem 348.

[30] Not that this is what Frege had in mind, but this result is the central theorem needed in the proof that an *inner model* for Fregean Arithmetic is definable in Fregean Arithmetic. The first-order domain consists of the natural numbers, together with Endlos. This result implies that the relativization of Hume's Principle to that domain is a theorem of Fregean Arithmetic, and it is easy to see that the restriction of each instance of comprehension is also a theorem.

satisfying certain conditions are isomorphic. Recall that the Isomorphism Theorem is:

$$\mathrm{Func}(Q) \,\&\, \neg\exists x(\mathcal{F}^{=}(Q)(a,x) \,\&\, \mathcal{F}(Q)(x,x)) \,\&\, \forall x(\mathcal{F}^{=}(Q)(a,x) \rightarrow \exists y.Qxy) \,\&\, \mathrm{Func}(R) \,\&\, \neg\exists x(\mathcal{F}^{=}(R)(m,x) \,\&\, \mathcal{F}(R)(x,x)) \,\&\, \forall x(\mathcal{F}^{=}(R)(m,x) \rightarrow \exists y.Rxy) \rightarrow \mathrm{Map}(\mathcal{F}^{\wedge}((R\pi Q);\ m,a))(\mathcal{F}^{=}(R)(m,\xi), \mathcal{F}^{=}(Q)(a,\xi) \,\&\, \mathrm{Map}[\mathrm{Conv}(\mathcal{F}^{\wedge}((R\pi Q);\ m,a))](\mathcal{F}^{=}(Q)(a,\xi), \mathcal{F}^{=}(R)(m,\xi))$$

The conditions in question are thus:

1. $\mathrm{Func}(P)$
2. $\neg\exists x[\mathcal{F}^{=}(P)(a,x) \,\&\, \mathcal{F}(P)(x,x)]$
3. $\forall x[\mathcal{F}^{=}(P)(a,x) \rightarrow \exists y.Pxy]$

Earlier, we called a series whose determining relation satisfies these conditions a simply endless series and said that the Isomorphism Theorem shows that all simply endless series are isomorphic. Better to understand the significance of this Theorem, let us write the conditions slightly differently. We write '0' for 'a' and introduce a concept Nξ, as in Theorem 263:

1. $\mathrm{Func}(P)$
2. $\neg\exists x[Nx \,\&\, \mathcal{F}(P)(x,x)]$
3. $\forall x[Nx \rightarrow \exists y.Pxy]$
4. $Nx \equiv \mathcal{F}^{=}(P)(0,x)$

Conditions (1)–(4) are axioms for arithmetic: The more familiar Dedekind-Peano Axioms are easily derived from these (and conversely).[31] What the proof of Theorem 263 shows is thus that any two structures satisfying *Frege's* axioms for arithmetic are isomorphic.

As was said, that Frege so intended the proof can not be shown by quoting him: He is not very good about explaining the significance of

[31] For, if some natural number precedes zero, then Px0 and $\mathcal{F}^{=}(P)(0,x)$, whence $\mathcal{F}(P)(0,0)$. And, if Pxz and Pyz, for *natural numbers* x, y, z, then, since $\mathcal{F}^{=}(P)(0,x)$, $\mathcal{F}^{=}(P)(0,y)$, and Pξη is functional, we have, by Theorem 243, that either $\mathcal{F}(P)(x,y)$ or $\mathcal{F}(P)(y,x)$, or x=y. But, if the former, since Pxz, by Theorem 242, $\mathcal{F}^{=}(P)(z,y)$; and, since Pyz, $\mathcal{F}(P)(y,y)$. Similarly, if $\mathcal{F}(P)(y,x)$, then $\mathcal{F}(P)(x,x)$. Hence x=y.
Conversely, the Dedekind-Peano Axioms imply these. The only one we must prove is (2). We proceed by induction. If $\mathcal{F}(P)(0,0)$, then, by Frege's Theorem 124, for some x, Px0. Suppose $\neg\mathcal{F}(P)(x,x)$, x a natural number, and Pxy. Suppose $\mathcal{F}(P)(y,y)$. Then, we have the following strengthening of Theorem 124, Theorem 141:

$$\mathcal{F}(Q)(a,b) \rightarrow \exists x[Qxb \,\&\, \mathcal{F}^{=}(Q)(a,x)]$$

Hence, for some z, Pzy and $\mathcal{F}^{=}(P)(y,z)$. But Pξη is one-one, so z=x.; so $\mathcal{F}^{=}(P)(y,x)$; but Pxy, so $\mathcal{F}(P)(x,x)$. Contradiction.

the theorems he proves. There is nevertheless reason to think that he did so intend it: Namely, the hypothesis that the proof is to show that any structures satisfying these axioms *for arithmetic* are isomorphic resolves an annoying puzzle about the structure of Part II of the *Grundgesetze*. What is puzzling is the status accorded to the proof that no natural number follows itself in the Pred-series: This theorem is granted its very own section, advertised as a major result. The modern reader, with the Dedekind-Peano Axioms firmly in mind, naturally reads this result as but an important lemma in the proof that every natural number has a successor. But Frege accords it pride of place: For it is among *his* axioms for arithmetic.

It is worth noting, too, that Frege's axioms are extremely intuitive. The natural numbers are the members of a series which begins with the number zero (Axiom 4, induction); each number in this series is followed by one (Axiom 3, that every natural number has a successor) and exactly one number (Axiom 1, that Pred(ξ,η) is functional). And, so to speak, each number in the series is followed by a *new* number, by one which has not previously occurred in the series: That is, the successor of a natural number never precedes it in the number-series. Formally:

$$\neg\exists x[Nx \,\&\, Pxy \,\&\, \mathscr{F}(P)(y,x)]$$

As is easily seen, however, in the presence of the other axioms, this is equivalent to Frege's Axiom 2, that no natural number follows after itself in the number-series. It is, of course, essential that Frege's axioms for arithmetic should be intuitive and connected with our ordinary applications of arithmetic: For Frege intends to show that *arithmetic* is a branch of logic, not that some formal theory which looks something like arithmetic can be developed within something which looks like logic.

The import of the second axiom is, indeed, explained in detail by Frege, in a passage which is worth quoting in full:

> The proposition ... says that no object which belongs to the number-series beginning with zero follows after itself in the number-series. Instead of that, we can also say: "No *finite* number follows after itself in the number-series". The importance of this proposition will be made more evident by the following considerations. If we determine the number belonging to a concept Φ(ξ), or, as one normally says, if we count the objects falling under the concept Φ(ξ), then we successively associate these objects with the number-words from "one" up to a number-word "*N*", which will be determined by the associating relation's mapping the concept Φ(ξ) into the concept "member of the

> series of number-words from 'one' to 'N'" and the converse relation's mapping the latter concept into the former. "N" then denotes the sought number; i. e., N is this number. This process of counting may be carried out in various ways, since the associating relation is not completely determined.
> The question arises whether, by another choice of this relation, one could reach another number-word 'M'. Then, by our assumptions, M would be the same number as N, but, at the same time, one of these two number-words would follow after the other, e.g., 'N' would follow 'M'. Then N would follow in the number-series after M, which means that it would follow after itself. That is excluded by our proposition concerning finite numbers (*Gg* I § 108).

That is to say: That the result of the process of counting – the number-word one reaches by counting – is well-determined is, according to Frege, a consequence of the fact that no natural number follows after itself. And so, this fact about the natural numbers has a special, and central, role to play in providing a justification of our ordinary application of arithmetic.

One might wonder, however, whether Isomorphism Theorem can possibly have the significance I have suggested it has. After all, one might say, Frege does not even write it down. – It should, however, be remembered that the *Grundgesetze* is a *mathematical* work and that Frege surely could have assumed that his readers could make certain inferences for themselves: The Isomorphism Theorem is, as we have seen, an *utterly trivial* consequence of Theorems 254 and 259 which no careful reader could miss. Indeed, in Frege's proof of Theorem 263, Theorem 259 is not needed in the general form in which Frege proves it; all he needs is that $\mathcal{F}$^(Q π Pred; a,0)(ξ,η) is the converse of $\mathcal{F}$^(Pred π Q;0,a)(ξ,η). The fact that he *does* prove Theorem 259 in its more general form should at least indicate that he intends the reader to realize that a proof like that of Theorem 263 could be given of a more general result, namely, the Isomorphism Theorem.

Besides, one should not put too much weight on the distinction between what Frege writes down and what he leaves for the reader to infer. Frege never explicitly records *mathematical induction* as a theorem either, and he can hardly be said to have been oblivious to its import.

7. *Closing*

If the foregoing account of Frege's proof of Theorem 263 and its significance is correct, then the evaluation of his work in the *Grundgesetze* must change. For the most part, the *Grundgesetze* has been ignored: When it has not been ignored, it has usually been dismissed, as if that were more charitable than the alternative, which is ridicule. As we have seen, however, not only does the *Grundgesetze* contain a derivation of axioms for arithmetic, in second-order logic, from Hume's Principle, it contains a proof, in Fregean Arithmetic, that *Frege's own* axioms for arithmetic determine a class of structures isomorphic to the natural numbers. Moreover, it contains a proof, in pure second-order logic, of the more general fact that all structures satisfying these axioms are isomorphic.

Now, I do not want here to compare the relative merits of Frege's and Dedekind's proofs of these theorems, and I am surely not going to suggest that Frege's work in these areas had very much impact on the working mathematicians of his day. However, the recent tendency, among some philosophers, to dismiss Frege's work as largely irrelevant simply can not be sustained, unless one is prepared, as I for one am not, to dismiss *Was Sind?* in the same breath.[32]

Appendix I

Proof that $\mathcal{F}^\wedge(R\pi Q;m,a)(\xi,\eta)$ is Order-Preserving

We prove that, if the R-series beginning with m and the Q-series beginning with a are simply endless, then $\mathcal{F}^\wedge(R\pi Q;m,a)(\xi,\eta)$ is order-preserving. The ordering relations which $\mathcal{F}^\wedge(R\pi Q;m,a)(\xi,\eta)$ respects are the weak ancestrals of $R\xi\eta$ and $Q\xi\eta$. To show that $\mathcal{F}^\wedge(R\pi Q;m,a)(\xi,\eta)$ and its converse are order-preserving, it is sufficient, in light of Theorem 259, to show that:

$$\mathcal{F}^\wedge(R\pi Q;m,a)(b,x) \ \&\ \mathcal{F}^\wedge(R\pi Q;m,a)(c,y) \rightarrow [\mathcal{F}(R)(b,c) \equiv \mathcal{F}(Q)(x,y)]$$

Suppose, then, that $\mathcal{F}^\wedge(R\pi Q;m,a)(b,x)$ and $\mathcal{F}^\wedge(R\pi Q;m,a)(c,y)$. By definition, $\mathcal{F}^=(R\pi Q)(m,a;b,x)$ and $\mathcal{F}^=(R\pi Q)(m,a;c,y)$. Hence, by Theorem 243: (i) $\mathcal{F}(R\pi Q)(b,x;c,y)$; (ii) $\mathcal{F}(R\pi Q)(c,y;b,x)$; or, (iii) b=c & x=y. Note that $\mathcal{F}^=(R)(m,b)$, by Theorem 232. Now, suppose $\mathcal{F}(R)(b,c)$; if (iii), then $\mathcal{F}(R)(b,b)$, contradicting the fact that the R-series beginning with m is simple. If (ii), then, by Theorem 232, $\mathcal{F}(R)(c,b)$, hence $\mathcal{F}(R)(b,b)$,

32 I should like to thank George Boolos, Michael Dummett, Mathieu Marion, and Jason Stanley for their advice and criticism.

by the transitivity of the ancestral, again contradicting the simplicity of the R-series beginning with m. Hence, (i) must hold: So, by Theorem 232, again, $\mathscr{F}(Q)(x,y)$.

By a parallel argument, if $\mathscr{F}(Q)(x,y)$, then (i) must hold; so $\mathscr{F}(R)(b,c)$, by Theorem 232.

Appendix II

Theorems Concerning the n-Ancestral

123: $\mathscr{F}_n(R)(\mathbf{a},\mathbf{b})$ & $\forall\mathbf{x}(R\mathbf{ax} \rightarrow F\mathbf{x})$ & $\forall\mathbf{x}\forall\mathbf{y}(F\mathbf{x}$ & $R\mathbf{xy} \rightarrow F\mathbf{y}) \rightarrow F\mathbf{b}$

127: $\forall F[\forall\mathbf{x}(R\mathbf{ax} \rightarrow F\mathbf{x})$ & $\forall\mathbf{x}\forall\mathbf{y}(F\mathbf{x}$ & $R\mathbf{xy} \rightarrow F\mathbf{y}) \rightarrow F\mathbf{b}] \rightarrow \mathscr{F}_n(R)(\mathbf{a},\mathbf{b})$

Both by definition.

124: $\mathscr{F}_n(R)(\mathbf{a},\mathbf{b}) \rightarrow \exists\mathbf{x}.R\mathbf{xb}$

(123_n), with $\exists\mathbf{x}.R\mathbf{x}\xi$. If $R\mathbf{ay}$, then, plainly, $\exists\mathbf{x}.R\mathbf{xy}$. Suppose $\exists\mathbf{x}.R\mathbf{xy}$ and $R\mathbf{yz}$. Obviously, $\exists\mathbf{x}.R\mathbf{xz}$.

128: $\mathscr{F}_n(R)(\mathbf{a},\mathbf{b})$ & $F\mathbf{a}$ & $\forall\mathbf{x}\forall\mathbf{y}(F\mathbf{x}$ & $R\mathbf{xy} \rightarrow F\mathbf{y}) \rightarrow F\mathbf{b}$

If $\forall\mathbf{x}\forall\mathbf{y}(F\mathbf{x}$ & $R\mathbf{xy} \rightarrow F\mathbf{y})$, then $\forall\mathbf{y}(F\mathbf{a}$ & $R\mathbf{ay} \rightarrow F\mathbf{y})$; hence, if $F\mathbf{a}$, then $\forall\mathbf{y}(R\mathbf{ay} \rightarrow F\mathbf{y})$. Apply ($123_n$).

129: $\mathscr{F}_n(R)(\mathbf{a},\mathbf{b})$ & $R\mathbf{ca} \rightarrow \mathscr{F}_n(R)(\mathbf{c},\mathbf{b})$

To apply (127_n), suppose $F\xi$ is hereditary and that $\forall\mathbf{x}(R\mathbf{cx} \rightarrow F\mathbf{x})$. Since $R\mathbf{ca}$, $F\mathbf{a}$. By (128_n), $F\mathbf{b}$.

131: $R\mathbf{ab} \rightarrow \mathscr{F}_n(R)(\mathbf{a},\mathbf{b})$

Suppose $F\xi$ is hereditary and that $\forall\mathbf{x}(R\mathbf{ax} \rightarrow F\mathbf{x})$. Since $R\mathbf{ab}$, $F\mathbf{b}$. Apply (127_n).

132: $\mathscr{F}_n^{=}(R)(\mathbf{a},\mathbf{b})$ & $R\mathbf{ca} \rightarrow \mathscr{F}_n(R)(\mathbf{c},\mathbf{b})$

Suppose $\mathbf{a}=\mathbf{b}$ and $R\mathbf{ca}$. Then $R\mathbf{cb}$, so $\mathscr{F}_n(R)(\mathbf{c},\mathbf{b})$, by ($131_n$). Hence, with ($129_n$):

$$[\mathbf{a}=\mathbf{b} \vee \mathscr{F}_n(R)(\mathbf{a},\mathbf{b})] \ \&\ R\mathbf{ca} \rightarrow \mathscr{F}_n(R)(\mathbf{c},\mathbf{b})$$

133: $\mathscr{F}_n(R)(\mathbf{a},\mathbf{b})$ & $R\mathbf{bc} \rightarrow \mathscr{F}_n(R)(\mathbf{a},\mathbf{c})$

By (127_n). Suppose $F\xi$ hereditary, $\forall\mathbf{x}(R\mathbf{ax} \rightarrow F\mathbf{x})$. Since $\mathscr{F}_n(R)(\mathbf{a},\mathbf{b})$, $F\mathbf{b}$. Since $R\mathbf{bc}$ and $F\xi$ is hereditary, $F\mathbf{c}$.

134: $\mathcal{F}^{=}{}_{n}(R)(a,b) \mathbin{\&} Rbc \rightarrow \mathcal{F}_{n}(R)(a,c)$

From (133_n), as (132_n) from (129_n).

136: $\mathcal{F}_{n}(R)(a,b) \rightarrow \mathcal{F}_{n}^{=}(R)(a,b)$
139: $a=b \rightarrow \mathcal{F}_{n}^{=}(R)(a,b)$
140: $\mathcal{F}_{n}^{=}(R)(a,a)$

All immediate from the definition of "$\mathcal{F}^{=}$" (and identity).

137: $\mathcal{F}_{n}^{=}(R)(a,b) \mathbin{\&} Rbc \rightarrow \mathcal{F}_{n}^{=}(R)(a,c)$

From (134_n) and (136_n).

141: $\mathcal{F}_{n}(R)(a,b) \rightarrow \exists x[\mathcal{F}_{n}^{=}(R)(a,x) \mathbin{\&} Rxb]$

(123_n), with $\exists x[\mathcal{F}_{n}^{=}(R)(a,x) \mathbin{\&} Rx\xi]$. Suppose Ray. Then, by (140_n), Ray and $\mathcal{F}_{n}^{=}(R)(a,a)$; hence, $\exists x[\mathcal{F}_{n}^{=}(R)(a,x) \mathbin{\&} Rxy]$. Suppose $\exists x[\mathcal{F}_{n}^{=}(R)(a,x) \mathbin{\&} Rxy]$ and Ryz. Since $\mathcal{F}_{n}^{=}(R)(a,x) \mathbin{\&} Rxy$, $\mathcal{F}_{n}^{=}(R)(a,y)$, by ($137_n$). So $\mathcal{F}_{n}^{=}(R)(a,y) \mathbin{\&} Ryz$, so $\exists x[\mathcal{F}_{n}^{=}(R)(a,x) \mathbin{\&} Rxz]$. Done.

144: $\mathcal{F}_{n}^{=}(R)(a,b) \mathbin{\&} Fa \mathbin{\&} \forall x\forall y[Fx \mathbin{\&} Rxy \rightarrow Fy] \rightarrow Fb$

Suppose a=b. Then, if Fa, Fb. Hence, with (128_n):

$$[a=b \vee \mathcal{F}_{n}(R)(a,b)] \mathbin{\&} Fa \mathbin{\&} \forall x\forall y[Fx \mathbin{\&} Rxy \rightarrow Fy] \rightarrow Fb$$

152: $\mathcal{F}_{n}^{=}(R)(a,b) \mathbin{\&} Fa \mathbin{\&} \forall x\forall y[\mathcal{F}_{n}^{=}(R)(a,x) \mathbin{\&} Fx \mathbin{\&} Rxy \rightarrow Fy] \rightarrow Fb$

Suppose the antecedent, use (144_n), with $\mathcal{F}_{n}^{=}(R)(a,\xi) \mathbin{\&} F\xi$. Since Fa, by ($140_n$), $\mathcal{F}_{n}^{=}(R)(a,a) \mathbin{\&} Fa$; i. e., a falls under $\mathcal{F}_{n}^{=}(R)(a,\xi) \mathbin{\&} F\xi$. Suppose, further, $\mathcal{F}_{n}^{=}(R)(a,x) \mathbin{\&} Fx$ and Rxy. Then, Fy, by hypothesis, and $\mathcal{F}_{n}^{=}(R)(a,y)$, by ($137_n$); so, $\mathcal{F}_{n}^{=}(R)(a,\xi) \mathbin{\&} F\xi$ is hereditary. Hence, by (144_n): $Fb \mathbin{\&} \mathcal{F}_{n}^{=}(R)(a,b)$.
Definition: $\mathrm{Func}_{n,n}(R\xi;\eta) \equiv\mathrm{df}\ \forall x\forall y\forall z(Rxy \mathbin{\&} Rxz \rightarrow y=z)$

242: $\mathrm{Func}_{n,n}(R\xi;\eta) \mathbin{\&} \mathcal{F}_{n}(R)(d,n) \mathbin{\&} Rda \rightarrow \mathcal{F}_{n}^{=}(R)(a,n)$

(123_n), with $\mathcal{F}_{n}^{=}(R)(a,\xi)$. We must show that, if $R\xi;\eta$ is n,n-functional and Rda, then:

(i) $\forall x[Rdx \rightarrow \mathcal{F}_{n}^{=}(R)(a,x)]$
(ii) $\forall x\forall y[Rxy \mathbin{\&} \mathcal{F}_{n}^{=}(R)(a,x) \rightarrow \mathcal{F}_{n}^{=}(R)(a,y)]$

(ii) follows immediately from Theorem 137_n. For the proof of (i), suppose that Rdx. Since $R(\xi;\eta)$ is n,n-functional and Rda, we have that x=a, whence $\mathcal{F}_{n}^{=}(R)(a,x)$, by Theorem 139. Done.

243: $\mathrm{Func}_{n,n}(R\xi;\eta) \mathbin{\&} \mathscr{F}_n^{=}(R)(\mathbf{a},\mathbf{b}) \mathbin{\&} \mathscr{F}_n^{=}(R)(\mathbf{a},\mathbf{c}) \rightarrow [\mathscr{F}_n^{=}(R)(\mathbf{c},\mathbf{b}) \vee \mathscr{F}_n(R)(\mathbf{b},\mathbf{c})]$

(144_n), with: $\mathscr{F}_n^{=}(R)(\xi,\mathbf{b}) \vee \mathscr{F}_n(R)(\mathbf{b},\xi)$. We must prove, that, if $\mathrm{Func}_{n,n}(R\xi;\eta)$ and $\mathscr{F}_n^{=}(R)(\mathbf{a},\mathbf{b})$:

(i) $\mathscr{F}_n^{=}(R)(\mathbf{a},\mathbf{b}) \vee \mathscr{F}_n(R)(\mathbf{b},\mathbf{a})$

(ii) $\forall\mathbf{x}\forall\mathbf{y}\{R\mathbf{xy} \mathbin{\&} [\mathscr{F}_n^{=}(R)(\mathbf{x},\mathbf{b}) \vee \mathscr{F}_n(R)(\mathbf{b},\mathbf{x})] \rightarrow [\mathscr{F}_n^{=}(R)(\mathbf{y},\mathbf{b}) \vee \mathscr{F}_n(R)(\mathbf{b},\mathbf{y})]\}$

(i) follows immediately, since $\mathscr{F}_n^{=}(R)(\mathbf{a},\mathbf{b})$. Suppose, then, that R**xy** and either $\mathscr{F}_n^{=}(R)(\mathbf{x},\mathbf{b})$ or $\mathscr{F}_n(R)(\mathbf{b},\mathbf{x})$. If the latter, then $\mathscr{F}_n(R)(\mathbf{b},\mathbf{y})$, by Theorem 133_n. Suppose, then, that $\mathscr{F}_n^{=}(R)(\mathbf{x},\mathbf{b})$. Then either **x=b** or $\mathscr{F}_n(R)(\mathbf{x},\mathbf{b})$. If **x=b**, then, since R**xy**, R**by**, whence $\mathscr{F}_n(R)(\mathbf{b},\mathbf{y})$, by Theorem 131_n. But, if $\mathscr{F}_n(R)(\mathbf{x},\mathbf{b})$, then since R**xy** and $R(\xi;\eta)$ is n,n-functional, $\mathscr{F}_n^{=}(R)(\mathbf{y},\mathbf{b})$, by Theorem 242.

Important Theorems in the Proof of Theorem 263

263: $\exists Q[\mathrm{Func}(Q) \mathbin{\&} \neg\exists x.\mathscr{F}(Q)(x,x) \mathbin{\&} \forall x(Gx \rightarrow \exists y.Qxy) \mathbin{\&} \exists x\forall y(Gy \equiv \mathscr{F}^{=}(Q)(x,y))] \rightarrow Nx{:}Gx = \infty$

262,η: $\mathrm{Func}(Q) \mathbin{\&} \neg\exists x.\mathscr{F}(Q)(x,x) \mathbin{\&} \forall x(\neg\exists y.Qx,y \rightarrow \neg\mathscr{F}^{=}(Q)(a,x)) \rightarrow Nx{:}\mathscr{F}^{=}(Q)(a,x) = Nx{:}\mathscr{F}^{=}(\mathrm{Pred})(0,x)$

259: $\forall y\forall z\{\mathscr{F}^{\wedge}(Q\pi R;\ a,m)(y,z) \equiv \mathrm{Conv}[\mathscr{F}^{\wedge}(R\pi Q;\ m,a)](y,z)\}$

258: $\mathscr{F}^{=}(R\pi Q)(m,a;n,y) \rightarrow \mathscr{F}^{=}(Q\pi R)(a,m;y,n)$

257: $\mathscr{F}^{=}(R\pi Q)(m,a;n,y) \mathbin{\&} \forall z\forall w(Fzw \rightarrow \forall u\forall v(Rzu \mathbin{\&} Qwv \rightarrow Fuv)) \mathbin{\&} Fma \rightarrow Fny$

256: $\mathrm{Func}(Q) \mathbin{\&} \forall y(\neg\exists z.Qy,z \rightarrow \neg\mathscr{F}^{=}(Q)(a,y)) \rightarrow \mathrm{Map}[\mathscr{F}^{\wedge}(\mathrm{Pred}\ \pi\ Q;\ 0,a)](\mathscr{F}^{=}(\mathrm{Pred})(0,\xi),\ \mathscr{F}^{=}(Q)(a,\xi))$

254: $\mathrm{Func}(R) \mathbin{\&} \mathrm{Func}(Q) \mathbin{\&} \neg\exists y(\mathscr{F}^{=}(R)(m,y) \mathbin{\&} \mathscr{F}(R)(y,y)) \mathbin{\&} \forall y(\neg\exists z.Qy,z \rightarrow \neg\mathscr{F}^{=}(Q)(a,y)) \rightarrow \mathrm{Map}[\mathscr{F}^{\wedge}(R\pi Q;\ m,a)](\mathscr{F}^{=}(R)(m,\xi),\ \mathscr{F}^{=}(Q)(a,\xi))$

253: $\mathrm{Func}(R) \mathbin{\&} \mathrm{Func}(Q) \mathbin{\&} \neg\exists y(\mathscr{F}^{=}(R)(m,y) \mathbin{\&} \mathscr{F}(R)(y,y)] \rightarrow \mathrm{Func}(\mathscr{F}^{\wedge}(R\pi Q;\ m,a))$

252: $\mathrm{Func}(R) \mathbin{\&} \mathrm{Func}(Q) \rightarrow \mathrm{Func}_{2,2}(R\pi Q)$

248: $\mathrm{Func}_{2,2}(R\pi Q) \mathbin{\&} \neg\exists x[\mathscr{F}^{=}(R)(m,x) \mathbin{\&} \mathscr{F}^{=}(R)(x,x)] \rightarrow \mathrm{Func}(\mathscr{F}^{\wedge}(R\pi Q;\ m,a))$

246: $\mathscr{F}_2^{=}(R\pi Q)(m,a;b,d) \rightarrow \mathscr{F}^{=}(R)(m,b)$

244: $\mathcal{F}_2(R\pi Q)(m,a;b,d) \rightarrow \mathcal{F}(R)(m,b)$

243: $\mathrm{Func}(R) \,\&\, \mathcal{F}^{=}(R)(m,t) \,\&\, \mathcal{F}^{=}(R)(m,n) \rightarrow [\mathcal{F}(R)(n,t) \vee \mathcal{F}^{=}(R)(t,n)]$

242: $\mathrm{Func}(R) \,\&\, \mathcal{F}(R)(d,n) \,\&\, Rda \rightarrow \mathcal{F}(R)(a,n)$

241: $\mathrm{Func}(\mathcal{F}^{\wedge}(R\pi Q;\, m,a)) \,\&\, \forall x[\mathcal{F}^{=}(Q)(a,x) \rightarrow \exists y.Qxy] \rightarrow \mathrm{Map}(\mathcal{F}^{\wedge}(R\pi Q;\, m,a))(\mathcal{F}^{=}(R)(m,\xi),\, \mathcal{F}^{=}(Q)(a,\xi))$

241,ζ: $\forall x[\mathcal{F}^{=}(Q)(a,x) \rightarrow \exists y.Qxy] \rightarrow \forall x\{\neg\exists y[\mathcal{F}^{\wedge}(R\pi Q,\, m,a)(x,y) \,\&\, \mathcal{F}^{=}(Q)(a,y)] \rightarrow \neg\mathcal{F}^{=}(R)(m,x)\}$

240,α: $\forall x[\neg\exists y.Qxy \rightarrow \neg\mathcal{F}^{=}(Q)(a,x)] \,\&\, \mathcal{F}^{=}(R)(m,n) \rightarrow \exists y.\mathcal{F}^{\wedge}(R\pi Q;\, m,a)(n,y)$

239: $\exists x.\mathcal{F}^{\wedge}(R\pi Q;\, m,a)(m,x)$

237,δ: $\forall y[\neg\exists z.Qyz \rightarrow \neg\mathcal{F}^{=}(Q)(a,y)] \rightarrow \forall y\forall z\{\exists x.\mathcal{F}^{\wedge}(R\pi Q;\, m,a)(y,x) \,\&\, Ryz \rightarrow \exists x.\mathcal{F}^{\wedge}(R\pi Q,\, m,a)(z,x)\}$

235: $\mathcal{F}^{\wedge}(R\pi Q;\, m,a)(c,d) \rightarrow \mathcal{F}^{=}(Q)(a,d)$

232: $\mathcal{F}(R\pi Q)(m,a;c,d) \rightarrow [\mathcal{F}(R)(m,c) \,\&\, \mathcal{F}(Q)(a,d)]$

231: $\mathcal{F}(R\pi Q)(m,a;n,b) \,\&\, \forall x\forall y\forall z\forall w[Fxy \,\&\, Rxz \,\&\, Qyw \rightarrow Fzw] \,\&\, \forall x\forall y[Rmx \,\&\, Qay \rightarrow Fxy] \rightarrow Fnb$

230: $\forall x\forall y(Fxy \rightarrow \forall z\forall w(Rxz \,\&\, Qyw \rightarrow Fzw)) \rightarrow \forall x\forall y(Fxy \rightarrow \forall z\forall w[(R\pi Q)(x,y;z,w) \rightarrow Fzw))$

212: $\forall y[\neg\exists z.Qyz \rightarrow \neg\mathcal{F}^{=}(Q)(a,y)] \,\&\, \mathcal{F}^{=}(Q)(a,d) \,\&\, \mathcal{F}^{\wedge}(R\pi Q;\, u,v)(y,d) \,\&\, Ryz \rightarrow \exists x.\mathcal{F}^{\wedge}(R\pi Q;\, u,v)(z,x)$

209: $\mathcal{F}^{=}(R\pi Q)(u,v;c,d) \,\&\, Rco \,\&\, Qdb \rightarrow \mathcal{F}^{=}(R\pi Q)(u,v;o,b)$

156: $\forall x[\mathcal{F}^{=}(\mathrm{Pred})(0,x) \rightarrow \exists y.\mathrm{Pred}(x,y)]$

145: $\neg\exists x.[\mathcal{F}^{=}(\mathrm{Pred})(0,x) \,\&\, \mathcal{F}(\mathrm{Pred})(x,x)]$

71: $\mathrm{Func}(\mathrm{Pred})$

Tree of Important Theorems in the Proof of Theorem 263

The following is a graphical representation of the dependencies among the main theorems used in Frege's proof of Theorem 263. Theorems in italics are general results concerning the ancestral. Uses of definitions or theorems which are immediate consequences thereof have been omitted. The chart and the list of theorems which follows should enable the reader to reconstruct Frege's proof of Theorem 263 without difficulty.

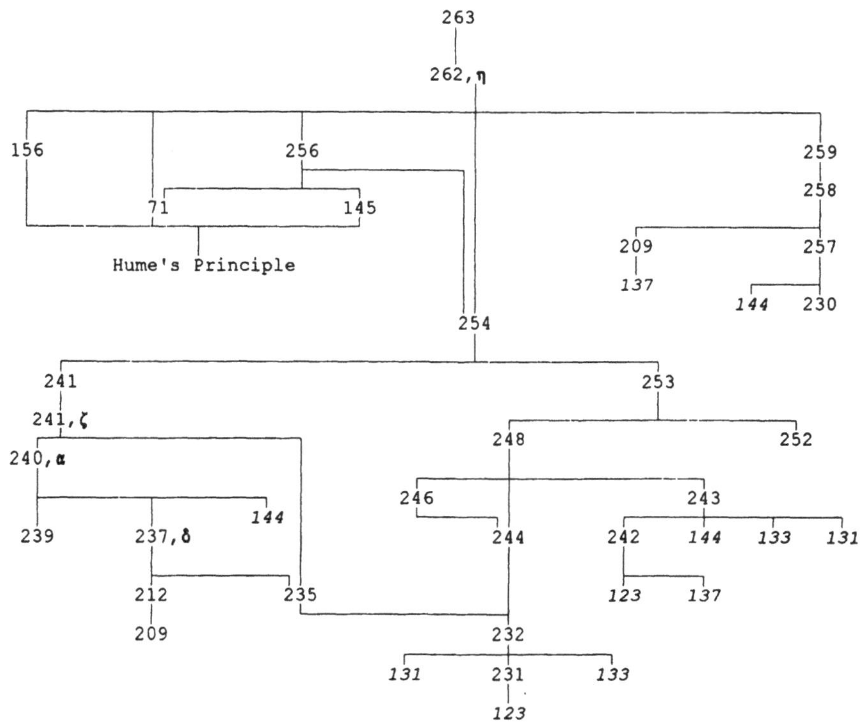

Whence the Contradiction?*

GEORGE BOOLOS

Chapter 17 of Michael Dummett's *Frege: Philosophy of Mathematics*[1] begins with the question: how did the serpent of inconsistency enter Frege's paradise? In the section of that chapter called 'How the serpent entered Eden' Dummett says, 'The second-order quantifier presents an altogether different problem, and it is to its presence in Frege's language that the contradiction is due'. Dummett regards Frege's remarks concerning the second-order quantifier as insufficient to stipulate the references of terms (in particular, the truth-values of sentences) formed by second-order quantification and writes that Frege's 'amazing insouciance concerning the second-order quantifier was the primary reason for his falling into inconsistency'. But in the very last chapter of the book he says, 'We may say that his mistake lay in supposing there to be a totality containing the extension of every concept defined over it; more generally, it lay in his not having the glimmering of a suspicion of the existence of indefinitely extensible concepts.' One might wonder whether it was Frege's insouciance or his naïveté that Dummett thinks is to blame for the error. But although it may seem as if Dummett is offering incompatible diagnoses of the contradiction, he has, I believe, a unitary account of its etiology to provide.

There are two pieces of unarguable mathematical fact on which Dummett's explanation of the contradiction rests. The first is that the first-order fragment of the system of Frege's *Grundgesetze* is con-

---

* First published in *Proceedings of the Aristotelian Society Supplementary Volume* LXVII (1993), 213–33. Reprinted by courtesy of the Editor of the Aristotelian Society.

[1] Michael Dummett, *Frege: Philosophy of Mathematics*, (Cambridge, MA: Harvard University Press, 1991). Let me say that although I disagree with several of its central contentions, I greatly admire Dummett's book.

sistent.[2] The second is that, as one might put it, one cannot assign different members of a set to different subsets of that set so that to every subset at least one member is assigned. (With the aid of the notion of the ordered pair, friends of plurals can make the appropriately general statement by saying: no matter what some objects may be, there are some things for which no item is such that the ordered pairs of that item and those things are all and only those objects.) A generalization of the second fact, pointed out by Dummett, is that the set of all functions from and to a given set can be injected into that set if and only if it contains exactly one member.

As the argument of Zermelo shows that $x \cup \{\{y \in x : \neg y \in y\}\}$ is always a proper superset of the set x, and as, on the von Neumann construction of the ordinals, $\bigcup x \cup \{\bigcup x\}$ is always a larger ordinal than any ordinal in the set x, so, according to Dummett, the argument of the Russell paradox shows that $'x\forall F(x = 'F \rightarrow \neg F(x))$ – let us call this item r – can never belong to the domain, can never be one of the objects over which the first-order variables range. [I shall modernize notation throughout, often writing ''F' instead of ''yF(y)' etc.] Note, though, that r is the extension of a concept whose name '$\forall F(\xi = 'F \rightarrow \neg F(\xi))$' begins with a second-order quantifier. Thus the introduction of second-order quantifiers forces an extension of the domain to comprise such new objects as r. If only first-order quantifiers are present, r cannot be defined, and it is possible to satisfy the first-order fragment over the natural numbers, as we shall see. But once second-order quantifiers are added, no domain is large enough to contain all extensions of concepts defined on that domain. Each domain gives rise to a more encompassing one, containing all extensions of concepts defined over the original domain. It was because Frege didn't have a glimmering of a suspicion of the way each domain must give rise to a properly wider one that he could be insouciant about the second-order quantifier.

Dummett's diagnosis is subtle: it is a striking but not particularly well known fact that the first-order fragment of Frege's system is consistent. And his account appears to be attractively commonsensical, for it looks like a reformulation in Fregean terminology of familiar and incontestable set-theoretic facts, such as that the power-set of a set is

[2] A particularly perspicuous description of this system and proof of its consistency have been given by Terence Parsons. See his 'On the Consistency of the First-Order Portion of Frege's Logical System, '*Notre Dame Journal of Formal Logic*, Vol. XXVIII, No. 1, January 1987. A version of Parsons' proof is presented below.

never injectable into that set. But powerful and unified as it is, I think it is too recherché. There is a much simpler account of what went wrong which lays the blame, not on the stipulations Frege made concerning the truth-conditions for sentences beginning with second-order quantifiers, but on those for identities of value ranges. On the account I want to argue for, the culprit is the obvious one, Basic Law V.

Before discussing Dummett's analysis of Frege's mistake about quantifiers, I want to express some doubt concerning Dummett's idea of an indefinitely extensible concept.[3]

'What the paradoxes revealed,' Dummett writes, 'was not the existence of concepts with inconsistent extensions, but of what may be called indefinitely extensible concepts'. Dummett defines a definite totality as one 'quantification over which always yields a statement determinately true or false'.[4] According to Dummett, 'No definite totality comprises everything intuitively recognisable as an ordinal number', and the concept of an ordinal number is for that reason a prototypical example of an indefinitely extensible concept. An indefinitely extensible concept has nothing that should be termed its extension, but has, rather, what may be called a sequence of extensions, of indeterminate ordinal length. Governing indefinitely extensible concepts like *ordinal* and *set* are principles of extendibility that take us from one of these extensions to the next in the sequence: the principles for *ordinal* and *set* presumably refer to operations such as $\alpha \mapsto \alpha \cup \{\alpha\}$ and $x \mapsto x \cup \{\{y \in x : \neg y \in y\}\}$.

Dummett's use of the term 'totality' rather than 'set' or 'class' leads me to suspect strongly that he believes that we cannot actually quantify over all the ordinals (or sets) there are; that whenever we quantify over some ordinals, there is at least one ordinal we have failed to quantify

[3] A superb discussion of this issue, with which I find myself in complete agreement, is found in Richard Cartwright's 1989 ASL lecture, 'Speaking of everything', published in *Noûs* 28 (1994), 1–20. I myself discussed the matter in an unpublished reply to Charles Parsons' essay 'Sets and classes', in his *Mathematics in Philosophy: Selected Essays* (Ithaca: Cornell University Press, 1983).

[4] There is a danger here of which Dummett seems unaware. Whether or not a totality is definite will depend, if Dummett's words are taken literally, on what the predicates of the language are. $\mathrm{Th}(<\mathrm{On},<>)$ is easily decidable (I am grateful to Ehud Hrushovski for explaining the proof of this folkloric result to me). Thus every statement in the language whose predicates are just $<$ and $=$, and in which quantifiers range over what Dummett calls 'the intuitive totality of all ordinals', will have a 'definite' or 'determinate' truth value, and the ordinals under less-than could not be reckoned an indefinite totality.

over, e. g., $\cup W \cup \{\cup W\}$, where W is the totality of all those ordinals we take ourselves to be quantifying over. Of course Dummett knows perfectly well that there is no set of all ordinals (an ordinal being a set), no set containing all sets, and no class containing all classes. Nevertheless, it would seem that he does think that there has to be a – what to call it – totality? collection? domain? containing all of the things we take ourselves at any one time to be talking about. He would seem to believe that whenever there are some things under discussion, being talked about, or being quantified over, for example some or all of the ordinals, there is a set-like item, a 'totality', to which they all belong. That is, he supposes that whenever we quantify, we quantify not over all the (ordinals or) sets that exist but only over some of them, and that, similarly, whatever sets we do on any occasion quantify over form a totality X which omits the item $\{x \in X : x \notin x\}$. Since $\{x \in X : x \notin x\}$ is a set (or a set-like item, an item 'intuitively recognisable as a set'), we have not managed to quantify over all the sets there are.

Benson Mates once stated: 'Any thing or things whatever constitute the entire membership of a class; in other words, for any things there are, there is exactly one class having just those things as members'.[5] And Nelson Goodman has written, 'Yet by the logicians' usage, any things whatever make up a class or set'.[6] Logicians (like Mates) who think so err. The following things don't constitute the entire membership of any one class: the classes that don't contain themselves.

It may be replied, and I suspect that Dummett would agree, that nevertheless, whenever we use quantifiers, there must always be some domain, some totality of objects, over which our variables of quantification range; so if we take ourselves to be quantifying over all classes, then we must assume that there is a totality or domain containing all classes. And it may be thought that it is part of what we mean by 'quantify over' that there must be some such domain.

Certain textbooks may reinforce this impression by telling us that to specify an interpretation we must first specify a non-empty set (class, collection, totality), the universe of discourse (or domain), over which our variables range, and then specify subcollections of the domain for each monadic predicate letter, etc.

---

[5] Benson Mates, *Skeptical Essays* (Chicago: University of Chicago Press, 1981), p. 43.

[6] Nelson Goodman, 'A world of individuals', in his *Problems and Projects* (Indianapolis: Hackett Publishing Company, 1972), p. 155.

But not all. If we look at the presentation of class theory found in Kelley's *General Topology*[7], we find that the theory presented there is a full-fledged theory of classes in which variables range over (pure) classes and in which 'set' is defined to mean 'member of a class'. Kelley writes, 'A remark on the use of the term "class" may clarify matters. The term does not appear in any axiom, definition, or theorem, but the primary interpretation [Kelley's footnote: Presumably other interpretations are also possible.] of these statements is as assertions about classes (aggregates, collections). Consequently the term 'class' is used in the discussion to suggest this interpretation'.

Kelley's axiom of extent (extensionality) reads, 'For each x and each y it is true that x=y if and only if for each z, z∈ x when and only when z∈ y'. His comment: 'Thus two classes are identical iff every member of each is a member of the other'.

Now it seems to me that insofar as we have any grip at all on the use of the phrase 'quantify over', we have to say that Kelley, in laying down his axiom of extent, is quantifying over all classes (aggregates, collections). I take it that when Kelley says 'each', he means it. How else are we to understand the axiom of extent, in view of Kelley's comment, except as saying that *any* classes x and y are identical iff x and y have the same members?

But why should we for a moment think that therefore there must be a collection of all the things Kelley was using his variables to range over? If one checks the exposition in *General Topology*, at any rate, one will find no suggestion at all that there must exist some sort of superclass, containing all the classes that the theory talks about. As Kelley's exposition makes plain, we can perfectly well explain or understand a standard formal language without having to suppose that there is a collection or other totality over which the variables of the language range: we can simply say: Our variables range over all classes (and '∈' applies to any x and y, in that order, just when x belongs to y).

Or: over all ('absolutely', if you insist, all) objects there ('really') are. If Frege thought his variables could so range, as of course he did, he was not in error. To his credit, Frege did not have the glimmering of a suspicion of the existence of indefinitely extensible concepts.

I've argued elsewhere that by using the plural number in translating sentences that begin with second-order quantifiers we can make *second*-order quantification intelligible, even if we suppose that the first-

[7] John L. Kelley, *General Topology* (Princeton: Van Nostrand, 1955).

order variables in our sentences range over all the objects there are. (Hartry Field once pointed out to me that my scheme cannot be used to translate sentences containing dyadic, or, more generally, polyadic, second-order variables.) I won't repeat the argument here, but turn instead to Frege's alleged amazing insouciance.

The account Dummett presents in chapter 17 of the cause of the inconsistency seems to me to offer a number of interpretative difficulties, some of which arise from the explanations Frege himself gives in the first thirty-three sections of *Grundgesetze*, especially with the notorious 'proof' that Frege offers in section 31.[8] Without being perfectly certain about the matter, I take it that in that chapter, Dummett is attempting to explain the origin of, or primary reason for, the inconsistency of Frege's system, and not merely the failure of Frege's argument in section 31 for the conclusion that inconsistency cannot occur. If only the latter, then I do not see that he has provided an explanation of the inconsistency. (A failed consistency proof for a consistent system would hardly be a serpent in Eden.) I also take it, though, that Dummett wishes to claim that if one examines the consistency proof and sees why it fails, one will see that the blame for the contradiction is to be ascribed to Frege's carelessness concerning the second-order quantifier. (But I am not sure, since at the end of the chapter Dummett states, correctly in my view, that the argument fails to show the consistency of even the first-order portion of the system.)

A second difficulty concerns causation, responsibility, blame. What caused the avalanche? The cry of the yodeller? Or the presence of snow on the mountainside? If there had been no snow, there certainly would have been no avalanche. Is second-order quantification the yodel or the snow? I will argue that Dummett has taken what should be thought of as a 'background condition' to be the cause of Frege's trouble.

A third worry concerns Dummett's rather surprising claim that Frege's explanation of the second-order quantifier, unlike the one he gives for the first-order quantifier, appears to be substitutional rather than objectual. Dummett says that there is meagre evidence for attributing to Frege the classical conception of the totality of all functions from and to the domain.[9] This last worry first.

---

[8] In section 31 Frege attempts to prove that every term of his formal system has a reference [*Bedeutung*], which of course must be unique.

[9] Richard Heck tells me that Dummett is now no longer entirely satisfied with the views he expressed on this matter.

A rather bad typographical error occurs on p. 218 of Dummett's book, doubtless caused by the difficulty of reproducing Frege's two-dimensional notation. What Frege actually said in the sentence Dummett quotes from section 20 is 'Now we understand by "$\forall F(\forall a F(a) \rightarrow F(\Gamma))$" the truth value of one's always obtaining a name of the value true whichever function-name one inserts in place of "F" in "$(\forall a F(a) \rightarrow F(\Gamma))$"'. Typo or no, one might well think that Dummett is probably right: Frege's talk here of inserting function-names in place of second-order variables certainly accords ill with an objectual interpretation of the second-order quantifier.

However, the evidence for the (standard) attribution of the classical conception strikes me as weighty and that for a substitutional reading of Frege's second-order quantifiers equivocal at best.

It is to be noted that Frege occasionally uses the language of substitutional quantification when discussing *first*-order quantification. In the passage from section 31 quoted by Dummett on p. 215, Frege says, 'Now "$\Phi(\xi)$" has a reference if, for every referential proper name "Δ", "$\Phi(\Delta)$" refers to something. If so, this reference is either always the value true (whatever "Δ" refers to) or not always'. We certainly should not be willing to ascribe to Frege a substitutional interpretation of the first-order quantifiers on the strength of that remark.

Dummett cites a passage from section 25 of *Grundgesetze*, where Frege writes, 'Let $\Omega_\beta(\phi(\beta))$ be a second-level function of one argument of the second kind, whose argument place is indicated by ϕ. Then $\forall F \Omega_\beta(F(\beta))$ is the value *true* only when for every suitable argument the value of our second-level function is the value *true*'. Dummett says, 'this is a comment, not a stipulation, since it is not laid down what $\forall F \Omega_\beta(F(\beta))$ is to be when the condition is not fulfilled; and no explanation is given of what constitutes a "suitable argument"'.

'Suitable' is Dummett's translation of *passend* (Furth renders it 'fitting'), and at the end of section 23, where he also defines arguments of the second kind as first-level functions of one argument, Frege does indeed define *passend*. He says, 'The objects and functions whose names are suitable for the argument places of the function are *suitable* [itals. Frege's] for the function'. The suitability of argument-places is defined syntactically, in the obvious way. Note that Frege says 'objects and functions', treating them alike.

Section 25, entitled 'Generality with respect to second-level functions. Basic Law IIb', is an explanation of the correctness of Basic Law IIb, '$(\forall F M_\beta(F(\beta)) \rightarrow M_\beta(\mathbf{F}(\beta))$', Universal Instantiation for functions.

Frege glosses this Law, 'What holds for all first-level functions of one-argument holds for any', a gloss which seems to me to support an objectual reading. The use of 'any' here, corresponding to the roman letter 'F', which is an unbound variable in Frege's symbolism, renders rather implausible the suggestion that he is talking only about functions that are definable in the language of the system. His next sentence is, 'Consequently $(\forall F \Omega_\beta(F(\beta)) \rightarrow \Omega_\beta(\phi(\beta)))$ is always the True, whatever first-level function of one argument $\phi(\xi)$ may be...', which again sounds as if he had an objectual reading of '$\forall F$' in mind. I take it that Frege, like Kelley when he says 'each', means it when he says 'all', 'any', or 'whatever'. Frege says that $\forall F \Omega_\beta(F(\beta))$ is true only when a certain condition is met since he is engaged in showing that then $\Omega_\beta(\Phi(\beta))$ will also be true: he is trying to justify IIb. Thus it would not be particularly to the point for him there to give sufficient conditions as well for the truth of $\forall F \Omega_\beta(F(\beta))$.

In any event, in section 24, Frege does give the requisite general account: 'If after a concavity with a Gothic function letter there follows a combination of signs composed of the name of a second-level function of one argument and this function letter, which fills the argument-places, then the whole denotes the True if the value of that second-level function is the True for every fitting argument; in all other cases it denotes the False'.

To tell whether Frege understands his second-order quantifiers substitutionally or objectually, one must look at how he translates formulas of his symbolism containing those quantifiers into natural language. If he regularly translates '$\forall F$' as 'for every function...', or something similar, then we may conclude that he understands them objectually; if as 'for every function name...', then substitutionally. I think the preponderance of examples favours an objectual interpretation.

In addition to those in sections 24 and 25, three other good objectual examples are found in sections 20, 34, and 45, where Frege discusses Leibniz's law, Frege's own surrogate for $\in$, and the ancestral, respectively.

I am thus inclined to think it likely that in section 25 Frege may have been supposing there that every concept is named by some function-name (and hence that he would have to accept that there are uncountably many function-names). Or more likely, lacking the notion of satisfaction, Frege may have simply expressed his meaning imprecisely in the sentence Dummett quotes. Despite his sophistication, care, and rigor, Frege did after all lack the notion of satisfaction (truth of a formula with respect to an assignment of appropriate items to its free variables), which would be introduced by Tarski about 40 years after the first volume of *Grundgesetze* appeared. I find it plausible that Frege would have accepted the natural Tarski-style definition of the satisfaction conditions for formulas of his language as a friendly amendment to his views.[10]

I don't want to discuss the matter at any greater length, though, because no matter how Frege's second-order quantifiers are to be construed, the concept crucial to Dummett's discussion of Frege's explanation of second-order quantification has a concept-name that it is very easy to construct: $\forall F(\xi = {}'F \rightarrow F(\xi))$.

If we abbreviate this formula as $H(\xi)$ and then consider the sentence $\forall F({}'H = {}'F \rightarrow F({}'H))$, which we shall call TT, we find, according to Dummett, that the 'stipulations intended to secure for it a determinate truth-value go round in a circle'.

Dummett's argument is that if we try to determine whether or not TT is true according to the stipulations Frege has made, we shall find that we are thrown back to determining whether or not TT is true and therefore cannot suppose that Frege has satisfactorily specified conditions that determine when TT is true. For, according to the specifications, TT is true if and only if the result of substituting each concept name for F in $({}'H = {}'F \rightarrow F({}'H))$ is true (or as I prefer to think Frege must have meant, if and only if for every concept F, the extension of the concept H is identical with that of F only if the extension of H falls under F). And if so, then $({}'H = {}'H \rightarrow H({}'H))$ is true, and therefore so is $H({}'H)$, which is identical to TT, the very sentence whose truth-value we are trying to determine. Dummett adds that if we had substituted $G(\xi)$, abbreviating $\forall F({}'F = \xi \rightarrow \neg F(\xi))$, for $H(\xi)$ 'we should, with a little help from Axiom V, have obtained the Russell contradiction'.

What this reasoning is supposed to show is unclear to me. At least this much is certain: from certain of Frege's specifications for the second-order quantifier and the signs for equality and the conditional, we cannot, by employing one very obvious line of deduction, determine that $H({}'H)$ is false. (We have after all substituted a concept name for a universal second-order quantifier, an odd way to proceed if we are trying to show $H({}'H)$ true.)

---

[10] I am indebted to Richard Cartwright, Richard Heck, and Jason Stanley for a number of observations in these last four paragraphs.

But we haven't shown that we can't show H('H) false by some other perhaps not so obvious line of argument, and we haven't shown that there is no way to show H('H) true.

The argument, curiously, makes no mention of, or appeal to, Basic Law V. It shows that Frege's specifications tell us that among the conditions necessary for the truth of TT is the condition that TT be true. To show that the specifications tell us that among the conditions necessary for the falsity of TT is the condition that TT be false, it would appear to be necessary to appeal to Basic Law V (which is the obvious license for the passage from H('H) to $\forall F(\text{'}H = \text{'}F \rightarrow F(\text{'}H))$). As it stands, however, the argument shows only that if H('H) is true, then Frege's specifications will require that H('H) be true. Supplementation by Basic Law V would seem to be needed to show that if H('H) is false, Frege's specifications will also require that it be false.

However, by means of a non-obvious argument due to Curry, we can in fact deduce that H('H) is false. Let C(ξ) abbreviate

$$\exists F(\xi = \text{'}F \wedge F(\xi)) \rightarrow \neg H(\text{'}H).$$

Suppose C('C). Then $\exists F(\text{'}C = \text{'}F \wedge F(\text{'}C)) \rightarrow \neg H(\text{'}H)$. But since 'C = 'C and C('C), $\exists F(\text{'}C = \text{'}F \wedge F(\text{'}C))$, and therefore $\neg H(\text{'}H)$. Thus we have shown that if C('C) holds, then H('H) is false. We must now show that C('C) holds. Suppose that for some F, 'C = 'F and F('C). By Basic Law V, $\forall x(C(x) \leftrightarrow F(x))$. But since F('C), C('C). So if $\exists F(\text{'}C = \text{'}F \wedge F(\text{'}C))$, then C('C), i. e., $\exists F(\text{'}C = \text{'}F \wedge F(\text{'}C)) \rightarrow [\exists F(\text{'}C = \text{'}F \wedge F(\text{'}C)) \rightarrow \neg H(\text{'}H)]$, and therefore $\exists F(\text{'}C = \text{'}F \wedge F(\text{'}C)) \rightarrow \neg H(\text{'}H)$, i. e., C('C) holds. Of course '¬H('H)' could have been replaced in this argument by 'H('H)' or 'p' or '⊥'. With the aid of Basic Law V we can prove whatever we please.

In section 31 of *Grundgesetze*, Frege says, 'By our stipulations [*Festsetzungen*], that "'eψ(e) = 'eφ(e)" is always to have the same truth-value as "∀a(ψ(a) = φ(a))"...'. I find it hard to understand why it is the stipulations Frege gave concerning the second-order quantifier that are to be held responsible for the contradiction *rather* than this stipulation concerning the truth-values of identities between value-range names.

Hume's principle is the statement that no matter which things the Fs and Gs may be, the number of Fs is the same as the number of Gs just in case the Fs and Gs are in one-one correspondence. Dummett calls Hume's principle 'the original equivalence'; Crispin Wright calls it $N^{=}$ (for numerical equality). If we let $F \approx G$ be some standard formula of second-order logic expressing the equinumerosity of the objects assigned to the variables F and G, and let # be a sign for a function from concepts to objects and read #F as: the number of Fs, then we may symbolize Hume's principle: $\forall F \forall G(\#F = \#G \leftrightarrow F \approx G)$.

Some years ago I showed that any proof of an inconsistency in the theory obtained by adjoining Hume's principle to second-order logic could be readily converted into the proof of an inconsistency in second-order arithmetic. Charles Parsons seems to have been the first person to realize[11] that the converse also holds (apart from Frege himself, to whom the ascription of such a recognition would be a charitable anachronism). It is instructive to consider the result of replacing the value-range operator ' by the cardinality operator # in Dummett's discussion. It would seem that if the second-order quantifier were responsible for Frege's difficulty, then by substituting 'cardinality', denoted: #, for 'value-range' in Dummett's argument, we should be able to show that Frege's specifications do not provide certain sentences about cardinality, analogous to TT, with truth-values. So let I(x) abbreviate $\forall F(x = \#F \rightarrow F(x))$ and let us try to determine the truth-value of I(#I).

We argue: for I(#I) to be true, $\forall F(\#I = \#F \rightarrow F(\#I))$ must be true, thus $\#I = \#I \rightarrow I(\#I)$ must be true, and therefore I(#I) must be true. So we seem to have gone round in a circle and it might occur to us to conclude that the argument shows that had Frege instead stipulated, '$\#e\psi(e) = \#e\phi(e)$' is always to have the same truth-value as '$\psi \approx \phi$' [abusing notation], he would not have secured a determinate truth value for I(#I).

But we have not used all the resources at hand. For, as Basic Law V was available to the real Frege so Hume's principle would have been at the disposal of our imaginary Frege. And Hume's principle enables us to see that I(#I) is in fact false. For assume I(#I), i. e., $\forall F(\#I = \#F \rightarrow F(\#I))$. Let G(x) be $x \neq x$. Then $\neg I(\#G)$, for otherwise I(#G), i. e. $\forall F(\#G = \#F \rightarrow F(\#G))$, whence $(\#G = \#G \rightarrow G(\#G))$ and G(#G), i. e. $\#G \neq \#G$, impossible. Therefore also $\#I \neq \#G$. And now let F(x) be $(x = \#G \vee [x \neq \#I \wedge I(x)])$. Then since #I does, but #G does not, fall under I, $I \approx F$, and by Hume's principle, $\#I = \#F$. Then by the definition of I(x), F(#I), i. e., $(\#I = \#G \vee [\#I \neq \#I \wedge I(\#I)])$. But that is absurd: $\#I \neq \#G$, as we just saw. And because Hume's principle is consistent, we cannot also show I(#I) true.

[11] See his 'Frege's theory of number', in his *Mathematics in Philosophy: Selected Essays*, op. cit.

The situation is the opposite for the other truth-teller-like formula similarly obtained from the formula $\exists F(x = {}'F \wedge \neg Fx)$ found in a variant proof of the inconsistency of Basic Law V: if J(x) is $\exists F(x = \#F \wedge Fx)$, then it turns out that we can prove J(#J) from Hume's principle.

It is certainly not the case that if Basic Law V is replaced with Hume's principle, then the truth-values of all sentences are determined thereby. Each of the four sentences 'The number of natural numbers is/is not identical with the number of numbers and the number of numbers is/is not identical with the number of (self-identical) objects' is consistent with the result of replacing V by Hume. Thus apart from the inescapable Gödelian incompleteness, there are some very fundamental questions about cardinality that Hume fails to resolve.

To recapitulate: Dummett's argument shows only that if one does not appeal to Basic Law V, one cannot readily deduce from Frege's stipulations concerning the second-order quantifier and the other usual logical symbols, what the truth-value is of H('H). Similarly for Hume's principle and I(#I). If we supplement Frege's explanations of the quantifiers with Hume's principle, we can show that I(#I) true (and its companion J(#J) false); we cannot show I(#I) false, since second-order logic plus Hume's principle is consistent. However, assuming Hume's principle hardly settles all questions, even all elementary questions. If we supplement Frege's explanations with Basic Law V, we find that we can deduce that H('H) is true and also deduce that it is false.

It is, in my view, not so much Frege's insouciance concerning second-order quantifiers that was responsible for his downfall as his adoption of a theory about a function from second- to first-order objects that could not possibly be true, facilitated by a lingering attachment to the idea that 'contextual definitions' like Hume's principle and Basic Law V, are, if not logically true, then near enough as could make no difference. If the difficulty were where Dummett takes it to be, the introduction of the cardinality operator should be as uncertain and dangerous as that of the operator assigning value-ranges to concepts. Some uncertainty there indeed is; but danger is not now conceivable, for the result of adjoining Hume's principle to second-order logic and second-order arithmetic ('analysis') are equiconsistent.

It is of interest to examine Terence Parsons' construction of a model for the first-order fragment of Frege's system, which works by inductively assigning denotations to value-range terms. Seeing why it cannot be extended to the full system brings out that it is not any ill-foundedness in Frege's stipulations concerning the second-order quantifier that is to blame for the contradiction.

In the model, the variables, which are all first-order, are to range over the natural numbers. Inductively define the *rank* of a value-range term 'αA(α), which may contain free variables, to be the least natural number greater than the rank of any value-range term contained in the formula A(α). Now add to the language of the system a constant **i** denoting i, for each natural number i. Order all *closed* value-range terms of the expanded language in an ω^2-sequence, with those of lower rank preceding those of higher. Let J be a standard pairing function. Now inductively assign a natural number as denotation to each closed value-range term 'αA(α), as follows: let m be the rank of 'αA(α). If for some term 'βB(β) that is earlier in the ω^2-sequence than 'αA(α), A(**i**) and B(**i**) have the same value (perhaps a truth-value) for all natural numbers i, then let 'αA(α) denote the same number as 'βB(β). (The definition is OK, since if there are two such terms 'βB(β) and 'γC(γ), we may may inductively assume that they have been assigned the same denotation.) Otherwise, let 'αA(α) denote the least number J(m,n) not yet assigned to any closed term of rank m. (Again, the definition is OK, since there are only finitely many terms *of rank m* that precede 'αA(α) in the sequence.)

The notion *the value of* A(**i**) is unproblematic, for the value of A(**i**) will depend only on the denotations of closed terms *of lower rank*, which may be assumed inductively to have been fixed, and a universal quantification ∀xD(x) is true if and only if for every i, D(**i**) is true.

Then since there cannot be an earliest closed value-range term 'αA(α) such that for some earlier closed value-range term 'βB(β), 'αA(α) and 'βB(β) have the same denotation iff for some i, (A(**i**) ↔ B(**i**)) is false, Basic Law V holds in the model.

As rereading it will make plain, Frege's argument in Section 31 can in no way be regarded as an anticipation of this proof of Parsons'. It is a natural question to ask where Parsons' proof fails, as it must, if second-order quantifiers are added to the language. Although, following Dummett, one might guess that an ill-foundedness in the truth-conditions will turn out to be responsible for the failure, that guess would be mistaken.

The trouble is that if second-order variables are added, the language will then contain the open terms 'αF(α), F a second-order variable. The rank of these terms is *0*. We handled first-order quantifiers by adding constants **i** denoting members of the domain to the language. Were we to try to handle (monadic) second-order quantifiers similarly, we should have to add to the language a new monadic predicate constant **S** for each set S of members of the domain. But then whenever S and T are different subsets of the domain we should have to assign $'\alpha\mathbf{S}(\alpha)$ and $'\alpha\mathbf{T}(\alpha)$ different members of the domain as their denotations, which cannot be done. Thus it is the Cantor-Russell *aporia* that screws up the attempt to construct a model for the Basic Law V in the full second-order language, and not any ill-foundedness in the truth-conditions for the second-quantifier.

A recent paper by Richard Heck[12] sheds considerable light on Basic Law V and the origin of the contradiction. His article seems to me to do in, once and for all, the idea that 'contextual definitions' like Hume's principle or Basic Law V, have, in general, any privileged logical status.

Heck observes that every sentence is equivalent to the existential quantification (over %) of some 'contextual definition', by which we here mean a sentence of form $\forall F \forall G(\%F = \%G \leftrightarrow E(F,G))$, where F and G are variables of the same type, usually not that of individuals, % is a sign for a function from the entities over which F and G range to individuals, and E(F,G) defines an equivalence relation on those entities. Basic Law V and Hume's Principle are contextual definitions, with $\forall x(F(x) \leftrightarrow G(x))$ and $F \approx G$ playing the role of E(F,G).

Heck's observation is proved thus: let a sentence ϕ be given. Then the relation E(F,G) on concepts F and G expressed by: $\phi \vee \forall x(F(x) \leftrightarrow G(x))$ is an equivalence relation on concepts. For either ϕ is false, in which case E(F,G) holds if and only if the same objects fall under F and G, or ϕ is true, in which case E(F,G) is the universal relation on concepts, i. e. holds no matter what F and G might be, Either way, E is an equivalence relation. Now suppose ϕ holds. Let a be some object in the domain. For all F, let $\%F = a$. Then, no matter what F and G may be, $\%F = \%G$ holds; but also, by our supposition, so does $\phi \vee \forall x(Fx \leftrightarrow Gx)$. Conversely, suppose $\forall F \forall G(\%F = \%G \leftrightarrow (\phi \vee \forall x(Fx \leftrightarrow Gx)))$ holds, but ϕ does not. Then $\forall F \forall G(\%F = \%G \leftrightarrow \forall x(Fx \leftrightarrow Gx))$ holds. Russell's argument then yields a contradiction. (Substitute % for '.)

Thus there is a contextual definition that implies any given non-logical truth ϕ; moreover ϕ implies the existential quantification of the definition. How to demarcate those contextual definitions that should count as logical truths in some extended sense of the expression from

[12] Richard Heck, 'On the Consistency of second-order contextual definitions', *Noûs*, Vol. XXVI, No. 4, 1992.

those that should not would seem to be a philosophical problem we have no hope of solving at present.

An observation quite analogous to Heck's, but concerning sentences of form $Tr(\ulcorner\psi\urcorner) \leftrightarrow \psi$ ('T-sentences', or 'Tarski biconditionals'), has been made by Vann McGee.[13] McGee shows that in any formal theory capable of proving the diagonal lemma (not a stringent requirement), any sentence whatsoever will be equivalent to some Tarski biconditional. (His argument is independent of the choice of the truth-predicate Tr() for the language of the theory.) The argument is brief. Let ϕ be an arbitrary sentence of the language. Let A(x) be the formula $Tr(x) \leftrightarrow \phi$. By the diagonal lemma, there is a sentence ψ such that the theory proves $\psi \leftrightarrow A(\ulcorner\psi\urcorner)$, i. e., $\psi \leftrightarrow (Tr(\ulcorner\psi\urcorner) \leftrightarrow \phi)$. Therefore, by the associativity (!) and commutativity of the biconditional, ϕ is equivalent in the theory to the Tarski biconditional $Tr(\ulcorner\psi\urcorner) \leftrightarrow \psi$.

McGee's observation renders thoroughly implausible the assumption, natural enough, that T-sentences, apart from a few special, well known, and easily isolable problem cases (e. g. $Tr(\ulcorner\psi\urcorner) \leftrightarrow \psi$, ψ having been obtained by applying the diagonal lemma to $\neg$Tr()), are 'partial definitions' of truth or possess only minimal content. One currently fashionable view of truth holds that all there is to the concept of truth is the infinite set of all Tarski biconditionals (minus a few troublemakers), each of which makes almost no claim at all, but which taken together make a significant claim about truth. McGee's result suggests that the problem of purging the undesirables is likely to be difficult at best.

With McGee's and Heck's examples in mind, it doesn't take much to come up with an instance of the comprehension scheme of set theory equivalent (in the presence of the null set axiom) to any sentence ϕ whatsoever: $\exists y \forall x(x \in y \leftrightarrow (\neg\phi \wedge x \notin x))$. Fortunately, we are past considering comprehension statements as having the slightest claim to our credence. Contextual definitions deserve the same regard.

Sentences $\forall F \forall G(\%F = \%G \leftrightarrow E(F,G))$ assert that the function denoted by %, which standard notational conventions insist be a function of 'mixed' type defined on all concepts, maps concepts into individuals in a way that respects the equivalence relation those concepts defined by E(F,G). Whether there is any such function will depend typically only on how many individuals there are. If E(F,G) is $\forall x(F(x) \leftrightarrow G(x))$,

[13] Vann McGee, 'Maximal Consistent Sets of Instances of Tarski's Schema (T)', *Journal of Philosophical Logic*, Vol. XXI, No. 3, 1992.

as in V, there will be no suitable function for % to denote regardless of what the domain is; if E(F,G) is F≈G, as in Hume, there will be a suitable function iff there are (Dedekind) infinitely many individuals; if E(F,G) asserts that (the values of) F is finitely different from G, then a suitable function will exist iff there are only finitely many individuals; if E(F,G) is $\forall x((F(x) \leftrightarrow F(x)) \wedge (G(x) \leftrightarrow G(x))$, then no matter what the individuals may be, there will always exist a suitable function.

Of course, for his derivation of arithmetic from Hume to work, Frege had to understand # as a function from concepts to objects (and not, say, as one from concepts to second-level concepts). Two difficulties then beset the thought that Hume is analytic: there must be infinitely many individuals and there must be a biggest number, the number of all the things there are. The second may be surmountable; the first, I have elsewhere argued, is not. Properly understood, Hume only seems to be analytic; seeing how it can be put to work reveals it as synthetic if either analytic or synthetic.

And Basic Law V is not only not analytic, it is simply a (higher-order) logical falsehood. For contraposing and making obvious abbreviations, we may write the bad half of Basic Law V:

$\forall F \forall G(F \neq G \rightarrow {}'F \neq {}'G)$. The smooth breathing sign ' is then seen to be a sign for a function allegedly mapping concepts one-one into objects, the non-existence of which was carefully proved by Frege in the appendix to *Grundgesetze*, after his derivation of the Russell paradox.

Here is a suggestion about the genesis of Frege's error in putting forth Basic Law V. Seeing no way to 'solve the Julius Caesar problem', Frege plumps in the *Grundlagen* for extensions and the principle 'extensions of concepts are the same iff the same objects fall under the concepts'. His qualms are assuaged by the evident structural similarity of this principle to Hume's principle, which both appears to be *thoroughly* obvious to Frege and strikes him as the preeminent analytic principle governing the notion of a number. He takes it that the principle about extensions is, similarly, the preeminent analytic truth governing the notion of an extension. After *Grundlagen*, he is struck by another analogy: concepts-as-functions. And if concepts are functions, there must be objects, value-ranges of functions, corresponding to extensions, and a corresponding principle, (the full) Basic Law V, governing functions and their value-ranges. Frege is now hooked: the principle about extensions strikes him as a *confirming instance* of Basic Law V, indeed as its single most important confirming instance.

But although we may guess at Frege's trains of thought, I think we cannot explain how the serpent entered Eden except to say: it is a brute fact that you cannot inject the power set of a set into that set (although you can, if the set is infinite, map its power set into it in such a way that equinumerous subsets are always taken to the same element). Perhaps hoodwinked by an analogy with an analytic-looking principle about numbers, Frege simply failed to notice that he was trying to do precisely that.

Confronted with the Russell paradox, Frege responded with a patch, which failed. It is not clear that had the patch worked, it would have counted as vindicating logicism, for it is by no means certain that Frege's revised notion of an extension and the new version of Basic Law V governing it could have counted as any more a logical notion and a logical truth than the notion of number itself and Hume's principle.

But there is a patch that works, and one not too distant in character from Frege's idea of concepts as 'two-sided' entities.[14] And it seems fair to say that it has as much claim as Frege's would have had (had it worked) to vindicate logicism about the natural numbers. But like Frege's revision of Basic Law V, and unlike Hume's principle and the original Basic Law V, its drawback is that there is no notion, and certainly no unquestionably logical notion, that the patched axiom can be thought to be analytic *of*.

So let V(x) be x=x. Call a concept F *bad* if $F \approx V \approx [x{:}\neg Fx]$, i. e., if there are just as many F objects, as non-F objects, as *objects*. Let E(F,G) hold iff either the same objects fall under F and G or F and G are both bad. E(F,G) is easily seen to define an equivalence relation.

Then the arithmetic of the natural numbers can be derived in the result of adjoining $\forall F \forall G(\%F = \%G \leftrightarrow E(F,G))$ to second-order logic; we may define zero as $\%[x{:}x \neq x]$ and successor in either the Zermelo or the von Neumann manner. Supplementation is required to obtain the theory of the real numbers. And it is not difficult to show $\forall F \forall G(\%F = \%G \leftrightarrow E(F,G))$ satisfiable.[15]

[14] Terry Parsons suggested to me this two-sided improvement of the repair given in my 'Saving Frege from Contradiction', *Proceedings of the Aristotelian Society*, N.S. Vol. LXXXVII, 1986–7, and inquired whether it is consistent.

[15] Thus: Ø is (von Neumann) 0; no ordered pair is a von Neumann ordinal. Let $D_0 = \{\emptyset\}$, $D_{n+1} = \{\langle z,i\rangle : z \subseteq D_n \wedge i = 0,1\}$, and $D = \cup D_n$. D is infinite. If F is a finite subset of D, then for some n, $F \subseteq D_n$, and $\langle F,1\rangle \in D_{n+1} \subseteq D$; if F is a cofinite subset of D, then for some n, $D\text{-}F \subseteq D_n$, and $\langle D\text{-}F,0\rangle \in D_{n+1} \subseteq D$. Otherwise F is infinite co-

The example shows that Frege's stipulations concerning the conditions under which an equality between value-range terms is true may be revised so that they assign each sentence at most one truth-value, without altering the class of well-formed expressions and without preventing the derivation of arithmetic or otherwise 'paralysing' the system, in a way that preserves a reasonable amount of the naive understanding of the notion of an extension. (Of course we have drawn on an understanding of sets sophisticated by decades of experience and theory in devising the repair.)

Dummett sketches a consistency proof for the first-order fragment of the system. It would be interesting to know exactly what the strength of the fragment is. Zero and the unordered pair can be defined, as $'x\ x{\neq}x$ and $'z(z{=}x \vee z{=}y)$. Hence so can each particular hereditarily finite set and successor (à la Zermelo): $\{x\}$, i. e. $'y\ y{=}x$. And one can define complements of these: V, i. e. $'x\ x{=}x$; $V\text{-}\{x,y\}$, i. e. $'z(z{\neq}x \wedge z{\neq}y)$, etc. On the other hand, one can certainly not define $\in$ (otherwise one could form the term $'x\ \neg x{\in}\ x$), and hence not $\subseteq$ (for $x{\in}\ y$ iff $\{x\}{\subseteq}y$) and hence neither $\cup$ nor $\cap$ (for $x{\subseteq}y$ iff $x{\cup}y = y$, iff $x{\cap}y = x$). Thus Dummett is certainly right to call the system 'paralysed'. But it seems to me that the greater the paralysis, the less tenable his opinion that it is second-order quantification that is the primary reason for the inconsistency; only if the first-order fragment had been strong enough to yield arithmetic or an interesting portion of it, would it be tempting to ascribe the inconsistency to the second-order quantifiers. For example, if some constructive or predicative second-order version of Frege's system could be defined and shown to be consistent and adequate for arithmetic, then Dummett's claim would acquire a force it now lacks.

One may compare the question why the *Grundgesetze* system is inconsistent with the somewhat curious question why (formal) arithmetic is incomplete. Quite evidently, one cannot either say that it is the presence of variables or that it is the presence of addition (and successor) even though variable-free arithmetic and addition with multiplication

---

infinite and therefore satisfies 'bad'. Set $\%F = \langle F,1\rangle$, $\langle D\text{-}F,0\rangle$, or $\varnothing$ according as F is finite, cofinite, or infinite coinfinite. And then $\langle D,\%\rangle \models \forall F\forall G(\%F = \%G \leftrightarrow (\forall x(Fx \leftrightarrow Gx) \vee$ F and G are bad$))$. For let F and G be arbitrary subsets of D. If F is cofinite, then $\%F = \langle D\text{-}F,0\rangle \in D$. Moreover, since F is cofinite, F and G are not both bad. And then $\%F = \%G$ iff $\%G = \langle D\text{-}F,0\rangle$, iff G is cofinite and $D\text{-}G = D\text{-}F$, iff $\forall x(Fx \leftrightarrow Gx)$. Similarly, if F is finite. Finally, if F is infinite coinfinite, then $\%F = \varnothing$. Then $\%F = \%G$ iff $\%G = \varnothing$, iff G is infinite coinfinite, iff G satisfies 'bad', iff F and G satisfy 'bad'. And if $\forall x(Fx \leftrightarrow Gx)$, then F and G satisfy 'bad'.

alone are complete. The reason, I think, is that the two theories are mutilated, or at least very much weakened, fragments of the original. On the other hand, it is perhaps slightly tempting to say, 'You know, surprisingly, it's the presence of multiplication in the language', the idea being that addition without multiplication, as opposed to variable-free arithmetic or arithmetic without addition, is an interesting and natural fragment of the whole arithmetic. (Since addition 'comes before' multiplication, arithmetic without addition is not 'natural'.) Were there some minor and salient feature of formal arithmetic whose removal would yield completeness (multiplication is hardly 'minor'), we might be tempted to ascribe incompleteness to the presence of that feature. I conclude that since there are revisions to the system of *Grundgesetze* that restore consistency, permit the development of arithmetic, and are significantly less drastic than the wholesale elimination of second-order quantification, it is not the presence of second-order quantifiers or Frege's specifications concerning them that are to be held responsible for the contradiction.

Reply to Boolos

MICHAEL DUMMETT

Professor Boolos and I are agreed that Frege's theory would be rendered consistent if either (i) Axiom V were deleted, or (ii) only first-order quantification were admitted. The substance of our disagreement is therefore restricted to the question which is the snow and which the yodel, in his metaphor, or which the match and which the matchbox.

What is the point of enquiring what led to the inconsistency in Frege's system? Not merely to discover how so clever a man could have come to make so colossal a blunder, but to locate the mistake in his reasoning. Frege aimed to answer the question what justifies us in assuming the existence of abstract objects, and, in particular, the objects of a fundamental mathematical theory. The question remains unresolved. Crispin Wright has argued that Frege's approach was fundamentally correct, without, however, enquiring why, in *Grundgesetze*, it went so disastrously wrong. It is therefore of interest to ask whether his attempt was utterly refuted by the catastrophe that overtook him, and, if not, how much of it can be sustained.

The basis of Frege's answer was the context principle, according to which reference to a range of abstract objects is justified if the senses of sentences involving it can be stipulated without presupposing the existence of those objects. The principle did not, as Boolos's remarks might suggest, express a predilection for contextual definition, which Frege came utterly to reject: it concerned how to explain what *cannot* be defined. If Frege's idea is not fundamentally correct, we have no inkling of an alternative one: it is therefore of interest to determine whether it was only his application of it that went so badly awry. For

First published under the title 'Basic Law V' in *Proceedings of the Aristotelian Society* XCIV (1994), 243–49. The paper was written in response to both George Boolos and Peter Clark. (The reply to Clark is on pp. 250–51.) Reprinted by courtesy of the Editor of the Aristotelian Society.

Frege, to fix the sense of a sentence was to determine its truth-conditions. The fundamental question is therefore why the method by which he believed that he had determined the truth-conditions of the sentences of his theory failed so signally to do so; why its partial embodiment in the axioms actually rendered them inconsistent is a subordinate, though more dramatic, question.

Frege would not have seen the task of laying down the truth-conditions for the sentences of his theory as that of specifying the intended model of it. That would have involved first stating which objects were to belong to the domain of the individual variables, and then interpreting the primitive symbols with respect to it; it would thus have presupposed a prior knowledge of the objects reference to which was supposed to be being justified *ab initio*. The context principle required the reference of the terms of the theory to be stipulated by laying down the values of functions, including concepts, taking their referents as arguments, the whole procedure being validated by a proof that a unique reference had been stipulated for every well-formed expression: Frege's consistency proof.

For Frege, and for anyone who believes his justification for speaking of abstract objects to have been in part correct, the problem is not so much what made his theory inconsistent as how, in the face of the semantics he devised for it, it *could* have been inconsistent. Boolos remarks that a failed consistency proof for a consistent system would hardly be a serpent in Eden. This dismissive observation would be just if the proof were a mere bright idea appended by Frege to his main exposition. It was not: it was integral to his entire conception of the manner in which to justify introducing a range of abstract objects.

Second-order quantification was essential for the inconsistency. What value-ranges there are depends on what first-level functions there are, that is over which functions the function-variables range. As Boolos mentions, I should now modify the firmness of my attribution to Frege of a substitutional interpretation of second-order quantification. I have not been converted to his view that an objectual interpretation is to be preferred: rather, I have come to think that no answer can be given to the question. Frege does not appear to have been conscious of the distinction between the two ways of understanding quantification. Now if we put an objectual interpretation, classically understood, on the second-order quantification of Frege's theory, there is, as I said in the book, a unique and obvious explanation for the inconsistency, namely that Axiom V, so understood, makes an inconsistent demand on the cardinality of the domain. Boolos refuses to consider any but an objectual interpretation, and accordingly regards that as the only explanation needed.

This diagnosis is, however, comparatively unilluminating, because, whatever was in Frege's mind, there is nothing in the theory that *compels* us to adopt an objectual rather than a substitutional interpretation of second-order quantification as Frege employs it, whereas the inconsistency remains if we adopt the latter, though it vanishes if we admit only first-order quantification. It is therefore more enlightening to enquire after the underlying reason for the inconsistency even against the background of a substitutional interpretation of second-order quantification.

In a first-order version of Frege's theory, there would be no domain of functions to consider the abstraction operator as defined over: there would therefore be no need to take the operator as defined over more than the meagre range of functions explicitly expressible. Since the expressive power of the language would be so weakened, it happens that Axiom V can in this case be satisfied. How, then, does Frege's consistency proof break down in the presence of second-order quantification? If we interpret it substitutionally, we cannot rely on considerations of cardinality. My strategy was to show first by example, without adverting to Axiom V, that Frege's inductive reasoning is grossly ill-founded. This does not make the theory inconsistent: it is what makes it *possible* for it to be inconsistent. I then observed that by using a very similar example, we obtain Russell's paradox by a modest appeal to Axiom V (with one of the function-variables held constant); we need to know only of one particular function that any function having the same value-range must be co-extensive with it. Cardinality does not enter the diagnosis: what leads to inconsistency is the circularity of Frege's stipulations. (Boolos offers a strange reason for setting aside the issue between objectual and substitutional quantification, namely that my example concerned a function expressible in the language; of course, if the topic is to be considered in the light of substitutional quantification, one can consider only such a function.)

Frege's consistency proof was not, indeed, sound even for the first-order fragment of his theory, since he in effect attempted to specify simultaneously the domain and the reference within it of value-range terms. His argument did not succeed, as he claimed, in fixing the truth-values of all sentences containing those terms; the stipulations intended to determine the truth-values of statements of identity between value-ranges were ill-founded in a manner analogous to that vitiating the proof of determinacy for sentences involving second-order quantification.

I differ from Boolos in preferring a diagnosis not assuming a classical interpretation of the function-variables, and, perhaps, in finding Frege's enterprise interesting, if bungled. There is a larger disagreement between us about domains of quantification in general.

When we quantify, must there be a domain over which we are quantifying? Boolos describes Frege as using his individual variables to quantify over all the objects there are, and commends him for doing so. The point bears on that just discussed. As Boolos also does, I remarked in my book (p. 227) that Frege's stipulations in *Grundlagen* do not determine the truth-values of all statements of identity between cardinal numbers, for instance that between Aleph-0 and the number of all objects. But from a precise criterion for the identity of colours we cannot derive the truth-value of the statement that two curtains have the same colour: reality determines that in the light of our criterion. And if our individual variables range over all the objects that there are, it may be asking too much of a criterion of identity for numbers that it should decide whether there are more than denumerably many of them: that, perhaps, is to be left to reality.

The view Boolos is advocating entails that we *can* quantify without restriction over all the objects there are, but is not to the effect that we always *ought to* do so: restriction of the range he holds to be quite in order. His aim is, rather, to controvert a view he tentatively attributes to me, saying that 'it would seem that he does think that there has to be a – what to call it – totality? collection? domain? containing all of the things we take ourselves at any one time to be talking about. He would seem to believe that whenever there are some things under discussion, being talked about, or being quantified over, ... there is a set-like item, a "totality", to which they all belong'; and again 'It may be replied, and I suspect that Dummett would agree, that whenever we use quantifiers, there must always be some domain, some totality of objects, over which our variables of quantification range'.

Boolos's repeated use of such phrases as 'Dummett would seem to think' and 'I suspect that he believes', and his difficulty in hitting on a word to express what he suspects me of supposing that there has to be for quantification to be possible, demonstrate that he is not *citing* any opinion I have expressed, but diagnosing what he takes to lie behind my explicit views. His diagnosis is quite mistaken; but I find an obstacle to discussing the matter on his terms. He evidently regards the question whether or not, when we quantify, we must assume that there is a what-do-you-call-it containing all the objects over which we are quantifying as *prior* to the question whether we are interpreting our statements truth-conditionally or constructively: we have to settle the former question first, and only then can we address the second. From my standpoint, we cannot deal with these questions sequentially: they are intertwined.

There are two salient ways of interpreting assertoric utterances: as statements or as claims (in the terminology I shall use here). With a statement is associated a condition for its truth: if the condition is satisfied, the assertion is correct, if not, the statement is false and its assertion incorrect. A claim is to the effect that a certain intellectual or linguistic feat can be performed, by the speaker or by someone he can call as a witness: if, when challenged, he performs it, or someone does so on his behalf, his claim was justified; if he is unable to do so or to produce anyone who can, it was unjustified. The evaluation of a claim as justifiable is as objective a matter as the evaluation of a statement as true: it can be justified or shown to be unjustified to everyone's satisfaction, without the intrusion of any subjective preference or judgement. The difference between the two types of assertoric utterance is that the condition for the truth of a statement is independent of the speaker's or anyone else's abilities or epistemic condition, except, of course, when the statement was about him, whereas the condition for a claim to be justified always turns on what the speaker or someone he can call on to act for him is able to do.

If the sentences of a mathematical theory are to be interpreted as all being determinately true or false and hence as capable of being used to express statements having definite truth-conditions, the quantifiers, when applied to predicates formed from sentences with determinate truth-conditions, must yield sentences which themselves have determinate truth-conditions. If we follow Frege in taking the predicate to be associated with a function mapping each object among those over which the individual variables range on to one of the values *true* and *false*, we may say that a proponent of a truth-conditional interpretation assumes that he can determine the truth-value of the quantified sentence as the outcome of the simultaneous action of this function on all the objects over which the individual variables range. It is of course perfectly possible to present a mathematical theory, as Kelley does, without enquiring into this question; that provides no reason for thinking that there is no question to be enquired into.

For such an infinitary operation to yield sentences with determinate truth-values, it must be quite definite over just *which* objects the variables of quantification range. What is necessary to secure this? Now what, from a realist standpoint, is needed for statements to the effect that something holds good of every oak-tree, say, or of some oak-trees, to have a definite truth-value is only that the term 'oak-tree' should have been given a clear criterion of application and a clear criterion of identity. Once we have supplied these, then, on a realist view of the external world, reality fixes what oak-trees there are. A completely resolute platonist may maintain that the same holds good of real numbers: once we have laid down which, among the mathematical objects there are, are to quality as real numbers, mathematical reality determines which real numbers there are.

Nevertheless, the platonist, however resolute, cannot keep this up indefinitely: that is why the set-theoretic paradoxes were such a shock. The concept *ordinal number*, as intuitively understood, admits as clear a criterion of application as does the concept *real number*: an ordinal number is the order-type of a well-ordering. But the platonist cannot regard the variables of quantification of a theory of ordinal numbers as ranging over everything satisfying that criterion, because *ordinal number* must be regarded as an indefinitely extensible concept by anyone, however devout a platonist. Boolos would dispute this, since he repudiates the very notion of an indefinitely extensible concept, saying, 'To his credit, Frege did not have the glimmering of a suspicion of the existence of indefinitely extensible concepts'. Plainly, he takes this denial to follow from his repudiation of the view that the objects over which the individual variables of a mathematical theory range form a collection, super-class or what-do-you-call-it. But suppose that a platonistically inclined mathematician has formulated a theory whose variables range over ordinal numbers, and has satisfied himself that its sentences have determinate truth-values in virtue of his having formed a definite conception of which objects fall among those over which his variables range – definite enough for it to be claimed that the infinitary operations yielding the truth-values of quantified sentences have in all cases a determinate outcome. Then the objects over which the variables range can be well-ordered by magnitude, and their order-type will satisfy the criterion for being an ordinal number, but cannot lie among the objects over which those variables range.

I have, in the foregoing discussion, avoided using the word 'range' as a noun, or any other noun such as 'domain', 'kind' or 'type', as it would otherwise have been natural to do, lest George Boolos should exclaim triumphantly, 'What did I tell you? A what-do-you-call-it!'. But the claim that, for quantified sentences of a mathematical theory to be capable of being construed as making statements with determinate values, *true* or *false*, there must be a means of determining over just which objects the variables of quantification range has nothing to do with any question whether those objects form a collection or super-class. It is true enough that, if we believe that we have such a means, it lies to hand to speak of the totality consisting of those objects, or of them as constituting the domain of quantification: but, in the general case, nothing hangs upon the use of such locutions. No thesis can be elicited to the effect that the objects over which the variables range form a collection or totality, which may then be challenged by enquiring what the word 'collection' or 'totality' is intended to mean.

What makes it natural to speak of the objects over which the individual variables of a theory range as forming a domain or totality is the indefinite extensibility of the concept *set* or *class*. If we have a theory of sets not admitting a universe set, and form a definite conception of which sets it is over which our variables range, nothing stands in the way of regarding them as forming a set in the intuitive sense of the term, although not one of those over which the variables of the theory range. Our doing so is not demanded by our taking those variables as ranging over them, or by our regarding the sentences of our theory as having determinate truth-values; but it is a consequence of our intuitive understanding of the concept *set*. If, with Boolos, we refuse to recognise the phenomenon of indefinite extensibility, we have no way of expressing this: we then find ourselves resorting, as he does, to speaking of a super-set or (if our theory admitted classes as well as sets) of a super-class, or totality, or collection, or set-like entity. Since such forms of expression are patently spurious, we then find ourselves denying the propriety of the step we have taken, as Boolos does; in denying it, we do violence to the intuitive notion of a set, and commit ourselves to that type of mystification which characterises every failure to grasp the lesson of the paradoxes.

Boolos mistakes my meaning when he says of me that he is led by my use of the term 'totality' to 'suspect strongly that he believes that we cannot actually quantify over all the ordinals (or sets) there are'. On the contrary, I hold, and in fact plainly said, that there is no bar to taking the variables of quantification of some theory as ranging over everything falling under some indefinitely extensible concept – over everything intuitively recognisable as an ordinal number, for example. There is a proviso, however, as I also plainly said: namely that no claim should be made that the sentences of that theory all possess determinate truth-values. Those sentences must be interpreted as serving, not to make statements in our restricted sense, but as expressing claims. This is because the quantifiers can no longer be explained in terms of infinitary operations. The statement 'Every ordinal has a successor' is true because it is intrinsic to the intuitive concept *ordinal number* that a means exists, given any object recognisable as falling under it, of defining another object also so recognisable and related to it as its successor; but an explanation along these lines will not secure a truth-value for every arbitrary quantified sentence. Boolos's misreading is due to his determination to state, and controvert, my view of what it is intelligible to quantify over in advance of any recognition of the distinction between the two ways of interpreting sentences, including the sentences of a mathematical theory: since my view is that the choice between these two ways of interpreting them bears crucially upon what we can intelligibly claim to be quantifying over, he cannot but misrepresent it.

Boolos is undoubtedly right to say that it is intelligible to quantify over all objects whatever; but sentences involving such quantification can be interpreted only as expressing claims, not as making statements, so that the view Boolos ascribes to Frege remains unjustified. This is because the (formal) concept *object* (or *identical with itself*) embraces all others; since some of the others are indefinitely extensible, it must be also. Completely unrestricted quantification is not illegitimate; what is illegitimate is a truth-conditional interpretation of sentences involving it. Boolos, however, took the distinction between the two ways of interpreting sentences to be of little consequence for the present issue: quite wrongly, if my argument has been sound.

On the Structure of Frege's System of Logic

CHRISTIAN THIEL

In this paper, I shall concern myself with some problems of the *Grundgesetze* system, the structure of which, I think, has not been made sufficiently perspicuous up to the present, in spite of continuous efforts made by Schweitzer 1935, Bartlett 1961, Resnik 1963, Hinst 1965, some minor probings of my own, and many others.[1]

Investigations of this kind, i. e., investigations of systematic difficulties with formal systems definitely belonging to the past, are not in the focus of interest today. They have not attracted as many scholars as have questions about Frege's proper place within or between movements in the philosophical tradition, or his place in relation to the Neo-Kantian schools in the late 19*th* and early 20*th* century, or Frege's influence on Wittgenstein. It seems that it has also been of greater interest to emphasize the alleged failure of Frege's semantics (like Hintikka), or even to claim Frege's narrow-mindedness and general mediocrity (like Baker & Hacker), or – recently – to suggest a parallel between the political convictions of the hoary Frege, drifting from a national liberalism towards the right end of the political spectrum, on the one hand, and his alleged conservatism in the domain of both mathematics and of methodology, on the other, as proclaimed by Toth in 1984, in Menzler-Trott's paper of 1989 and in his forthcoming book, and in Mehrtens's book of 1990.

Whereas the topics listed so far concern the *history* of logic and foundations, scholars interested in the systematic issues raised by Frege have further investigated "classical" subjects like Frege's philosophy of (or opinions on) language, or the context principle, or the axiological component of Frege's concept of *Bedeutung*, or either his discovery or total ignorance of what Angelelli has baptized "modern abstraction". In contrast to all that, the problems or themes to which I wish to draw your attention are rather of a "technical" nature.

---

[1] References by author's name and year are to the bibliography at the end of this paper.

Avoiding the very general subject of the interplay of syntax and semantics in Frege's *Grundgesetze*, I will select some of these topics for consideration. I will try to give my judgment on the present state of the discussion about them, which is a risky enterprise since I have been somewhat out of contact with the community of Frege scholars after my centenary edition of the *Grundlagen*. The probability is that I have missed important contributions to the literature and solutions of problems that I still take to be open.

The first problem I take to be open is the *inconsistency* problem. Frege's system of the *Grundgesetze* is inconsistent, as we all know. But there is no unanimity yet as to the "real" reasons for the failure. Two of the most prominent devices of Frege's logistic programme are under suspicion: *Wertverläufe* on the one hand, the *horizontal* on the other.

How does Frege introduce *Wertverläufe* into the system of the *Grundgesetze*? If we do not artificially restrict the meaning of the term "introduction", there are *three* different ways in which Frege introduces *Wertverläufe*.

First, *talk about Wertverläufe* is introduced already in § 3, as follows (Montgomery Furth's translation, see Frege 1964, 36):

> I use the words
>
> "the function $\Phi(\xi)$ has the same *course-of-values* as the function $\Psi(\xi)$"
>
> generally to denote the same as [allgemein als gleichbedeutend mit] the words
>
> "the functions $\Phi(\xi)$ and $\Psi(\xi)$ have always the same values for the same argument".

This stipulation of a synonymy which very likely – and in spite of Dummett's strong denial – is not only a *Bedeutungsgleichheit* but even a *Sinngleichheit*, might be called the "semantical" introduction. I will deal later with the question of what the meaning of *other* propositions about *Wertverläufe* is to be (i. e., of propositions that are not equations).

The second and the third ways of introducing *Wertverläufe* are, in contradistinction to the first, "syntactical" and therefore ought to be called ways of introducing *names* for *Wertverläufe*, *Wertverlaufsnamen*.

The second way simply mirrors the semantic introduction of *Wertverläufe* by stipulating as a "law of logic" ("ein logisches Gesetz", GGA I, 14)

$$\vdash (\grave{\varepsilon} f(\varepsilon) = \grave{\alpha} g(\alpha)) = (\smile\!\!\!\!\!^{\mathfrak{a}}\; f(\mathfrak{a}) = g(\mathfrak{a}))$$

which is assigned the number V already in § 20, in anticipation of the systematic numbering of the axioms or "Grundgesetze" in § 47 (GGA I, 61).

And thirdly, *Wertverläufe* are syntactically introduced in § 31 as the values of a second-level function $\grave{\varepsilon}\varphi(\varepsilon)$, one of the eight "Urfunktionen", the "primordial functions".

Since basic law V introduces *equations* between names of *Wertverläufe*, i. e., introduces *two* such names at a time, and since it is the *only* basic law importing *Wertverläufe*, it seems that in the system of *Grundgesetze*, propositions with only *one Wertverlaufsname*, like

$$\grave{\varepsilon}(-\,\varepsilon) = (-\!\!\overset{\mathfrak{a}}{\smile}\!\!-\, \mathfrak{a} = \mathfrak{a}) \qquad (\Psi)$$

cannot be derived – at least not "without going through a contradiction", to use an obscure and puzzling phrase of James Michael Bartlett ("ohne über einen Widerspruch zu gehen").

I should have used this latter restriction when I published in 1978 a short note under the title "Die Unvollständigkeit der Fregeschen 'Grundgesetze der Arithmetik'", maintaining the non-derivability of the proposition Ψ just mentioned, by an argument derived from § 10 (cf. Thiel 1978). As Kutschera was quick to point out, the *Grundgesetze* system, being inconsistent, is complete (more complete than one would wish). Indeed, Frege's theorem Ia symbolizes the *ex contradictione quodlibet*, and taking the two halves of Frege's version of the Zermelo-Russell antinomy as antecedents, we can, by detachment, derive *every* well-formed proposition, e. g. Ψ.

Can Ψ also be derived *without* using theorem Ia? As far as I know this is an open question, but one which it should not be too difficult to answer. Łukasiewicz proposed, in 1938, an axiom system for classical propositional logic which lists Frege's theorem Ia as a separate axiom. If this axiom system is indeed independent (which I have not checked so far, but which I assume), then one could take it as a model for a modified Fregean system for which we could either show Ψ to be derivable, or else prove by metalogical considerations that it is underivable. (My conjecture is that it is underivable).

I hasten to add that even this result would not be beyond doubt since we have used a weakened propositional calculus for *deduction*, but allowed *substitutions* of well-formed expressions with *Wertverlaufsnamen*, thereby going beyond any propositional calculus. Several investigations into Frege's logic simply ignore the fact that the system of *Grundgesetze* does not contain a classical propositional calculus as a *subsystem* – in the sense that one cannot single out those axioms of Frege's *Grundgesetze* that correspond to purely propositional laws of logic, *without* restricting the formation rules for well-formed expressions, too.

I will not go deeper into this question (which is not at all easy) and I will also skip the complex of "Frege's Way Out". I would rather like to discuss another possible source of inconsistency in Frege's *Grundgesetze*. His explanation of the horizontal (GGA I, 9; I abbreviate "das Wahre" as "*W*", and "das Falsche" as "*F*")

$$-\Delta \leftrightharpoons \begin{cases} W & \text{if } \Delta = W \\ F & \text{otherwise} \end{cases}$$

has reminded modern readers of the use of a truth (meta-)predicate "*T*" and an equivalence

$$T(\text{“}p\text{”}) \leftrightarrow p.$$

But $T(\text{“}p\text{”})$ is not $T(p)$, and v. Kutschera (1989, 25n.) has aptly pointed out that $-\xi$ applied to a *name of* the True yields the False, and not the True, as value. Nevertheless, in papers of 1979 and 1980, Peter Aczel has suggested that, because of Frege's use of the horizontal, Tarski's antinomy might be derivable in the *Grundgesetze* – a suggestion that makes sense only if the derivation does not take its way through the Zermelo-Russell or any other antinomy.

I was presumably rather too hasty in publishing a sketch of such a derivation within *Grundgesetze* in 1983, without checking beforehand whether the necessary means for it are really available in the *Grundgesetze*. The expressibility of the Gödelizing function and the diagonalization function as well as of related *Wertverläufe* within the *Grundgesetze* system was immediately doubted by Michael M. Richter, and recently in print by v. Kutschera *(loc.cit.)*. But to the best of my knowledge, the question remains undecided.

Carlos Dufour (unpublished) claims to have found a reconstruction of Aczel's idea which gets around the problem of strength of the proof-theoretic means of Frege's *Grundgesetze*, but I must admit that I have not yet understood his argument. On the other hand, it may be that Peter Aczel does *not* claim the derivability of Tarski's antinomy in Frege's system, but only the undefinability of a truth predicate in it. Nevertheless, Aczel explicitly says that the horizontal "is responsible for Russell's paradox" (1989, 17) and "gives an internal definition of the collection of truths" (ibid.). But think again: Since $-\xi$ is definable by $\xi = (\xi = \xi)$, $\xi = \zeta$ is likewise undefinable within the system, as indeed stated by Aczel. Why then is it the horizontal that is responsible for Russell's paradox, and not the equality sign?

A large number of open problems surrounds the so-called permutation argument in § 10. In my view, these are not pseudo-problems, in spite of the inconsistency of the *Grundgesetze* system. I have come to ascribe this inconsistency not (or not only) to basic law V (i. e., not to the logical or syntactical form of a particular axiom), but to an insufficient delimitation of the function names admissible in it (*v. infra*). Admitting that I have not yet investigated the possible kinship of this approach with Nino Cocchiarella's "realist theory of predication along Fregean lines" (Cocchiarella 1974, 80, cf. also chs. 2 and 4 in Cocchiarella 1987), I will turn to the basic idea of my own version in a moment, after an outline of the two other *desiderata*.

In *Grundgesetze* § 10, tacitly taking up a question which had already arisen in the *Grundlagen* concerning *Begriffsumfänge*, Frege proceeds to assign a reference also to expressions other than equations between *Wertverlaufsnamen*, e. g. to

$$\grave{\varepsilon}(-\varepsilon) = (\smile^{\mathfrak{a}}\!\!-\, \mathfrak{a} = \mathfrak{a}).$$

To achieve this, he reduces this case to that of equations between *Wertverlaufsnamen*, by identifying the True *W* with the *Wertverlauf* $\grave{\varepsilon}(-\varepsilon)$, and the False *F* with the Wertverlauf $\grave{\varepsilon}(\varepsilon = \top\!\smile^{\mathfrak{a}}\!\!-\, \mathfrak{a} = \mathfrak{a})$. As *W* and *F* are the *only* objects not referred to by *Wertverlaufsnamen* earlier, the identification settles the hitherto open cases – or so Frege hopes.

He tries to prove by a *permutation argument* that *any Wertverlauf* can be identified with the True and *any other* with the False, without contradicting the stipulation of basic law V. Schroeder-Heister believes to have a counter-example to this general claim (1987, 77). Adrian W. Moore and Andrew Rein, however, maintain that the model-theoretic proof of Schroeder-Heister is inadequate for Frege's procedure, because any adequate interpretation must first interpret Frege's second-level function $\grave{\varepsilon}\varphi(\varepsilon)$, and that Frege's identifications are completely correct. It seems to me that in spite of Terry Parsons's lucid exposition of the issue, the controversy has not yet been settled. So let me list its final solution as another *desideratum*.

Likewise, Michael Resnik's beautiful study on "Frege's proof of referentiality" (Resnik 1986) may turn out to be only the penultimate word on Frege's unsuccessful attempt at proving the referentiality of all well-formed expressions of the *Grundgesetze* system. It is conceivable that this attempt at a proof – unparallelled in other Fregean writings – will never be fully disentangled. It is clear that Frege aims at giving some kind of inductive proof, but, due perhaps to the *gap-formation principles* ("Lückenbildungsprinzipien"), the induction process cannot proceed without loops and it becomes possible that an expression is reached by the process in two different ways and gets two different references on this occasion (see Figure 1).

So it would be desirable to have a more penetrating analysis of Frege's attempt along the lines of Michael Resnik's investigation. I can agree with practically everything in his paper of 1986, except his argument for the presumably true statement that referentiality and consistency are independent. Resnik's simple model-theoretic argument for this thesis seems to me to show only that every well-formed expression has been *given* a reference, but not that it has been given *only one* reference (although I may be mistaken here). At any rate, it is *not* true that the conclusion from referentiality to consistency is invalid *because* two derivable expressions "A" and "$\top A$" need not be names of the two different truth-values (if "A" is a name of a *Wertverlauf* distinct from W and from F, then $\top A = W$, but not $A = F$). This case could be excluded only if all derivable expressions were names of truth-values, and concerning this there is a fatal indeterminacy in Frege's *Begriffsschrift*. The expression "2" is not a name of a truth-value; it will later be shown to be a genuine *Wertverlaufsname*. So "$\top 2$" is a name of the True, and Frege can write "$\vdash\!\top 2$" (GGA I, 11). But whereas Frege can state his amalgamation laws,

$$--\Delta = -\Delta;\ -\top\Delta = -\top-\Delta = \top-\Delta,$$

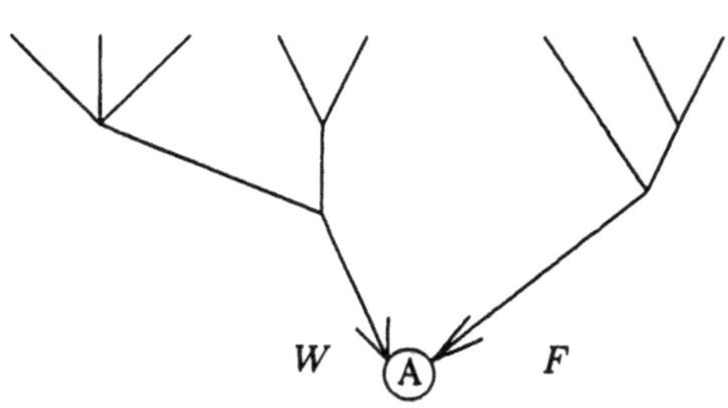

Figure 1: Clashing branches in a pedigree

there is no law Δ = (– Δ): the formula "Δ = (– Δ)" says only that Δ is a truth-value, and hence it is not *generally* valid. This, however, leaves us with on open question about the proper treatment of a premiss

in an application of the detachment rule. Should the *succedens B* be detached as "*B*" or as "– *B*"? We do not get an answer; the *Begriffsschrift* notation does not allow us to distinguish between

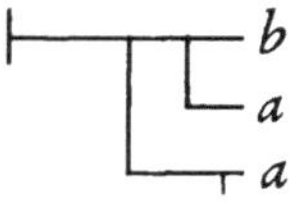

and

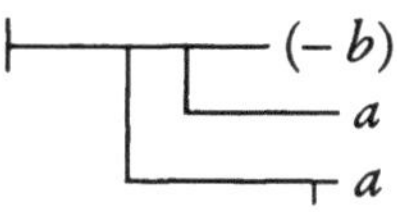

except by bracketing.

The third *desideratum* which I want to mention here is an elucidation of the construction and import of theorem χ in Frege's *Nachwort* to vol. II of *Grundgesetze*. In the *Nachwort*, Frege gives two different derivations of the antinomy, the first involving a particular expression "∀", the second using Frege's elementhood symbol "∩", but both making use of *Wertverlaufsnamen*. Immediately after that, however, in order to strengthen his point against Vb,

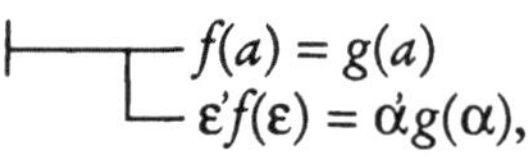

Frege presents a confutation of this law by making his argument independent of *Wertverläufe* (which "are after all under suspicion": Frege 1964, 133, cf. GGA II, 257 b), and he arrives at the falsehood of Vb as his result.

Of course, Frege cannot mean to derive the *negation* of Vb, since this law contains in its very formulation two *Wertverlaufsnamen*. What he does, is to derive a theorem χ,

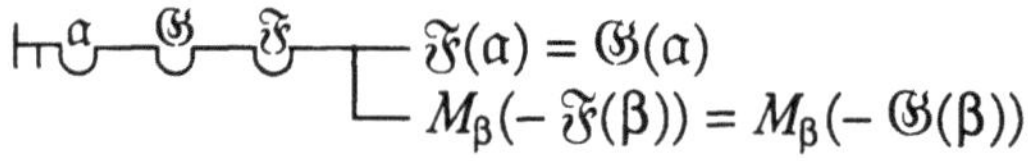

This is equivalent to

$$\vdash\!\overset{\mathfrak{G}}{\smile}\!-\!\overset{\mathfrak{F}}{\smile}\!\!\begin{array}{l}\overset{\mathfrak{a}}{\smile}\ \mathfrak{F}(\mathfrak{a}) = \mathfrak{G}(\mathfrak{a}) \\ \llcorner\ M_\beta(-\ \mathfrak{F}(\beta)) = M_\beta(-\ \mathfrak{G}(\beta)),\end{array}$$

and by taking $\acute{\varepsilon}\varphi(\varepsilon)$ as the schematically indicated second-level function M_β, we get the existence of two functions $\Phi(\xi)$ and $\Psi(\xi)$ such that $\acute{\varepsilon}(-\ \Phi(\varepsilon)) = \acute{\alpha}\ (-\ \Psi(\alpha))$, but *not* $\Phi(\Gamma) = \Psi(\Gamma)$ for some Γ, i. e. $\vdash\!\overset{\mathfrak{a}}{\smile}\!-\ \Phi(\mathfrak{a}) = \Psi(\mathfrak{a})$). Frege claims that this violates Vb regarding the concepts $-\ \Phi(\xi)$ and $-\ \Psi(\xi)$: *if* we were to adjoin Vb to the basic laws of *Grundgesetze* minus V, we would immediately encounter inconsistency.

Actually, Frege's formal derivation on pp. 259 ff. does not yield exactly what he states verbally as his result, since in the antecedens of χ, $\mathfrak{F}$ and $\mathfrak{G}$ appear with a prefixed horizontal, and in the succedens without it (and of course $\Phi(\Gamma) \neq \Psi(\Gamma)$ does not preclude $-\ \Phi(\Gamma) = -\ \Psi(\Gamma)$). One must indeed go back to pp. 257 ff. in order to find that $\mathfrak{F}$ and $\mathfrak{G}$ have really been constructed as *concepts* for which $\mathfrak{F}$ and $-\mathfrak{F}$, and $\mathfrak{G}$ and $-\ \mathfrak{G}$ coincide.

But if theorem χ is valid, then Frege is right in saying about the truncated *Grundgesetze* system that we cannot formulate in it an intuitively acceptable notion of set (or class, or *Begriffsumfang*). It is astonishing that this result, unlike Frege's attempted "Way Out" in the same *Nachwort*, has not stimulated further investigations before Boolos 1986/87. Although neither that paper nor its most recent successor, Boolos 1993, is directly concerned with theorem χ, they are both relevant for its assessment.[2] The context is dominated by the possibility to ignore Frege's logistic definition of number and to derive arithmetic directly from "Hume's Principle", i. e., the statement that the number belonging to a concept $\Phi(\xi)$ is equal to the number belonging to a concept $\Psi(\xi)$ if and only if $\Phi(\xi)$ and $\Psi(\xi)$ are equinumerous in the sense of a one-to-one correlation of the objects falling under them, respectively. The importance of this possibility, which may have been obvious to Frege and whose "discovery" is not quite so sensational as some have recently tried to make us believe, rests upon the fact that the formal derivation may proceed without resort to Frege's extensions of (second-order) concepts and his fundamental law V, and that a necessary but much weaker law can consistently be adjoined to second-order logic (which we take to be consistent). The proposal of Boolos 1986/87

[2] I am grateful to George Boolos and Michael Resnik for pointing this out to me in the discussion of my paper at the Frege conference in Munich in 1991.

to replace Frege's extensions by "subtensions" which are abstract objects with respect to an equivalence relation that takes into account the (formally defined) "smallness" of the concepts $\Phi(\xi)$ and $\Psi(\xi)$ is extremely interesting and indeed "a minimal departure from the project of the *Grundgesetze*" (Boolos 1986/87, 149) which might even have appealed to Frege had he known that his Way Out would not lead to an escape-door. On the question whether this new approach would simply amount to abandoning logicism altogether, or whether it could perhaps bear the name of a "new logicism", I have not arrived at a firm opinion. I still think it worth-while to investigate the possibility of formally retaining Frege's fundamental law V while putting restraints on the function names admitted, a suggestion to which I will turn shortly.

Frege concluded that with theorem χ "the extension of the concept in the traditional sense has been abolished" (GGA II, 260), and that "the usual understanding of the words 'extension of a concept' stands in need of correction" (GGA II, 256a: "dass die bisherige Auffassung der Worte 'Umfang eines Begriffes' einer Berichtigung bedarf"). This directs our attention to the fact (a fact important for the *historian* of logic and of the foundations of mathematics), that in Frege's time classes were clearly identified with the extensions of concepts (*Begriffsumfänge*), whereas today we identify them with sets and keep logic and set theory apart as disciplines, close as their cooperation may be. But for traditional logic, and obviously for Frege also, the doctrine of the extension of concepts is a genuine part of the logic of the concept (*Begriffslogik* or *Begriffslehre*), and it is in this sense that Frege can call his *Wertverläufe* logical objects, and basic Law V a fundamental law of logic.

But are we really forced to revise our notion of "the extension of a concept" and, accordingly, our characterization of the corresponding function $\grave{\varepsilon}\varphi(\varepsilon)$ by an additional axiom like Frege's fundamental law V or its later substitute $V^{\prime}$ (GGA II, 262b)? Let me turn, as promised, to my alternative analysis of the problem, outlined more than twenty years ago (Thiel 1972) and supplemented by details of the necessary revisions in subsequent years. This represents also a return to the topic referred to in the title of this paper, since the main idea behind my proposal is the recognition that the structure of a formal system is determined not by its axioms and deduction rules alone, but just as well by the formation rules for the admissible expressions – i. e., in the *Grundgesetze* system, by the "primordial functions" together with the rules for the formation of well-formed names from others already constructed. Only names taken from this explicitly delimited domain of well-formed names may be inserted into fitting argument places of function names of the proper kind. Much depends on the constraints we impose on this domain, and under different choices the inconsistency problem, the identifiability thesis of § 10, and the problems raised by theorem χ (all obviously inextricably interwoven) may appear in a different light.

All of the eight primordial names, from which the system of *Grundgesetze* starts, are function names: the horizontal, the negation sign, the condition sign, the equality sign, and the definite description operator as first-order function names (the last of which we will skip in the sequel since it is not involved in the derivation of the antinomy, nor in the formation of the "critical" expressions "∀" and "R"), the first-order universal quantifier and the name "$\grave{\varepsilon}\varphi(\varepsilon)$" of the *Wertverlauf* function as second-order function names, and the second-order universal quantifier as a third-order function name (with second-order functions as arguments, and a bound variable for first-order functions). Except for the *Wertverlauf* function (and the definite description function), all of them take only truth values as values. At the beginning, i. e., among the primordial names, there is no object name. All object names are derivative names constructed by inserting some admissible argument name into the argument place of a function name. This operation of insertion will be referred to as "(ι)". In Frege's system, all values of functions are objects; no function can be the value of some other function for some argument.

Removing an argument name from its place within an object name is, as it were, the inverse operation to insertion; it transforms the object name to which it is applied into a function name by leaving ("creating", "forming") a gap in it – an argument place open for object names generally. As we can apply gap-formation to object names of multiple composition, it may sometimes yield new function names, i. e. function names not appearing in the construction process leading to the complex object name subjected to our "gap-formation" process $\Phi(\Delta) \Rightarrow \Phi(\xi)$. In like manner, we may transform a compound one-place first-order function name into a two-place one by forming a second gap, $\Phi(\xi,\Delta) \Rightarrow \Phi(\xi,\zeta)$, and an object name containing another object name as part, into a second-order function name by removing the whole first-order function name which is the complement of this part: $\Phi(\Delta) \Rightarrow \varphi(\Delta)$. I will refer to these three kinds of operation as "(λ_0)", "(λ_1)", and "(λ_2)", respectively. Frege explains them, as well as (ι), in § 26 on pp. 43–44 of GGA I.

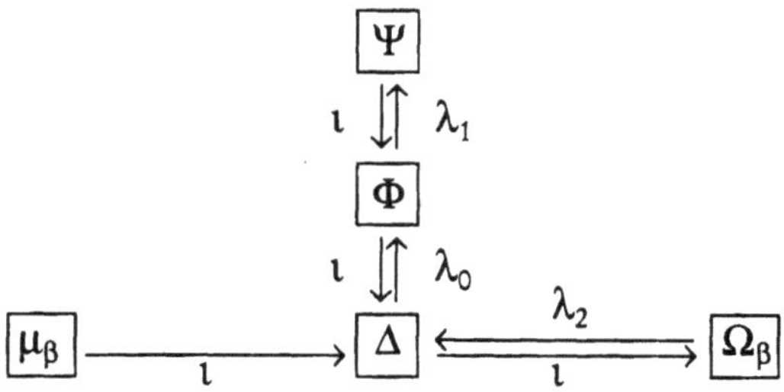

Figure 2: The general interplay

Figure 2 illustrates the interplay of the four operations generally. In the boxes, "Δ" stands for an object name, "Φ" for a one-place first-level function name, "Ψ" for a two-place first-level function name, "Ω_β" for a (one-place) second-level function name, and "μ_β" for second-level universal quantification (which is a third-level function, indicated schematically). Arrows run from the box with the function name to which the operation given beside the arrow is applied, to the boxes with the operation's result. Figure 3 takes us to the point of departure of Frege's system by including the names of its primordial functions, positioned next to "their" boxes:

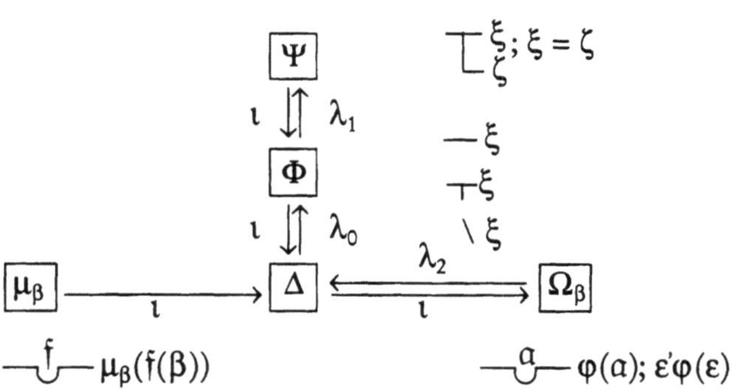

Figure 3: Interplay of the primordial names

From the names of the primordial functions, the operations (ι), (λ_0), (λ_1), and (λ_2) yield all the derivative names of Frege's *Grundgesetze*. Among them are names of the truth-values, and some of these names are sentences in the usual sense. As "basic laws" of his system (among which we may ignore basic law VI since we have skipped Frege's description function \ξ), Frege selects the sentences shown in Figure 5.

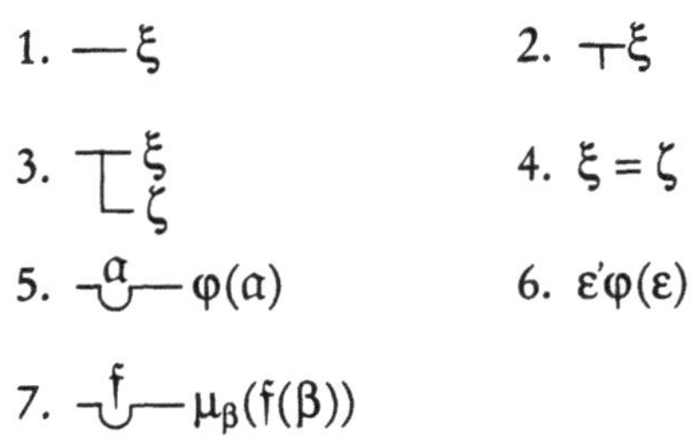

Figure 4: Frege's primordial functions

Here, "officially", the use of small Roman letters is an abbreviation in order to avoid the clumsy universal quantification of sentences with the corresponding Gothic (or rather, "German") letters, but in practice Frege does not treat small Roman letters in a different manner from "real" schematic letters like "*M*" in basic law IIb. Rules of deduction, listed among the rules of the system in § 48 (pp. 61–64), govern the derivation of the derivative sentences ("theorems") of the system.

Of course, every well-formed name of the system should have a reference, and one reference only. Frege provides for this by specifying the values for every argument admissible into the empty place(s) of all names of primordial functions – with the exception of the *Wertverlauf*

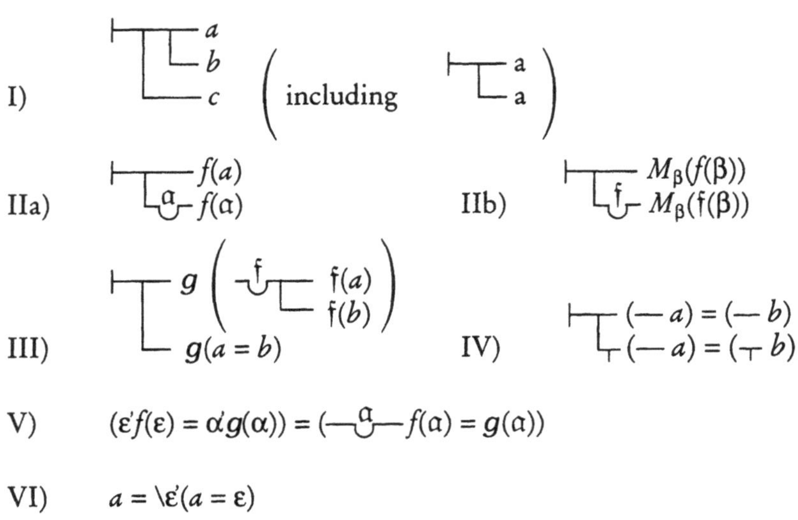

Figure 5: Basic Laws I–VI

function $\acute{\varepsilon}\varphi(\varepsilon)$. Nowhere in the *Grundgesetze* does Frege say that the value of the function $\acute{\varepsilon}\varphi(\varepsilon)$ for a one-place first-level function as argument should be the *Wertverlauf* of this latter function – presumably because he wants to avoid the miserable appeal (once offered in *The Foundations of Arithmetic*) to the reader's already established "acquaintance" with *Wertverläufe*. It seems that Frege relied on the boundary conditions imposed on operations with *Wertverlaufsnamen* by his basic law V, and on the information supplied by the following *criteria* for the referentiality of well-formed names, the successful passing of which shows a name to have (viz., to have been given by the explanations and stipulations of the system) exactly one reference, as required:

$$\begin{array}{lrcl}
\text{Ⓐ} & \otimes\,\Phi(\xi) & \Leftrightarrow & \bigwedge_{\Delta}(\otimes\,\Delta \rightarrow \otimes\,\Phi(\Delta)) \\
\text{Ⓑ} & \otimes\,\Delta & \Leftrightarrow & \bigwedge_{\Phi}(\otimes\,\Phi(\xi) \rightarrow \otimes\,\Phi(\Delta)) \wedge \\
 & & & \bigwedge_{\Psi}(\otimes\,\Psi(\xi,\zeta) \dot{\rightarrow} \otimes\,\Psi(\Delta,\zeta) \wedge \otimes\,\Psi(\xi,\Delta)) \\
\text{Ⓒ} & \otimes\,\Psi(\xi,\zeta) & \Leftrightarrow & \bigwedge_{\Gamma,\Delta}(\otimes\,\Gamma \wedge \otimes\,\Delta \dot{\rightarrow} \otimes\,\Psi(\Gamma,\Delta)) \\
\text{Ⓓ} & \otimes\,\Omega_{\beta}(\varphi(\beta)) & \Leftrightarrow & \bigwedge_{\Phi}(\otimes\,\Phi(\xi) \rightarrow \otimes\,\Omega_{\beta}(\Phi(\beta))) \\
\text{Ⓔ} & \otimes -\!\overset{\mathfrak{f}}{\smile}\!- \mu_{\beta}(\mathfrak{f}(\beta)) & \Leftrightarrow & \bigwedge_{M}(\otimes\, M(\varphi(\beta)) \dot{\rightarrow} \otimes -\!\overset{\mathfrak{f}}{\smile}\!- M_{\beta}(\mathfrak{f}(\beta)))
\end{array}$$

Figure 6: Frege's criteria of referentiality

It is obvious that these criteria are closely related to Frege's formation principles for well-formed expressions. There is no reason to expect that their fulfilment alone would guarantee a unique reference to each one of a given collection of names (let alone confer it upon them). But Frege had hoped that conformation to the criteria would make it possible to confer a unique reference on an expression which we wish to adjoin to a collection of names whose referentiality has already been established. He was of course thinking of *Wertverlaufsnamen*, to be adjoined to the collection of truth-value names in the system of *Grundgesetze*.

His hopes were frustrated, and although I must refer the reader to Resnik 1986 for a comprehensive discussion of the reasons for this failure, we can get a clear impression of the deficiencies of Frege's idea by considering, first, the name "$\top\xi$" of the negation function. According to Frege's first criterion, it is referential if and only if its value for any referential expression admissible into its argument places is referential:

$$\otimes \top\xi \Leftrightarrow \bigwedge_{\Delta}(\otimes\Delta \rightarrow \otimes\top\Delta).$$

This is indeed the case, because it has been guaranteed by Frege's explanation of the negation function as

$$\top\Delta \leftrightharpoons \begin{cases} F & \text{if } \Delta = W \\ W & \text{otherwise.} \end{cases}$$

It is understood that the succedens $\otimes\Delta$ of the quantified conditional in the criterion is true; if it were not, i. e. if "Δ" were not referential, "$\top\Delta$" would not be referential either. This easy arrival at a positive answer is due to the fact that the expression tested is a function name and, moreover, a primordial name, for which Frege has given us a complete explanation: even if in a particular case we do not know whether "Δ" is a name of the True or not (so that we do not know which of the two subcases of the explanation obtains), we can be sure that "$\top\Delta$" is referential.

This is not always so. Let us take the compound expression

$$\vdash\!\top\!\overset{\mathfrak{g}}{\smile}\!\!\begin{array}{l} \top\, \mathfrak{g}(\forall) \\ \llcorner\, \grave{\alpha}(-\,\mathfrak{g}(\alpha)) = \forall \end{array} \qquad\qquad (\mathrm{R})$$

by which Frege expresses the circumstance that the class $\forall$ does not belong (as an element) to itself. Is R referential? This question could be answered in the affirmative, by means of the criteria of referentiality stated by Frege in § 29 of *Grundgesetze*, only if it could be shown that all expressions of the form $\Psi(\forall)$ are referential. This however, we cannot prove since this is also the form of the expression R which we want to show to be referential: we have moved in a circle (cf. Figure 7).[3]

The close connection of this fact with the inconsistency of the system of *Grundgesetze* becomes clear if we try to discover the reference which Frege's stipulations must have conferred upon R if the system does what Frege had designed it to do. As R is an object name, and its explicit form starts with a horizontal, its reference must be either the True or the False. But Frege himself, in his *Nachwort* to the second volume of *Grundgesetze* (GGA II, 256 f.) has shown that these two referential specifications lead to a contradiction, viz. the Zermelo-Russell antinomy.

This failure demonstrates that some big mistake (or several?) must have occurred in the construction of the system of *Grundgesetze*.

---

[3] See Thiel 1965, p. 82 f., or p. 76 f. of Blakeley's translation Thiel 1968.

Frege's spontaneous reaction was to declare his basic law V false and to admit that his arguments in § 31 do not suffice to secure a reference for his combinations of signs in all cases.[4]

That this is so, cannot be doubted, but Frege scholars differ widely in their diagnoses and their recommended therapies, from Bartlett 1961 through Resnik 1963, Hinst 1965, Ch. Parsons 1965, Thiel 1965, 1975, and 1979, Resnik 1986, Boolos 1986/87 and 1993. Although I believe that the three last-mentioned papers mark the present state of our insight, it seems to me that they do not make some further deliberations superfluous. I emphasized (*v. supra*, p. 5) that even though basic law V may be "false", this may be due not to its formal structure but to an inappropriate liberality in admitting *Wertverläufe* like ∀, hand in hand with the function names out of which they were formed (by insertion of a first-order function name into the argument place of "ε'φ(ε)"). In Thiel 1979 (p. 484), I have dubbed expressions of this kind "Trojan expressions" for obvious reasons, and suggested that their exclusion might prove to be an alternative solution to Frege's inconsistency problem. I have elaborated this also in Thiel 1972 and in great detail in Thiel 1975, and do not consider it necessary to repeat, nor to summarize these expositions here.

What has remained in need of clarification is the rôle of "impredicativity" in this approach. As already noted in Bartlett 1961, expressions of the kind we had to analyse are "impredicative" in the broad sense usual in discussions of the foundations of mathematics. E. g.,

$$\vdash\!\overset{\mathfrak{g}}{\smile}\!\begin{array}{l} \mathfrak{g}(\xi) \\ \grave{\alpha}(-\,\mathfrak{g}(\alpha)) = \xi \end{array}$$

is a function name banished by the restrictions in Russell 1908 for the reason that "whatever contains an apparent variable must not be a possible value of that variable".[5] The formulation signals a kind of circularity that, from a "constructivist" point of view, is obviously involved in the use of a quantifier in an expression which (as it were, after the completion of its construction) turns out to belong to the range of that same quantifier. The analogous circularity in the context of Frege's ref-

[4] See Frege's letter to Russell, 22 June, 1902, in: G. Frege, *Wissenschaftlicher Briefwechsel*, ed. G. Gabriel et al. (Meiner: Hamburg 1976), 212–215; or *Philosophical and Mathematical Correspondence* (Blackwell: Oxford 1980), 131–133.

[5] B. Russell, Mathematical logic as based on the theory of types. *American Journal of Mathematics* 30 (1908), 222–262, quoted 237.

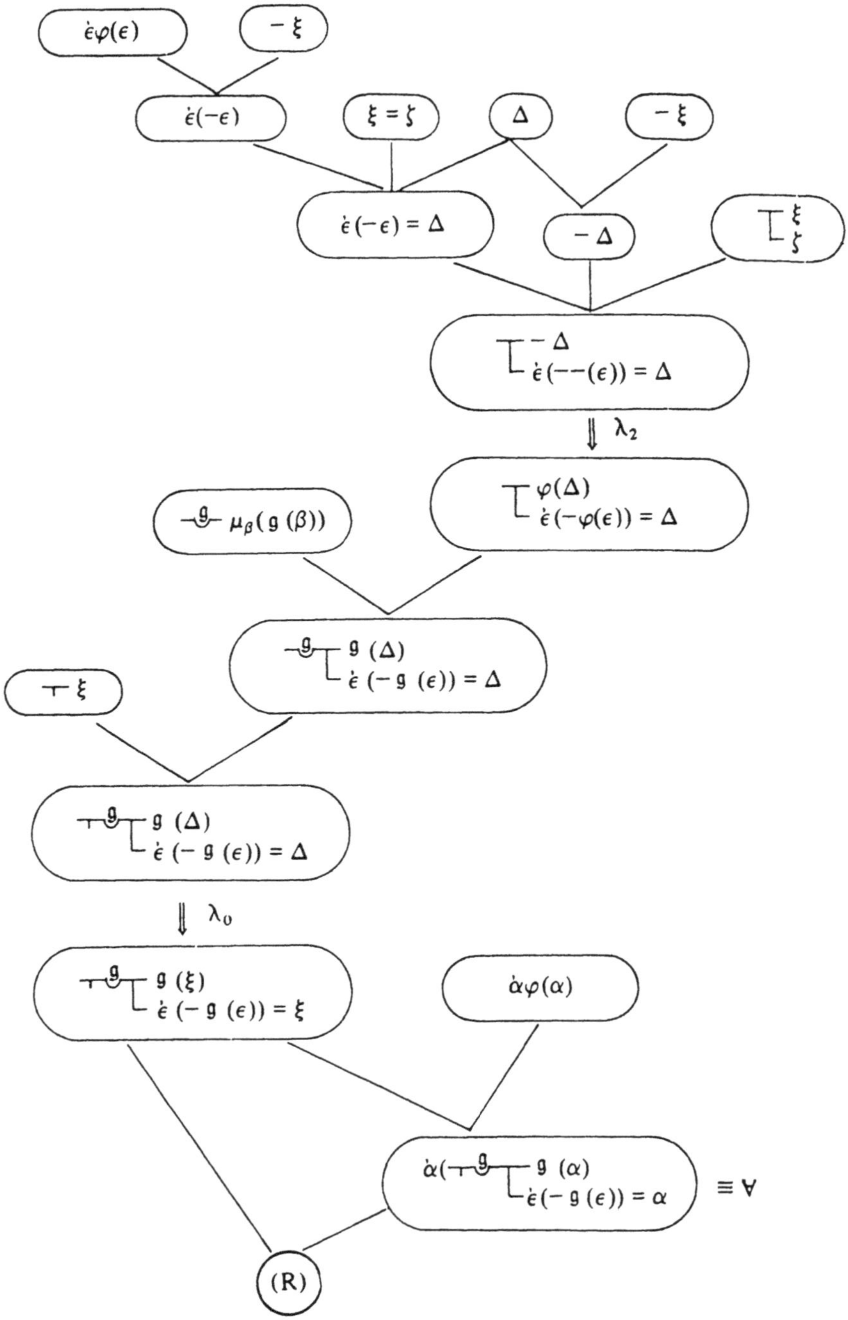

Pedigree of formula R

erentiality proof is that which was pointed out to exist between R, "∀" and the function name (call it "$P(\xi)$") obtained by replacing "∀" by "ξ" in R: R originates from inserting "∀" into a function name occurring "farther up" in its own pedigree, and our attempt at proving R referential has to refer back to the referentiality of "$P(\xi)$" the referentiality of which rests on the referentiality of R. Yet, the criteria of referentiality do *not* mirror the construction steps of the well-formed expressions – a somehow trivial statement in view of the fact that in any application of a gap-formation principle we lose the thread of referentiality which we keep as long as we create compound expressions by insertion: the referentiality of the function names we obtain depends on the function's values for *all* the arguments admissible into its empty places, and about these we know nothing.

Bibliography

Aczel, Peter: 1979, The structure of the formal language of Frege's Grundgesetze. *6th International Congress of Logic, Methodology and Philosophy of Science, Hannover, August 22 – August 29, 1979, Abstracts of Sections 13/14*, 16–17.

Aczel, Peter: 1980, Frege Structures and the Notions of Proposition, Truth and Set. In: J. Barwise / H. J. Keisler / K. Kunen (eds.), *The Kleene Symposium* (North Holland Publishing Company: Amsterdam/New York/Oxford 1980), 31–59.

Bartlett, James Michael: 1961, *Funktion und Gegenstand. Eine Untersuchung in der Logik von Gottlob Frege*. Phil. Diss. München 1961.

Boolos, George: 1986/87, Saving Frege from contradiction. *Proceedings of the Aristotelian Society* n. s. 87 (1986/87), 137–151.

Boolos, George: 1993, Whence the contradiction? *The Aristotelian Society. Supplementary Volume* 67 (1993), 213–233.

Cocchiarella, Nino B.: 1974, A new formulation of predicative second order logic. *Logique et Analyse* n. s. 17, no. 65–66 (1974), 61–87.

Cocchiarella, Nino B.: 1987, *Logical Studies in Early Analytic Philosophy*. Ohio State University Press: Columbus, Ohio 1987.

Frege, Gottlob: 1893, *Grundgesetze der Arithmetik. Begriffsschriftlich abgeleitet.* Band I. Hermann Pohle: Jena 1893.

Frege, Gottlob: 1903, *Grundgesetze der Arithmetik. Begriffsschriftlich abgeleitet.* Band II. *ibid.* 1903.

Frege, Gottlob: 1964, *The Basic Laws of Arithmetic. Exposition of the System.* Translated and edited, with an introduction, by Montgomery Furth. University of California Press: Berkeley/Los Angeles 1964.

Hinst, Peter: 1965, *Syntaktische und semantische Untersuchungen über Freges "Grundgesetze der Arithmetik"*. Phil. Diss. München 1965.

Kutschera, Franz von: 1989, *Gottlob Frege. Eine Einführung in sein Werk.* Walter de Gruyter: Berlin/New York 1989.

Mehrtens, Herbert: 1990, *Moderne Sprache Mathematik. Eine Geschichte des Streits um die Grundlagen der Disziplin und des Subjekts formaler Systeme.* Suhrkamp: Frankfurt a. M. 1990.

Menzler-Trott, Eckart: 1989, Ich wünsche die Wahrheit und nichts als die Wahrheit ... Das politische Testament des deutschen Mathematikers und Logikers Gottlob Frege. Eine Lektüre seines Tagebuchs vom 10. 3. bis 9. 5. 1924. *Forum* [Wien] 36. Jg., Nr. 432 (20. Dezember 1989), 68–79.

Menzler-Trott, Eckart: 1995?, *Gottlob Frege. Das politische Testament. Freges Tagebuch vom 10. März bis 9. Mai 1924.* Königshausen & Neumann: Würzburg [erstmals angezeigt 1990].

Moore, Adrian W. / Rein, Andrew: 1986, Grundgesetze, Section 10. In: L. Haaparanta / J. Hintikka (eds.), *Frege Synthesized. Essays on the Philosophical and Foundational Work of Gottlob Frege.* D. Reidel: Dordrecht etc. 1986, 375–384.

Moore, Adrian W. / Rein, Andrew: 1987, Frege's permutation argument. *Notre Dame Journal of Formal Logic* 28 (1987), 51–54.

Parsons, Charles D.: 1965, Frege's Theory of Number. In: Max Black (ed.), *Philosophy in America* (Cornell University Press: Ithaca, N. Y., and Allen & Unwin: London, 1965), 180–203; repr. with a Postscript in: Charles Parsons, *Mathematics in Philosophy, Selected Essays* (Cornell University Press: Ithaca, N.Y. 1983), 150–175.

Parsons, Terence: 1987, On the consistency of the first-order portion of Frege's logical system. *Notre Dame Journal of Formal Logic* 28 (1987), 161–168.

Resnik, Michael David: 1963, *Frege's Methodology. A Critical Study*. Ph. D. Thesis, Harvard University, Cambridge, Mass. 1963.

Resnik, Michael David: 1986, Frege's proof of referentiality. In: L. Haaparanta / J. Hintikka (eds.), *Frege Synthesized. Essays on the Philosophical and Foundational Work of Gottlob Frege.* D. Reidel: Dordrecht etc. 1986, 177–195.

Schroeder-Heister, Peter: 1984, Freges Permutationargument. Zu § 10 der "Grundgesetze der Arithmetik". In: G. Wechsung (ed.), *Frege Conference 1984. Proceedings of the International Conference held at Schwerin (GDR), September 10–14, 1984* (Akademie-Verlag: Berlin 1984; *Mathematical Research / Mathematische Forschung,* Band 20), 182–188.

Schroeder-Heister, Peter: 1987, A model-theoretic reconstruction of Frege's permutation argument. *Notre Dame Journal of Formal Logic* 28 (1987), 69–79.

Schweitzer, Hermann: 1935, Die Fregesche Einführung der Wertverlaufsnamen in seine Logik. In: H. Scholz / H. Schweitzer, *Die sogenannten Definitionen durch Abstraktion. Eine Theorie der Definitionen durch Bildung von Gleichheitsverwandtschaften.* Leipzig 1935 (*Forschungen zur Logistik und zur Grundlegung der exakten Wissenschaften*, Heft 3), Anhang I.

Thiel, Christian: 1965, *Sinn und Bedeutung in der Logik Gottlob Freges.* Anton Hain: Meisenheim am Glan 1965; English translation by Thomas J. Blakeley: *Sense and Reference in Frege's Logic* (D. Reidel: Dordrecht 1968).

Thiel, Christian: 1972, Gottlob Frege: Die Abstraktion. In: J. Speck (ed.), *Grundprobleme der großen Philosophen. Philosophie der Gegenwart I.* Vandenhoeck & Ruprecht: Göttingen 1972, [3]1985, 9–44 and 9–46.

Thiel, Christian: 1975, Zur Inkonsistenz der Fregeschen Mengenlehre. In: Ch. Thiel (ed.), *Frege und die moderne Grundlagenforschung. Symposium, gehalten in Bad Homburg im Dezember 1973* (Anton Hain: Meisenheim am Glan 1975), 134–159.
Thiel, Christian: 1978, Die Unvollständigkeit der Fregeschen "Grundgesetze der Arithmetik". In: J. Mittelstraß / M. Riedel (eds.), *Vernünftiges Denken. Studien zur praktischen Philosophie und Wissenschaftstheorie* (Walter de Gruyter: Berlin /New York 1978), 104–106.
Thiel, Christian: 1979, Bedeutungsvollständigkeit und verwandte Eigenschaften der logischen Systeme Freges. In: Franz Bolck (ed.), *"Begriffsschrift". Jenaer Frege-Konferenz 7.–11. Mai 1979* (Friedrich-Schiller-Universität: Jena 1979; *Wissenschaftliche Beiträge der Friedrich-Schiller-Universität Jena*), 483–494.
Thiel, Christian: 1983, Die Revisionsbedürftigkeit der logischen Semantik Freges. *Anuario Filosófico* 16 (1983), 293–301.
Toth, Imre: 1984, Three Errors in the Grundlagen of 1884: Frege and Non-Euclidian Geometry. In: G. Wechsung (ed.), *Frege Conference 1984. Proceedings of the International Conference held at Schwerin (GDR), September 10–14, 1984* (Akademie-Verlag: Berlin 1984; *Mathematical Research / Mathematische Forschung*, Band 20), 101–108.

The Horizontal

PETER SIMONS

1. Frege's Logics in their own Right

The inconsistency of Frege's logical system in *Die Grundgesetze der Arithmetik* has had more than just the obvious negative consequences for the study of his logic. The most natural consequence is that no one is inclined to develop the system further, but the effect has been more lasting on the reception of Frege's thought than just that. In the first instance, of course, the efforts of those sympathizing with something like Frege's views were naturally bent in the direction of developing a suitable alternative or replacement for the shattered system of Frege. With a little oversimplification, we can say that this task was well in hand by 1908, when Russell and Zermelo published their seminal papers offering different ways around the problem of the inconsistency.[1] This meant, again quite naturally, that Frege's logics were neglected for a long time in favour of those that replaced them. The rediscovery – or perhaps we should say, the first proper discovery – of Frege which has blossomed since the 1950s has concentrated largely on his philosophy of language and to some extent on his philosophy of mathematics, but largely neglected the particularities of the work which was most characteristic of him, namely his logic. But Frege's logic – despite its inconsistency – is more than a mere forerunner to that of Russell or Church or anyone else. It has its own oddities and beauties, and can be studied for its own sake both from an historical and from a logico-aesthetic point of view. There have been very few studies of this kind: perhaps the most significant are those on the insufficiency of "Frege's Way Out" of the inconsistency;[2] those concerning the source or sources of the inconsistency,[3] discussions of the permutation argument in *Grundgesetze* § 10,[4] Frege's attempt to show all the well-formed names of his

[1] Russell 1908, Zermelo 1908.
[2] Sobociński 1949, Quine 1955, Geach 1956, Resnik 1964.
[3] Thiel 1975, Aczel 1980, Dummett 1991, Ch. 17, Boolos 1993, Dummett 1994.
[4] Thiel 1976, Schroeder-Heister 1987, Moore & Rein 1986, 1987.

system are referential,[5] and the questions of the consistency and completeness of the first-order fragment of the *Grundgesetze* system.[6] These studies – to an extent which is determined mainly by their aims – examine the interstices of the system. This paper continues in that vein, and turns up what are I think suprising facts.[7]

A second strand of Frege research that has certainly been underemphasized is the study of his development. There has been a tendency – perhaps a little encouraged by Frege's own practice – to treat his views from back to front, reading later views back into earlier years, or taking earlier views as embryonic or inexact versions of later ones. This tendency has been dampened more recently by the admission of Frege's principal commentator, Michael Dummett, that he did not previously sufficiently appreciate the development in Frege's views.[8] There are now excellent studies of the development of Frege's views in the philosophy of mathematics by Gregory Currie[9] and on a larger scale embedded also in Dummett's *Frege: Philosophy of Mathematics*,[10] and the development of the crucial concept of *extension* has been charted by Tyler Burge.[11] This paper looks at the development surrounding the sign written '–' in Frege's logic: the horizontal.

In the foreword to *Grundgesetze* Frege lists the most important changes in his formal logic or concept-script which differentiate the *Grundgesetze* logic of 1893 from the *Begriffsschrift* logic of 1879. The changes, which span roughly the period 1885–1890, were sufficiently important to force Frege to abandon an almost completed manuscript and recast his theories in the form in which we know them.[12] They are so thoroughgoing that it is more correct to speak of 'Frege's logics' in the plural than 'Frege's logic' in the singular. The introduction of courses of values, a generalization of the extensions of concepts Frege had found necessary in the *Grundlagen*, is one major difference. But the differentiation of sense from signification (reference), the conception of the signification of sentential clauses as truth-values, and the

---

[5] Resnik 1986.
[6] Thiel 1978, Schroeder-Heister 1987, Parsons 1987.
[7] A paper coming close to the topics I consider here is Martin 1984, which discusses both the importance of the horizontal and the many-valued aspect of Frege's semantics.
[8] See Dummett 1993, p. 172.
[9] Currie 1986.
[10] Dummett 1991.
[11] Burge 1984.
[12] Frege 1893, p. ix, 1964, pp. 4–5.

resulting radical modification to the understanding of concepts and relations go hand in hand with this. The resulting changes to the graphical form of the concept-script and the derivations made with it are, as Frege notes, much less radical than the change in the way in which the signs are understood, and the chief device enabling this large change in meaning to accompany a minimal change in superficial form is the reinterpretation of the horizontal stroke sign.

The horizontal is introduced – or reintroduced – in *Grundgesetze* with such little pomp that it is easy not to see how elegantly it fits into the system of primitive concepts and facilitates the further definitions. It is indeed part of his genius that choices with complicated motivations should be made to seem so natural and straightforward by Frege. Only when one attempts to modify the system, to tinker with some of its parts, does this easy elegance emerge. Frege himself had neither the time nor, later, the inclination to investigate the by-ways of his own logical system, but from an historical point of view this is justified, since it reveals aspects which a simple retracing of his path do not: we come to appreciate better how Frege thought. For instance, he knew well that at least two of his primitive concepts, the horizontal and negation, were definable in terms of the others, but chose not to eliminate them. In fact the conditional is eliminable too, and one can go even a little further. To understand why Frege did not avail himself of these reductions when in other ways he was relentlessly sparing in his use of definitions we now take for granted (like the existential quantifier) is to see a little more clearly his attitude to economy and elegance within logic itself.

2. *The Content Stroke in Begriffsschrift*

In Frege's *Begriffsschrift (BS)* is a logical system which we can now find relatively familiar. There are what one may call propositional connectives, namely negation N and material implication C, the universal quantifier Π, and identity $\equiv$, given a metalinguistic interpretation, but otherwise obeying familiar laws.[13] We find recognizable axioms, rules and theorems of predicate calculus: the axioms are the following

[13] When talking about functions by using their special symbols, I shall for the most part not follow Frege's later practice of marking the argument places using Greek place-holder consonants. So I shall talk of C rather than $C\xi\zeta$ etc. While admirable in principle, Frege's insistence is annoying in practice, and I do not think any crass mistakes have arisen here by ignoring it, or by being cavalier about quotation marks.

formulas with their original numeration and notated here in Łukasiewicz's style:[14]

1. $\vdash CaCba$
2. $\vdash CCcCbaCCcbCca$
8. $\vdash CCdCbaCbCda$ (Łukasiewicz later showed this to be not independent)
28. $\vdash CCbaCNaNb$
31. $\vdash CNNaa$
41. $\vdash CaNNa$
52. $\vdash C(c \equiv d)Cf(c)f(d)$
54. $\vdash (c \equiv c)$
58. $\vdash C\Pi x f(x) f(c)$

The stated rule is modus ponens; Frege also implicitly uses substitution and a quantifier distribution rule. Notice the neat separation of axioms into groups:

the first three are purely implicational
the next three add negation to implication
the next two add principles of identity
the last one adds universal instantiation.

This is not quite a presentation of modern first-order predicate logic with identity for three reasons. Firstly, quantification is, as we might now say, typically ambiguous. In derivations in Part III of *BS* (formulas 77, 91, 93), Frege allows a function letter to stand over the concavity of his quantifier symbol, as he explicitly says in § 11 it can do, and as he needs in order to derive the theorems he needs about the ancestral. So the logic is in fact second-order. Secondly, the meaning of the Italic or free variables in Frege is officially that they replace quantifiers with maximum scope. So for deriving laws involving the Italic letters, strictly the quantifier law $\vdash \Pi F \Pi y C \Pi x F(x) F(y)$ (the "official" version of 58) is needed. Frege's practice is however more felicitous than his official explanation. He uses the Italics as schematic letters, and the derivations he gives at the beginning allow him to replace schematic letters by other, sometimes more complex schemata. Thirdly, unlike standard predicate

[14] Glorious though Frege's symbology is, and I do think it is both beautiful and perspicuous, it is still, in these days of computer graphics, costly in effort and paper, so I have linearized using an eclectic mixture of bracketless prefix and bracketed infix notation which stays reasonably close to Frege's own mixture of bracketless (stroke) and infix notation.

logic, Frege allows the identity sign to be flanked by expressions for any content whatsoever. Disregarding the metalinguistic interpretation of identity, this effectively distinguishes identity from the connectives *N* and *C* in that these may may only take judgeable contents as arguments. So there is an implicit restriction of some arguments, but not others, to judgeable contents.

Frege explains the difference between judgeable and unjudgeable contents early on, in very brief terms. Some contents, such as 'house', are not judgeable. I cannot judge that house, but I can judge that opposite magnetic poles attract one another, so that opposite magnetic poles attract one another is a judgeable content. To express the mere thought of something's being the case as distinct from the judgement that it is, Frege omits the small vertical judgement stroke from his assertion sign '⊢' leaving what he calls his content stroke, '–'. His explanation of its role as that of tying the symbols which follow it into a whole[15] and his requirement that only a sign for judgeable content stand to the right of it conspire to render it theoretically superfluous. For, as David Bell has pointed out,[16] if a symbol or symbol combination to the right of it is not already a whole or is not judgeable, then there is no magic the content stroke can work which will unify it or make it a sign for a judgeable content: '– help and' is just as nonsensical and non-unified as plain 'help and', and '– house' is just as unjudgeable as plain 'house'. For all that, the horizontal stroke notation is not completely without a use in Frege's notation. For where schematic letters occur to the right of it, its presence serves to *show* that the letters stand for judgeable contents. Frege is thus spared the need to somehow *say* a content is judgeable, which would require each time a jump to the metalevel, or else distinguish judgeable from unjudgeable contents by a syntactic device such as different types of variables, which would require the laws of identity to be stated for each type of variable separately. Further, the content stroke was obviously of heuristic value to Frege in giving him something on which literally to hang the negation strokes, condition strokes and quantifier concavities of his symbolism, thus providing a perfect and natural way to recognize and show scope. Thus, despite its obscure theoretical status, the heuristic and historical importance of the content stroke is considerable.

---

15 *BS*, § 2.
16 Bell 1979, p. 86.

The use of identity and of a single run of schematic variables flanking identity symbols show that Frege was prepared to some extent to allow judgeable and unjudgeable contents to consort together in one class. This suggests that the distinction was not as firm and absolute as, say, the later distinction between objects and functions. Further, surprising evidence comes from the Preface to *BS*, where Frege says the two axioms 31 and 41 of double negation can be replaced by the single axiom $\vdash(NNa \equiv a)$. But now, if a is *not* a judgeable content, NNa is not admissible, since for Frege, Na is the same as $N(-a)$: Frege explicitly says the short horizontals on either side of his little negation stroke are to be considered as content strokes.[17] So $NNa \equiv a$ can only be a judgeable content if a is. Here the fact that a must be judgeable does not show itself by a's being to the right of a horizontal (either a content stroke alone or one adjoining a connective or quantifier). The restriction must be discerned by more reasoning. A more perspicious axiom would have been $\vdash(NNa \equiv -a)$, which, by Frege's merger rule for horizontals, is the same as $\vdash(NN{-a} \equiv -a)$, and the content letter is shown to be judgeable by standing in each case to the right of a horizontal line.

3. *From the Content Stroke to the Horizontal*

Between *BS* and the *Grundgesetze (GG)*, Frege's notation, while remaining outwardly very similar, underwent a massive reinterpretation. The reasons for the changes are complex and I cannot discuss them here. Although it was certainly not the most important change, the one with the greatest interpretative repercussions is to the symbol '–', which Frege stopped calling 'the content stroke' and called instead simply 'the horizontal'. Every formula in Frege's logic begins with '⊢', which again consists of the assertion or judgement stroke '|' followed by the horizontal '–': '|–' is simply written without gaps. Since '–' has a new interpretation, it follows that every formula containing '–' has a new interpretation, and that is quite simply every formula. All symbols incorporating a horizontal, namely those for negation, implication, and quantification are also reinterpreted, as, for different reasons, is identity, now written '=' instead of '≡'. So the only symbol in the whole notation which can even plausibly claim to have the same significance

---

[17] *BS*, § 7.

as in *BS* is the judgement stroke '|'. Rarely can calligraphic constancy have disguised a more thorough change of sense.

The transitional stages from content stroke to horizontal are not clearly represented in the corpus, and I shall not speculate here what they might have been. That there was a transitional stage of some sort is clear from Frege's statement in the Foreword to *GG* where he says that an inner transformation of the conceptual notation forced him to abandon a draft almost complete in manuscript. It is very unlikely that this was written as early as *BS*, since it would need to incorporate the *Grundlagen* view of numbers as concept extensions. Frege says that while the old signs have hardly changed their appearance and their inferential roles, or "algorithm", as he calls it, they have been given, as he puts it, different explanations.

The clear distinctions between sense and reference, between function and object, between functions of different types and orders, the acceptance of identity as an object language predicate, and the view of truth-values (hereafter, for short: TVs) as objects, all contribute to this transformation.

4. The Horizontal

The upshot of the transformations is that the horizontal is now given substantial work to do. This is first presented in the essay "Function and Concept", where Frege, in addition to presenting his new firm concept of function, shows the world for the first time the basic functions of his new logic, with the exception of the description function. The identity function is assumed understood, then Frege introduces in turn the course of values (hereafter: CV) operator, then the horizontal, negation, the universal quantifier, and implication. The order of introduction is slightly different from that in *GG*, but otherwise there are no significant differences.

In *GG*, as in "Function and Concept", Frege takes pains to explain as far as possible what his basic signs mean. They are *all* names of object-valued functions, that is, they always refer to objects when saturated by referential expressions of the right type. With one notable exception, namely the CV operator, Frege elucidates his basic functions by specifying exhaustively which values they take for which arguments. In outlining this in the order in which he does, I shall use a number of symbols he does not.

Firstly, I shall use '⊤' and '⊥' as proper names for the True and the False respectively. Although Frege introduces names for these later on in *GG*, they are complex course of value names, whereas Frege's elucidation of the horizontal and other logical functions presupposes in his German metalanguage the terms 'das Wahre' and 'das Falsche'.

Secondly, when Frege writes an equation like '$2^2 = 2 + 2$', he does not understand it as we do. This point is very easy to overlook in interpreting Frege because he runs together the functions of being a sentence and being an expression of an assertion. For us, an equation is a sentence, an expression of a proposition with a truth-value, whether it is asserted or not. For Frege on the other hand (in *GG*) it is not a sentence but a name of a truth-value, and he takes care to express such things using a nominal expression like, "2^2's being equal to 2 + 2" or "the truth-value thereof, that 2^2 is equal to 2 + 2", (where in *BS* he would have said "the circumstance or proposition that 2^2 is equal to 2 + 2", which is at least implicitly "sentential" or judgeable).

Because I want to continue in my language to use '=' as a sentence-forming functor, I shall appropriate this sign and replace this sign in Frege's formulas by the name-forming functor '≈'. Roughly speaking, when I assert '$a = b$' Frege would assert '$\vdash(a \approx b)$'. But I shall be speaking on the metalevel.

Otherwise I shall use my replacements for Frege's symbols autonymously, and shall, like Frege, give the intended reference of the signs case by case.

5. *The Basic Functions in Grundgesetze: Semantic Determination*

(1) The Horizontal

$$-a = \top \text{ if } a = \top$$
$$-a = \bot \text{ if } a \neq \top$$

(2) Negation

$$Na = \top \text{ if } a \neq \top$$
$$Na = \bot \text{ if } a = \top$$

(3) Identity

$$a \approx b = \top \text{ if } a = b$$
$$a \approx b = \bot \text{ if } a \neq b$$

(4) Universal Quantifier (first level only given, other cases analogous)

$$\Pi x\Phi(x) = \top \text{ if for all } a\text{: } \Phi(a) = \top$$
$$\Pi x\Phi(x) = \bot \text{ if for some } a\text{: } \Phi(a) \neq \top$$

(5) Course of Values Operator.

I shall write $\wedge x\Phi(x)$ instead of Frege's $\acute{\varepsilon}\Phi(\varepsilon)$. Frege's use of Greek and Gothic vowels for variables is not systematic, as I have shown elsewhere,[18] and will not be followed. Here, as the sole exception to his rule, Frege gives no elucidation except to point to Basic Law V, so that

$$\wedge x\Phi(x) = \wedge z\Psi(z) \text{ if for all } a\text{: } \Phi(a) = \Psi(a)$$
$$\wedge x\Phi(x) \neq \wedge z\Psi(z) \text{ otherwise.}$$

But this is, notoriously, insufficient to fix the referents of CV names. Frege's likewise well-known but inadequate attempt to rectify this problem is to stipulate what the values of functions are for CVs as arguments, as the cases arise. In the process, he stipulates as a convention that

$$\top = \wedge x(-x)$$
$$\bot = \wedge x(x \approx (N\Pi z(z \approx z)))$$

so that, recalling the elucidations of the functions $-$, $\approx$, Π and N, we get that

$$\top = \wedge x(x \approx \top)$$
$$\bot = \wedge x(x \approx \bot),$$

that is, the TVs are their own "singleton" CVs.

Obviously, if these were meant to be *definitions* of the TVs, they would be circular. But Frege thought that no definition of truth can be given.

(6) Description

Frege now introduces his "substitute for the definite article" \. This is determined as follows:

$$\backslash a = b \text{ if } a = \wedge x(x \approx b)$$
$$\backslash a = a \text{ if for all } b\text{, } a \neq \wedge x(x \approx b).$$

[18] Cf. Simons 1988, p. 37.

Writing '$\{b\}$' for '$\wedge x(x \approx b)$', this becomes

$$\backslash a = b \text{ if } a = \{b\}$$
$$\backslash a = a \text{ if for all } b, a \neq \{b\}.$$

We can consider $\backslash a$ in computing terms as a "get member of singleton" function. If the argument a is a singleton, \ "gets" its member and "returns" it as value. If a is not a singleton, \ "returns" a itself as default value. (Of course, this is procedural talk, which is not appropriate to Frege's platonistic view.) Such things are familiar nowadays: in Frege's day it was an innovation.

Description is, apart from the CV operator, the only basic function in *GG* that is not a *concept*, i.e., whose values are not always TVs. It is interesting to consider when its values *are* TVs. Obviously $\backslash a$ is a TV if a is, but what if a is a singleton CV whose member is a TV, i.e. $\{\top\}$ or $\{\bot\}$? By Frege's stipulation of the TVs, we have that

$$\{\top\} = \top$$
$$\{\bot\} = \bot.$$

So again, by the definition of \, $\backslash\{\top\} = \top$ and $\backslash\{\bot\} = \bot$, so $\top = \backslash\top = \backslash\backslash\top = \dots$ and $\bot = \backslash\bot = \backslash\backslash\bot = \dots$

Frege uses the description function only once, to define the function $\frown$, and thereafter he works with this function rather than using definite descriptions again. Defining a null CV as follows:

$$\wedge := \wedge x(N(x \approx x))$$

we can give the determination of $\frown$ as

$$\text{if for some function } \phi, b = \wedge x\phi(x), \text{ then } a \frown b = \phi(a)$$
$$\text{otherwise } a \frown b = \wedge.$$

(7) The Conditional

Finally Frege introduces the conditional or implication function C:

$$Cab = \bot \text{ if } a = \top \text{ and } b \neq \top$$
$$Cab = \top \text{ otherwise}$$

and then he goes on to mention higher-order functions and their associated quantifiers.

6. *Primitives*

On the way, Frege mentions almost in passing, a fact expressible as an equation which could be taken as a definition of the horizontal, namely[19]

$$\text{for all } a\text{: } -a = (a \approx (a \approx a))$$

so – is in fact genuinely, logically redundant in the system!

Frege did of course use definitions, quite a few in fact, in his system, but he did not bother with definitions for standard logical constants like conjunction or disjunction or existential quantification, though of course he knew that such definitions could be given. So it looks as though he was keen to keep the number of defined logical constants down to a minimum. Following this, had Frege upheld the same methodological principles in logic as did the Poles, especially Leśniewski, after him, he would have been obliged to actually remove – from his set of primitives, defining it as above in terms of ≈. The reason why he did not do so is perfectly obvious: to do so would have destroyed the graphical form of his notation for logic.

Frege could have introduced further economies. An obvious one would have been

$$Na := a \approx \bot$$

In correspondence, Russell mentioned to Frege the possibility of defining negation as

$$Np := Cp\Pi r.r$$

This seems to have genuinely surprised Frege, who wrote back on 13 November 1904: "this would save a primitive sign; but would probably need some new primitive laws, e. g.: ⊢ $CCCp\Pi a.a\ \Pi a.a\ p$" [I have changed the notation: Frege gives both his and Russell's]. To this, Frege adds a remark which sheds interesting light on his views as to what may count as a basic law: "this is really too complicated for a primitive law; and I do not know whether it can be reduced to something simpler."[20]

So Frege knew he could reduce the number of his primitives from seven to five (leaving out higher-order quantifiers). A somewhat more tricky task, which Frege probably never considered, is eliminating *C*. It

---

[19] *GG*, § 10.

[20] Frege 1976, p. 248, Frege 1980, p. 166.

can be done. In 1923, Tarski showed in his doctoral dissertation how to define conjunction in Leśniewski's protothetic in terms of material equivalence and universal quantification.[21] The analogue of Tarski's simplest definition will work here:

$$Kab := \Pi f(a \approx (f(a) \approx f(b))).$$

For clearly, if $f(a) \approx f(b)$ is $\top$ or $\bot$, Kab cannot be $\top$ unless a is a TV. But if a is $\bot$ then $a \neq b$. So if $Kab = \top$ and $a = \bot$ then $\top = \Pi f(\bot \approx (f(a) \approx f(b))$, i. e. $\top = \Pi f(N(f(a) \approx f(b))$. But by the existence of constant functions, $f(a) = f(b)$ for some f. So $a \neq \bot$, so since a must be a TV, $a = \top$, so $\top = \Pi f(f(a) \approx f(b))$, whence, substituting $a \approx \xi$ for $f(\xi)$ we have $a = b$, i. e. $Kab = \top$ iff $a = b = \top$, and since the right-hand side of the definition is a concept, $Kab = \bot$ otherwise. But now defining $Cab := NKaNb$ this gives us the required values for the required arguments. So Frege could have pared his primitives down to four: $\approx$, Π, \ and ^ (and higher-order quantifiers),[22] eliminating completely the "propositional" functions –, N and C. Nothing illustrates more clearly how much Frege's logic had lost the hierarchical structure of *BS* and had become "all of a piece" in *GG*.

7. *Frege's Many-Valued Logic*

In explaining the meanings of his constants case by case, Frege most often considers the TVs as one kind of object, and all others, what I shall call *third objects*, as a different kind of case. The logical constants are in fact given in large part by what their values are for $\top$ and for other objects than $\top$. Consider then the symbol 'O' to stand for anything which is a third object, i. e. not a TV, and look at the following semantic tables for –, N and C:

a	$-a$
$\top$	$\top$
O	$\bot$
$\bot$	$\bot$

a	Na
$\top$	$\bot$
O	$\top$
$\bot$	$\top$

		b		
	Cab	$\top$	O	$\bot$
	$\top$	$\top$	$\bot$	$\bot$
a	O	$\top$	$\top$	$\bot$
	$\bot$	$\top$	$\top$	$\top$

Indeed, subsuming O and $\bot$ together as U we have just

[21] Cf. Tarski 1983, p. 6.

[22] I imagined that this is as far as the economies could go, but Terence Parsons has shown that by reidentifying the TVs with two other CVs, first-order universal quantification can be eliminated also.

a	$-a$
T	T
U	U

a	Na
T	U
N	T

Cab	T	U
T	T	U
U	T	T

which is of course much more "classical" and shows how for many purposes ⊥ and third objects are treated alike.

Throwing all third objects into O makes Frege's logic look quasi-three-valued. For many semantic purposes, we can specify what the referents of expressions are using just the three "values" ⊤, O and ⊥. Being of kind O can of course be expressed in Frege's system: we define what it is to be a truth-value, as he does,[23] by

$$\mathrm{TV}(a) := a \approx -a$$

hence

$$\mathrm{O}(a) := N(\mathrm{TV}(a)).$$

The horizontal thus has the role of eliminating third objects from consideration as values by assimilating them to the False, reducing considerations to the two TVs, and though it shares this feature with other primitives except \ and ^, it exhibits this feature most prominently. From a formal point of view, if we consider being a third object as a third value, then the horizontal is like the three-valued functor of *strong assertion*[24]

p	Sp
*1	1
2	3
3	3

It is this choice of functor which allows Frege's *Verschmelzungsgesetze*, the laws of merging horizontals with the horizontal parts of other connectives,[25] to work so well. Any sign having a horizontal at right or left can be merged with a horizontal considered as function, *salva veritate*. Other neutralizers than – do not work nearly as well.

Comparison with the corresponding 3-valued tables for the complex functions Frege employs as conjunction and disjunction

---

[23] *GG*, § 5.
[24] Cf. Rescher 1969, p. 30, where it is called "external assertion".
[25] Cf. *GG* § 48.1.

NCaNb	T	O	⊥
T	T	⊥	⊥
O	⊥	⊥	⊥
⊥	⊥	⊥	⊥

*CN*ab	T	O	⊥
T	T	T	T
O	T	⊥	⊥
⊥	T	⊥	⊥

shows a certain asymmetry owing to the fact that – and *N* both neutralize the difference between falsity and thirdness, but for this reason the de Morgan laws are fulfilled. The neutralization is however more problematic for the complex connective *NCCabNCba*, corresponding to material equivalence, which has the sematic table

Eab	T	O	⊥
T	T	⊥	⊥
O	⊥	T	T
⊥	⊥	T	T

whereas identity by comparison has the partial table

$a \approx b$	T	O	⊥
T	T	⊥	⊥
O	⊥	?	⊥
⊥	⊥	⊥	T

where the reason for the question mark is obvious.

8. Some New Definitions

Frege shows, at least in the published works, what from our modern, more experimental point of view is a surprising incuriosity as to alternatives and the odd logical corners of his system (had he done so, he might have come upon the contradiction himself). Frege's determination to use the system to demonstrate logicism led him to inventions and innovations galore, but this determination and his conscientiousness as a teacher left him little time for exploring these byways in the way later logicians, especially in Poland, were wont to do.

So how, for instance, within Frege's system, can we have standard 3-valued material equivalence given by

$a \equiv b$	T	O	⊥
T	T	⊥	⊥
O	⊥	T	⊥
⊥	⊥	⊥	T

and other suitable or interesting connectives? There are a number of possibilities for defining 3-valued connectives, but since 'O' is not strictly speaking a third value we cannot count these in the same way for a standard 3-valued logic. Here are some connectives with their semantic tables.

(1) Strong Negation

Consider a negation-type functor with the following table

a	Ma
$\top$	$\bot$
O	$\bot$
$\bot$	$\top$

This can be defined as

$$Ma := a \approx \bot$$

and it has the interesting property that only left-merging works: $-Ma = Ma$ but $M{-a} = Na$. We could define being a truth-value in terms of the two negations as

$$\mathrm{TV}(a) := CNaMa.$$

(2) Neutral Negation

The most interesting definition is one of a function which switches the truth-values but leaves third objects alone. Its schematic table for all third objects b is then

a	Ua
$\top$	$\bot$
b	b
$\bot$	$\top$

To define this it seems we need the description function:

$$Ua := \backslash\, \hat{}x[NCC(a \approx x)(x \approx Na)NCN(x \approx -a)(x \approx a)]$$

There are three cases: if $a = \top$ then only $\bot$ falls under the concept in square brackets, if $a = \bot$ then only $\top$ does, while if a is a non-TV, only a itself falls under the concept. Note that $Ua = Na = Ma$ if a is a TV, $Ua = a$ if a is not a TV. There are no merging laws for U: $-Ua = U{-a} = Na$.

This is an instance of definition by cases, which is actually quite easy in Frege's system because of the function \. For instance, suppose we want to define a two-place function $k(a,b)$ as taking the value of $g(a,b)$ if another function f is such that $f(a,b) = c$, and taking the value of $h(a,b)$ if $f(a,b) = d$. Then a suitable definition is

$$k(a,b) := \backslash\, \hat{}x[NCC(f(a,b) \approx c)(x \approx g(a,b))NC(f(a,b) \approx d)\\ (x \approx h(a,b))]$$

(Such definitions are a lot more perspicuous in Frege's two-dimensional graphical notation.) Consider the cases: if $f(a,b) = c$ and $f(a,b) \neq d$ then

$$\begin{aligned} k(a,b) &= \backslash\, \hat{}x[NCC\top(x \approx g(a,b))NC\bot(x \approx h(a,b))] \\ &= \backslash\, \hat{}x[NC(x \approx g(a,b))N\top], \\ &= \backslash\, \hat{}x[NC(x \approx g(a,b))\bot], \end{aligned}$$

and this is $\top$ iff $x = g(a,b)$.
Case 2, where $f(a,b) = d$ and $f(a,b) \neq c$, is similar. Frege's definition of \ already automatically caters for
Case 3: $f(a,b) \neq c$ and $f(a,b) \neq d$. Then $k(a,b) = \backslash\, \hat{}x[NC\top\top] = \backslash\, \hat{}x[\bot] = \bot$.
Case 4: $f(a,b) = c$ and $f(a,b) = d$. Then $c = d$. If $g(a,b) = h(a,b)$ then $k(a,b) = g(a,b)$. If $g(a,b) \neq h(a,b)$ then nothing falls under the bracketed concept, so $k(a,b)$ is the CV $\hat{}x[NCC\top(x \approx g(a,b))NC\top(x \approx h(a,b))]$. At any event, the function $k(a,b)$ is always defined.

A simpler kind of case is given by a 1-place function X which is to have the value of f for arguments fulfilling a certain condition D and any old value otherwise. We can define

$$X(a) := \backslash\, \hat{}x[CD(a)(x \approx f(a))].$$

Case 1: $D(a) = \top$. Then the bracketed expression denotes $\top$ only for the value $f(a)$ of x, so $X(a) = f(a)$.
Case 2: $D(a) \neq \top$. Then the bracketed expression denotes $\top$ for all values of x, so $X(a) = \mathrm{V}$, the universal CV definable as

$$\mathrm{V} := \hat{}x(x \approx x).$$

We see that Frege's obsession with defining functions for all arguments pays off by yielding default values for unintended cases.

To return to the logical connectives Frege could have defined: in terms of neutral negation U we can define

(3) Weak Assertion

$$Wa := NUa$$

with table

a	Wa
⊤	⊤
O	⊤
⊥	⊥

and in terms of this

(4) Strong Negation

$$La := NWa.$$

These functions do not admit of unrestricted merging with the horizontal.

To get a

(5) Weak Disjunction

$a \vee b$	⊤	O	⊥
⊤	⊤	⊤	⊤
O	⊤	⊤	⊤
⊥	⊤	⊤	⊥

dual to Frege's conjunction we switch a and b to Ua and Ub in conjunction then negate, i. e.

$$a \vee b := NCWaUb$$

Similarly

(6) Weak Conjunction

$a \,\&\, b$	⊤	O	⊥
⊤	⊤	⊤	⊥
O	⊤	⊤	⊥
⊥	⊥	⊥	⊥

dual to Frege's disjunction can be obtained from Frege's disjunction as $NCWaNb$. For the

(7) Strong Material Equivalence

whose table was given above we use the definition of TV and define by cases:

$$a \equiv b := \ \backslash\, {}^\wedge x[NCCNCNTV(a)TV(b)(x \approx \top)NCCNTV(a)TV(b)(x \approx (a \approx b))].$$

All the connectives defined so far are concepts (TV-valued). There are however connective-like functions which we might want to resemble Łukasiewicz's connectives in 3-valued logic, but which are not concepts. For example,

(8) Łukasiewicz Material Implication

should be defined to satisfy

$a \supset b$	$\top$	O	$\bot$
$\top$	$\top$	O	$\bot$
O	$\top$	$\top$	O
$\bot$	$\top$	$\top$	$\top$

One way is:

if $a = \bot$ or $b = \top$ then $a \supset b = \top$
if $a = \top$ then $a \supset b = b$
if $b = \bot$ then $a \supset b = a$

if neither a nor b is a TV then $a \supset b = (a \equiv b) = \top$, which gives

$a \supset b$	$\top$	O	$\bot$
$\top$	$\top$	b	$\bot$
O	$\top$	$\top$	a
$\bot$	$\top$	$\top$	$\top$

The following definition by cases will work:

$$a \supset b := \backslash{\wedge}x[NCNCCCNCCN(\mathrm{TV}(a))(\mathrm{TV}(b))(a \approx \bot)(b \approx \top)$$
$$(x \approx \top)NC(a \approx \top)NC(a \approx \top)(x \approx b)NC(b \approx \bot)(x \approx a)]$$

Likewise conjunctions and disjunctions resembling those of Łukasiewicz can be given using the semantic tables required and obtaining the definitions using similar case by case specifications.

9. *Conclusion and Summary*

I have not been concerned to criticise Frege's view of TVs or CVs as objects – there are plenty of works where this kind of criticism can be found – but to take these for granted, and work within the system in a way which Frege did not. Since I have been concerned with definitions rather than with theorems or truth, the inconsistency of the system has not impinged on my considerations. Terence Parsons has shown that the first-order fragment of the *GG* system, with CVs but without higher-order quantification, is consistent, and apart from the Tarski-style definition of implication, none of the definitions I employ use higher-order quantification, so they go through in the consistent but still interesting fragment.

My main conclusions (not all of them new) are as follows:

(1) The content stroke in *BS*, although it does not receive a coherent explanation, plays a background semantic role in allowing only judgeable contents to stand as its arguments
(2) The *GG* system is radically different, except for its calligraphic form, from the *BS* system.
(3) The horizontal function $-\xi$, though it received an adequate explanation only in *GG*, was known by Frege to be redundant, as (later) was negation.
(4) The conditional is also definable using Frege's primitives.
(5) Frege had understandable reasons for not making the economies such definitions allowed.
(6) Whereas *BS* is constructed hierarchically, introducing new primitives in stages, *GG* is much more "of a piece".
(7) Among Frege's primitives, only the description function and the CV operator are not concepts.
(8) Except for the case of the CV operator, all of Frege's primitives are given exhaustive semantic elucidations.
(9) Frege's treatment of TVs as objects lends his system features which bring it into the vicinity of known many-valued, especially 3-valued, propositional logics.
(10) Frege had the resources to formulate definitions by cases.
(11) Provided syntactic restrictions on definition are fulfilled, definitions of objects and first-order functions are so designed that they should always provide default values (though because of the inconsistency this idea is not in fact realized).
(12) One can give numerous definitions of connectives exploring paths Frege did not bother to take.
(13) The overwhelming impression one retains, despite the inconsistency of *GG*, is one of Frege's superb logical craftsmanship, especially in the choice of primitives to facilitate smoothly functioning rules of inference.

References

Aczel, P. 1980: "Frege Structures and the Notions of Proposition, Truth and Set." In: J. Barwise, H. J. Keisler & K. Kunen, eds., *The Kleene Symposium*. Amsterdam: North-Holland, pp. 31–59.
Bell, D. A. 1979: *Frege's Theory of Judgement*. Oxford: Clarendon Press.
Boolos, G. 1993: "Basic Law (V)". *Aristotelian Society Supplementary Volume* **67**, 213–233.
Burge, T. 1984: "Frege on Extensions of Concepts, from 1884 to 1903." *The Philosophical Review* **93**, 3–34.
Currie, G. 1986: "Continuity and Change in Frege's Philosophy of Mathematics". In: Haaparanta & Hintikka, eds. 1986, pp. 345–373.
Dummett, M. A. E. 1991: *Frege: Philosophy of Mathematics*. London: Duckworth.
Dummett, M. A. E. 1993: *Origins of Analytical Philosophy*. London: Duckworth.
Dummett, M. A. E. 1994: "Chairman's Address: Basic Law V". *Proceedings of the Aristotelian Society* **94**, 243–251.
Frege, G. 1879: *Begriffsschrift, eine der arithmetischen nachgebildete Formelsprache des reinen Denkens*. Halle/S: Nebert. Reprinted in *Begriffschrift und andere Aufsätze*, Hildesheim: Olms, 1964, pp. 1–88. English translations: I (by T. W. Bynum) "Conceptual Notation, a formula language of pure thought modelled upon the formula language of arithmetic" in Frege 1972, pp. 101–203, and II (by S. Bauer-Mengelberg "*Begriffsschrift*, a formula language modeled upon that of arithmetic, for pure thought", in van Heijenoort, ed., 1967, pp. 1–82. Abbreviated in the text as *BS*.
Frege, G. 1893: *Grundgesetze der Arithmetik, begriffsschriftlich abgeleitet*. Vol. I. Jena: Pohle. Reprinted Darmstadt: Wissenschaftliche Buchgesellschaft, 1962. Partial English translation: Frege (1964). Abbreviated in the text as *GG*.
Frege, G. 1964: *Basic Laws of Arithmetic. Exposition of the System*, ed. & tr. M. Furth. Berkeley: University of California Press.
Frege, G. 1972: *Conceptual Notation and Related Articles*. Ed. & tr. T. W. Bynum. Oxford: Clarendon Press.
Frege, G. 1976: *Wissenschaftlicher Briefwechsel*. Hamburg: Meiner. English translation: Frege 1980.
Frege, G. 1980: *Philosophical and Mathematical Correspondence*. Oxford: Blackwell.
Geach, P. T. 1956: "On Frege's Way Out". *Mind* **65**, 408–409.
Haaparanta, L. & Hintikka, J. eds., 1986: *Frege Synthesized*. Dordrecht: Reidel.
Martin, J. N. 1984: "The Semantics of Frege's *Grundgesetze*". *History and Philosophy of Logic* **5**, 143–176.
Moore, A. W. & Rein, A. 1986: "*Grundgesetze*, Section 10". In: Haaparanta & Hintikka (eds.) 1986, pp. 375–384.
Moore, A. W. & Rein, A. 1987: "Frege's Permutation Argument". *Notre Dame Journal of Formal Logic* **28**, 51–54.
Parsons, T. 1987: "On the Consistency of the First-Order Portion of Frege's Logical System". *Notre Dame Journal of Formal Logic*. **28**, 161–168.

Quine, W. V. 1955: "On Frege's Way Out". *Mind* **64**, 145–159.
Rescher, N. 1969: *Many-valued Logic*. New York: McGraw-Hill.
Resnik, M. 1964: "Some Observations Related to Frege's Way Out". *Logique et analyse* N.S. **7**, 138–144.
Resnik, M. 1986: "Frege's Proof of Referentiality". In: Haaparanta & Hintikka, eds. 1986, pp. 177–195.
Russell, B. A. W. 1908: "Mathematical Logic as Based on the Theory of Types". *American Journal of Mathematics* **30**, 222–262.
Schroeder-Heister, P. 1987: "A Model-Theoretic Reconstruction of Frege's Permutation Argument". *Notre Dame Journal of Formal Logic* **28**, 69–79.
Simons, P. M. 1988: "Functional Operations in Frege's *Begriffsschrift*". *History and Philosophy of Logic* **9**, 35–42.
Sobociński, B. 1949: "L'analyse de l'antinomie russellienne par Leśniewski. II: La correction de Frege". *Methodos* **1**, 220–228.
Tarski, A. 1983: *Logic, Semantics, Metamathematics*. 2nd ed. Philadelphia: Hackett.
Thiel, C. 1975: "Zur Inkonsistenz der Fregeschen Mengenlehre". In: C. Thiel, ed., *Frege und die moderne Grundlagenforschung*. Meisenheim am Glan: Hain, pp. 134–159.
Thiel, C. 1976: "Wahrheitswert und Wertverlauf: Zu Freges Argumentation im § 10 der 'Grundgesetze der Arithmetik'". In: M. Schirn, ed., *Studien zu Frege I: Logik und Philosophie der Mathematik*. Stuttgart-Bad Cannstatt: Frommann-Holzboog, pp. 287–299.
Thiel, C. 1978: "Die Unvollständigkeit der Fregeschen 'Grundgesetze der Arithmetik' ". In: J. Mittelstraß & M. Riedel, eds., *Vernünftiges Denken: Studien zur praktischen Philosophie und Wissenschaftstheorie*. Berlin: de Gruyter, pp. 104–106.
van Heijenoort, J., ed, 1967: *From Frege to Gödel. A Source Book in Mathematical Logic, 1879–1931*. Cambridge, Mass.: Harvard University Press.
Zermelo, E. 1908: "Untersuchungen über die Grundlagen der Mengenlehre I." *Mathematische Annalen* **65**, 261–281. Translation in van Heijenoort, ed., 1967, pp. 199–215.

Frege and Natural Deduction

FRANZ VON KUTSCHERA

It is generally agreed that systems of natural deduction originate with Gerhard Gentzen, who first formulated them in his paper *Untersuchungen über das logische Schließen* (1934). My aim is not anything like a revision of this opinion, I only want to point out that Frege, in the first volume of his *Grundgesetze der Arithmetik* (1893, abbreviated here by GGA), has used a system closely related to the calculus of sequents that Gentzen developed for formalizing natural deduction – Gentzen calls it *Schlußweisenkalkül.* Gentzen's procedure might, therefore, have been suggested by Frege's system, but apparently it was not, since he only mentions Frege in his introductory remarks as one of those whose formalization of logic is very far from the kind of deduction actually employed in mathematical proofs.

Gentzen's aim was to stay as close as possible to this "real" deductive praxis, and although Frege would not have attributed any normative value to that, in GGA he designed his logic system to facilitate proofs. He used only 6 axioms (8, counting subcases), but 12 rules.[1] I shall restrict myself here to that part which contains elementary predicate logic, use modern symbols, distinguish terms and sentences and employ sentence schemes, so that substitution rules can be dispensed with. Then Frege's system, which I shall call F, looks like this:

Axiom schemes

A1a)	$A \supset A$	A2) $\bigwedge x A[x] \supset A[a]$
b)	$A \supset (B \supset A)$	A3) $\neg(A \equiv \neg B) \supset (A \equiv B)$.

Frege states his theorems in the form of implicational formulae $A_1 \supset (A_2 \supset \ldots (A_n \supset B) \ldots)$. $A_1, \ldots, A_n$ he calls *Unterglieder* (lower members), since in his symbolism they stand beneath the conclusion B, while B is the *Oberglied* (upper member). In view of our modern representation I

[1] The system is stated in GGA, pp. 61–64.

shall speak of antecedents and the succedent. Now we can also take $A_1,\ldots,A_i$ as the antecedents and $A_{i+1} \supset \ldots (A_{i+1} \supset \ldots (A_n \supset B) \ldots)$ as the succedent, for any i such that $1 \leq i < n$.[2] Therefore we have to distinguish between the implicational formula itself and its different analyses into antecedents and succedents. We can write such analyses in the sequential form $A_1, \ldots, A_i \rightarrow A_{i+1} \supset (A_{i+2} \supset \ldots (A_n \supset B) \ldots)$. Frege's inference rules – as Gentzen he calls them *Schlußweisen* – are formulated as rules for the manipulation of antecedents and the succedent in implicational formulae. They apply then to such formulae in an analysis. Therefore we can also conceive of them as rules for the derivation of sequents from other sequents. Then, however, we have to add the deduction rule:

R1* a) $\Delta, A \rightarrow B \vdash \Delta \rightarrow A \supset B$ (Importation)
b) $\Delta \rightarrow A \supset B \vdash \Delta, A \rightarrow B$ (Exportation).

Δ (and later on Γ) is to be a sequence of sentences. In this reconstruction applications of R1* may have to be inserted before and after those of the other rules in Frege's proofs. A sequent $\Delta \rightarrow A$ represents an implicational formula only if Δ is not empty, but we way understand $\rightarrow A$ as representing A.

Rules

Frege's first rule, the *Verschmelzungsregel*, can be omitted here since we do not use the horizontal, which transforms terms into sentences. Instead we have our R1*.

R2: $A_1, \ldots, A_n \rightarrow B \vdash A_{\pi(1)}, \ldots, A_{\pi(n)} \rightarrow B$ for all permutations π of $1,\ldots,n$. (Interchange of antecedents)

R3: $\Delta, A, \Delta' \rightarrow B \vdash \Delta, nB, \Delta' \rightarrow nA$ (*Wendung*, contraposition), where $nA = \neg A$, if A does not have the form $\neg B$, and $n\neg B = B$.

Frege says: "In a sentence an antecedent may be exchanged with the succedent, if the truth values of both are inverted simultaneously."[3]

We can resolve this rule, as Frege does, into

R3a) $\Delta, A, \Delta' \rightarrow B \vdash \Delta, \neg B, \Delta' \rightarrow \neg A$,
b) $\Delta, \neg A, \Delta' \rightarrow \neg B \vdash \Delta, B, \Delta' \rightarrow A$
c) $\Delta, \neg A, \Delta' \rightarrow B \vdash \Delta, \neg B, \Delta' \rightarrow A$
d) $\Delta, A, \Delta' \rightarrow \neg B \vdash \Delta, B, \Delta' \rightarrow \neg A$

---

[2] Cf. GGA, p. 22.
[3] GGA, pp. 27, 61.

R4: $\Delta,A,\Delta',A,\Gamma \rightarrow B \vdash \Delta,A,\Delta',\Gamma \rightarrow B$
(Contraction of antecedents)

R5: $\Delta \rightarrow A[a] \vdash \Delta \rightarrow \bigwedge xA[x]$,
where a does not occur in the formulae of the conclusion. (Generalization of succedent)

R6: $\rightarrow A; \Delta,A,\Delta' \rightarrow B \vdash \Delta,\Delta' \rightarrow B$
(*Schließen (a)*, Cut.)
This is a special case of

R7: $\Gamma \rightarrow A; \Delta,A,\Delta_1 \rightarrow B \vdash \Gamma,\Delta,\Delta' \rightarrow B$
(*Schließen (b)*, Cut.)

R8: $\Delta,A,\Delta' \rightarrow B; \Delta,\neg A,\Delta' \rightarrow B \vdash \Delta,\Delta' \rightarrow B$
(*Schließen (c)*, Excluded middle formulated as a second kind of cut rule).

The rest of Frege's rules concern substitution, change of bound variables and proceeding from a definition to an equivalence.

Let us now compare this Fregean calculus **F** with a modern formulation of a calculus of sequents, which we shall call **G**. Here I also use sequents with an empty succedent and define $\Delta\rightarrow$ by $\Delta \rightarrow \neg(A\supset A)$. Ω is to be a sequence of formulae that contains at most one member.

Axiom schema

RF: $A \rightarrow A$

Structural rules of inference

I:	$\Delta,A,B,\Delta' \rightarrow \Omega \vdash \Delta,B,A,\Delta' \rightarrow \Omega$	(Interchange)
AC:	$\Delta,A,A \rightarrow \Omega \vdash \Delta,A \rightarrow \Omega$	(Contraction)
AT:	$\Delta \rightarrow \Omega \vdash \Delta,A \rightarrow \Omega$	(Thinning in antecedent)
ST:	$\Delta \rightarrow \vdash \Delta \rightarrow A$	(Thinning in succedent)
C:	$\Delta \rightarrow A; \Delta,A \rightarrow \Omega \vdash \Delta \rightarrow \Omega$	(Cut)
TND:	$\Delta,A \rightarrow \Omega; \Delta,\neg A \rightarrow \Omega \vdash \Delta \rightarrow \Omega$	(Excluded middle)

Logical rules of inference

SI: $\Delta,A \rightarrow B \vdash \Delta \rightarrow A\supset B$ **AI:** $\Delta \rightarrow A; \Delta,B \rightarrow \Omega \vdash \Delta,A\supset B \rightarrow \Omega$

SN: $\Delta,A \rightarrow \vdash \Delta \rightarrow \neg A$ **AN:** $\Delta \rightarrow A \vdash \Delta,\neg A\rightarrow$

SG: $\Delta \rightarrow A[a] \vdash \Delta \rightarrow \bigwedge xA[x]$, where a does not occur in the conclusion.

AG: $\Delta,A[a] \rightarrow \Omega \vdash \Delta, \bigwedge xA[x] \rightarrow \Omega$.

The logical A-rules for the introduction of complex formulae into the antecedent are equivalent with the inversions of the corresponding S-rules, i. e. with

SI⁻: $\Delta \rightarrow A \supset B \vdash \Delta, A \rightarrow B$
SN⁻: $\Delta \rightarrow \neg A \vdash \Delta, A \rightarrow$
SG⁻: $\Delta \rightarrow \wedge x A[x] \vdash \Delta \rightarrow A[a]$.

Now the point is not the equivalence of the two systems **F** and **G** – after all it is well-known that both are complete systems of predicate logic – it is rather their close similarity: The rules of **G:** AI and SI^- are our R1*, SG is R5 and I,AC,C,TND are special cases of Frege's rules R2,R4,R7,R8. The rest can be easily obtained: AT from A1b and R7, SN from R1*,R3,R7, SN^- by R3, SG^- by A2 and R7. Inversely, Frege's rules are identical with those of **G** or generalizations that are easily seen to be admissible in **G**.

The essential difference, then, is that Frege states elimination rules for the succedent instead of introduction rules for the antecedent. The restriction to introduction rules is a prerequisite for Gentzen's *Hauptsatz*, cut elimination, which is not only useful for consistency proofs but also yields a purely mechanical procedure of proof-construction, as Kurt Schütte has shown. Thus, even if we would generously attribute the first calculus of sequents to Frege, its employment for proof-theoretical purposes would still be a very important and quite original contribution of Gentzen.

References

Frege, G.: Grundgesetze der Arithmetik, Bd. I, Jena 1893.
Gentzen, G.: Untersuchungen über das logische Schließen, Mathematische Zeitschrift 39 (1934), 176–210, 405–31.
Schütte, K.: Beweistheorie, Berlin 1960.

Part II:
Epistemology

Frege's Anti-Psychologism

EVA PICARDI

There is hardly a piece of writing by Frege – be it book or letter, article or review – where he misses the opportunity to stigmatize the evil of psychologism. An unmistakable mark that a philosopher has fallen prey to this infection is his tendency to blur certain distinctions, for example, between the laws of thinking and the laws of thought, between the sense of linguistic expressions and that which they can be used to talk about, between sense and mental representations. As far as mathematicians are concerned, even those (few) who refrain in their writings from psychologistic phraseology betray symptoms of psychologism in their careless employment of the word "sign": they claim to pay attention only to the sign but in reality switch freely from the sense of a sign to that which it stands for and even to the calculation techniques commonly associated with it. In a spirit similar to Frege's Georg Cantor in his *Mitteilungen zur Lehre vom Transfiniten* of 1887 denounces the *akademisch-positivistische Skepsis* (personified by Helmholtz and Kronecker) as being guilty of conceiving numbers as mere signs and moreover "not as signs of concepts which refer to sets, but as signs of the individual things counted in the subjective process of counting". ("Zeichen für die beim subjektiven Zählprozeß gezählten Einzeldinge").[1] A source of this error resides in the genetic fallacy of confusing the meaning and import of mathematical propositions with the mental acts whose performance is allegedly required for apprehending them (i. e. acts of psychological abstraction, of construction, of pattern recognition). But – Frege urges – the content of a sentence (what the sentence says) and our forming a judgement about it are two utterly different things: the use of the word "judgement" ("Urteil") to refer both to the content and to the act of judgement and the careless employment of the word "idea" ("Vorstellung") are but two examples of the conceptual confusions which mar all psychologistic accounts of logic.

[1] G. Cantor, *Abhandlungen mathematischen und philosophischen Inhalts*, ed. by E. Zermelo, Hildesheim, Olms, 1932, repr. 1966, pp. 378–439, pp. 382–3.

Frege's anti-psychologism has many targets and facets, some of which I have explored elsewhere by way of comparing Frege's conception of logic with certain aspects of the logical doctrines of Erdmann, Wundt, Sigwart and Kerry.[2] Frege's most sustained discussions of psychologism are to be found in his *Preface* to GGA, in his review of Husserl's *Philosophie der Arithmetik*, and in the First Logical Inquiry, "Der Gedanke". In 1893 the scapegoat of psychologism is Benno Erdmann, whereas in 1918 no particular miscreant is mentioned. Frege's target might very well have been a corpus of doctrines and not a specific author. However, internal textual evidence has led me to arrive at a conjecture concerning a specific author who may have been at the back of Frege's mind.

With the exception of this philological curiosity, which I will take up in the Appendix, I do not here intend to plunge deeper into the historical background of Frege's work, but I want rather to explore what I believe Frege took the main fault of psychologism to be, i. e. a mistaken picture of language which turns the objectivity of sense and the communication of thoughts into a mystery. The confusion between logical laws and psychological laws of thinking, which Frege discusses in the Preface to the first volume of GGA, is yet another facet of that mis-

---

[2] Cf. Picardi, "The Logics of Frege's Contemporaries or 'der verderbliche Einbruch der Psychologie in die Logik'", in D. Buzzetti, M. Ferriani (eds.), *Speculative Grammar, Universal Grammar and Philosophical Analysis of Language*, Amsterdam, Benjamins, 1987, pp. 173–204; *Kerry und Frege über Begriff und Gegenstand*, *History and Philosophy of Logic*, XV (1994), pp. 9–32, and *La chimica dei concetti. Linguaggio, logica, psicologia 1879–1927*, Bologna, Il Mulino, 1994, chs. 1 and 2.
I have mentioned Kerry, Wundt and Sigwart because they are among the authors whose work we know to have been actually studied by Frege. As is well known, Frege's article *Über Begriff und Gegenstand* is a reaction to the criticism which Benno Kerry had levelled against Frege's account of the notions of a concept and an extension of a concept in *Grundlagen der Arithmetik*. Between 1885 and 1891 the Austrian philosopher Benno Kerry (1858–1889) had published in the *Vierteljahrsschrift für wissenschaftliche Philosophie*, a review edited by R. Avenarius, a series of 8 articles under the title "Über Anschauung und ihre psychische Verarbeitung". In these articles Kerry discusses in detail the philosophical and mathematical doctrines put forward by G. Cantor, F. A. Lange, L. Kronecker and B. Bolzano. Even apart from their intrinsic value, Kerry's articles constitute a mine of information as regards the philosophical atmosphere in which Frege's work was written and received. Wilhelm Wundt's *Logik* is mentioned by Frege in two places of his *Nachlaß*, whereas as far as Sigwart is concerned we have only the testimony of Heinrich Scholz; it appears that Frege had taken notes or excerpts, now lost, from the *Logik* of Christoph Sigwart. Scholz's list is published in vol. I of M. Schirn (ed.), *Studien zu Frege/Studies on Frege*, Bad-Cannstatt, Frommann-Holzboog, 1976, as an appendix to A. Veraart's article which describes the history of Frege's *Nachlaß*.

taken picture. The assimilation of logical laws to psychological laws can be seen as the result of an extreme form of naturalism:[3] according to Frege, at the root of this confusion is the lack, on the part of psychologistic philosophers, of an adequate concept of truth. To the psychologistic philosopher only a relativistic notion of truth is available: truth is made to coincide with universal agreement and is implicitly relativized to the mental equipment of the "normal" representatives of the species *homo sapiens* at a given stage of its psychological and social evolution.[4] The chief defect of a naturalistic approach to logic is not just that it disregards the claims of *a priori* knowledge without offering any alternative account, but that, by embodying a relativistic notion of truth, it issues in a form of extreme subjectivism as regards meaning. It is Frege's conception of truth which, in my opinion, provides the link between his anti-psychologism in logic and his anti-psychologism in the account of meaning. The thesis I am attributing to Frege is a strong and controversial one: it amounts to the claim that nothing short of the classical notion of truth can give us a correct account of the meaning we attach to our utterances. Frege held that truth and meaning stand in the closest connection to each other: the account of meaning which he had outlined in his essays "Über Sinn und Bedeutung", "Über Begriff und Gegenstand", "Über Funktion und Begriff" provides in my opinion the most effective remedy against psychologism.

If what I have said is correct, then it is not only unpromising but positively misleading to tackle the issue of Frege's anti-psychologism by appealing to his conception of epistemology – a conception of which we can at best surmise its bare outlines. Philip Kitcher, for instance, has suggested that Frege implicitly relied on Kant's account of knowledge as far as the ultimate justification of logical principles is concerned, thereby himself succumbing to a form of psychologism, for, in Kitcher's opinion, Kant's theory of knowledge is psychologistic.[5]

---

[3] "Naturalism" is not Frege's term, but Heidegger's; cf. M. Heidegger, *Die Lehre vom Urteil im Psychologismus* (1913), in *Gesamtausgabe*, vol. I, *Frühe Schriften*, Frankfurt a. M., Klostermann, 1978. Heidegger's criticism of psychologistic theories of judgement follows by and large the arguments outlined by E. Husserl in his *Prolegomena zur reinen Logik* (1900) which precede the *Logische Untersuchungen*.

[4] This is the gist of Frege's criticism of Benno Erdmann's *Logik* in the Preface to GGA. Husserl's critical survey of Sigwart's and Erdmann's conception of logic in §§ 38–40 of his *Prolegomena zur reinen Logik* closely resembles the line of argument employed by Frege in GGA. Both Frege and Husserl identify the chief error of psychologism in the endorsement of a relativistic conception of truth. Both Sigwart's and Erdmann's books on logic were influential in Frege's times.

[5] P. Kitcher, "Frege's Epistemology", *Philosophical Review*, 88 (1979), 235–62.

Kitcher is not the first to advance the suggestion that Frege had slipped inadvertently into psychologism. The same charge, on very different grounds, was made by Russell and Jourdain. Before examining Kitcher's claim, it may be appropriate to review Frege's own rebuttal of the charge brought against him by Jourdain and Russell.

A new form of psychologism?

It must have been galling for Frege to read in Russell's letter of 12. 12. 1904: "We do not assert the thought, for this is a private psychological matter [eine psychologische Privatsache], we assert the object of the thought, and that is to my mind a certain complex";[6] and equally galling it must have been for him to be queried some ten years later by Philip Jourdain whether he "now regard[ed] assertion (⊢) as merely psychological" and whether "in view of what seems to be a fact, namely, that Russell has shown that propositions can be analysed into a form which only assumes that a name has a 'Bedeutung', & not a 'Sinn'" he would now hold that "'Sinn' was merely a psychological property of a name".[7]

This is by no means a minor clash of mentalities between the founding fathers of logicism; on the contrary, it is a sign of a deep philosophical difference. The charge that assertion is purely psychological is very likely Wittgenstein's. It seems quite probable that already in 1913–14 Wittgenstein had come to this conclusion and talked to Jourdain about it. Unfortunately the letters exchanged between Frege and Wittgenstein concerning the *Logisch-philosophische Abhandlung* never arrived at Wittgenstein's remark 4.441, where he expresses his view on assertion.

Had Russell (and Wittgenstein) uncovered a form of psychologism which had escaped the most vehement advocate of anti-psychologism? Had Frege himself fallen prey to the very same *Zeitkrankheit* which he had so ruthlessly exposed in other philosophers and mathematicians?

In his reply to Jourdain Frege touches neither on the subject of sense nor on that of assertion but only points out certain obscurities

6 G. Frege, *Wissenschaftlicher Briefwechsel* (=WB), G. Gabriel, H. Hermes, F. Kambartel, C. Thiel, A. Veraart (eds.), Meiner, Hamburg, 1976, p. 251; Engl. edition abridged by B. McGuinness and translated by H. Kaal, *Philosophical and Mathematical Correspondence* (=PMC), Blackwell, Oxford, 1980, p. 169.

7 Jourdain to Frege, 15. 1. 1914, WB, p. 126 PMC, p. 78.

which in his opinion beset Russell's notion of a variable. An earlier draft of the letter, however, is entirely devoted to these matters. Here Frege explains once more the reasons why he believes that the notions of sense and assertion, far from involving any concession to psychologism, are an effective antidote against it. One may wonder why in the letter Frege actually sent to Jourdain neither the notion of sense nor that of assertion are mentioned. Perhaps remembering his utter failure in 1904 to get across to Russell what he meant by sense, thought, cognitive value, Frege simply deemed it a waste of time to repeat the experiment with Jourdain.

I shall quote two passages from the draft of the letter to Jourdain for they are in my opinion the best and most concise account of what Frege took the bearing of the notion of sense on the issue of psychologism to be. The first argument concerns the contribution of the sense of a proper name to the sense of a sentence in which it occurs, in harmony with the principle of compositionality; the second concerns sense as the mode of presentation of the referent and its relevance to explaining the cognitive significance of identity statements. The first argument runs as follows:

> The possibility of our understanding propositions which we have never heard before rests evidently on this, that we construct the sense of a proposition out of the parts that correspond to the words. If we find the same word in two propositions e. g. 'Etna', then we also recognize something common to the corresponding thoughts, something corresponding to this word. Without this language in the proper sense would be impossible. We could indeed adopt the convention that certain signs were to express certain thoughts like railway signals [...]; but in this way we could not form a completely new proposition, one which would be understood by another person though no special convention had been adopted beforehand for this case. [...] Without meaning [*Bedeutung*] we could indeed have thought, but only a mythological or literary thought, not a thought that could further scientific knowledge. Without a sense we would have no thought and hence also nothing that we could recognize as true.[8]

After having described the Aphla-Ateb example – that is, the example of one and the same mountain designated by different names by different people living in its vicinity – and after having remarked that the thought expressed by "Atleb is a least 5000 meter high" differs from that of "Aphla is at least 5000 meter high" since one can judge the

[8] WB, pp. 127–28; PMC, pp. 79–80.

former true without judging the latter true, and vice versa, – as two acts of knowledge ("Erkenntnisthaten") are involved here, and not just one – Frege concludes:

> Now if the sense of a name was something subjective then the sense of the proposition in which the name occurs and hence the thought would also be something subjective and the thought one man connects with this sentence would be different from the thought another man connects with it; a common store of thoughts, a common science would be impossible. It would be impossible for something one man said to contradict what another man said, because the two would not express the same thought, but each his own. For these reasons I believe that the sense of a name is not something subjective [crossed out: in one's mental life], that it does not therefore belong to psychology and that it is indispensable".

Similar arguments can be given in support of the non-psychological character of the sense of predicates and relational expressions, but for reasons which I will explore later proper names constitute a delicate test-case for our intuitions. I do not here want to discuss the question whether Frege's notion of sense as something which can be shared and communicated, and which stands in the closest relation with truth, is ultimately defensible as it stands. I simply want to point out that he took the threat of psychologism to consist in an intolerable form of subjectivism, which in GGA he had labelled "idealism" and "solipsism". However, if all that Frege had said about sense had consisted in the formulation of principles of compositionality underlying our understanding of sentences, such an insight, though valuable, would give us no clue how to answer the question what it is for a hearer to recognize a sentence as bearing a given sense, or, indeed, as bearing the same sense that the speaker who addresses him confers upon it. To answer the latter question we must appeal to the concept of truth: the sense of a linguistic expression can then be characterized as its contribution to the determination of the semantic value of the sentence in the context of which it occurs. It is via the notion of truth and reference that we attain a conception of what it is for a sentence to express a thought, i. e. to possess a definite content which we may reasonably aim at communicating to others and to establish as true or not true to the satisfaction of all parties concerned. In default of such an account of meaning, of which I have here given a very rough sketch,[9] we are

[9] I am here following Dummett in his interpretation of Frege's conception of semantics. Cf. e. g. M. Dummett, *The Interpretation of Frege's Philosophy*, London, Duckworth, 1981.

powerless against psychologism, and no amount of Kantian transcendentalism could restore our faith in objectivity. Or so it seems to me.

In his book on Frege, Hans Sluga has claimed that "it would require a strong argument to show that Kantian transcendentalism is insufficient to account for the desired anti-psychologism, objectivism and apriorism. No such argument can be found in Frege's writings"[10]. Since, as I construe them, Frege's complaints against psychologism are essentially semantic in nature, I am at a loss to see how a mere appeal to Kantian transcendentalism could have secured the desired objectivity of content. Had Frege thought that Kantian transcendentalism was sufficient, he would perhaps have said so, and very probably he would not have bothered to devise his own a theory of sense and reference.

As both Cantor and Windelband recognized, psychologism can be seen as an offspring of a naturalistic reading of Kant. Windelband's assessment of the German philosophical scene between 1870 and 1890 is worth quoting:

> It was the fate of the revival of Kantianism that because of the interests of the scientific way of thinking it was at first restricted to the theory of knowledge, whose extreme developments toward a decidedly empiricist position led to positivistic transformations on the one hand and to the dissolution of philosophical problems into psychological problems on the other. Thus, just as during the time before Kant, the disastrous predominance of psychologism became widespread, for a few decades, particularly at German universities.[11]

Given the many different trends of Kantianism in German philosophy, the claim that Kantian transcendentalism of itself provides a barrier against psychologism stands in need of qualification and circumstantial evidence. Of course, I am not denying that Frege (young Frege especially) made use of concepts and formulations which derive from Kant. For instance, his celebrated characterization of sense as "die Art des Gegebenseins eines Gegenstandes", his claim in "Der Gedanke" that perceptual experience involves a nonsensible element, his ascription of conceptual priority to the sentence used to express a judgement over its component parts, all these ring a familiar (Kantian) bell. But adverting to these features helps us not at all in understanding Frege's conception of logic and his anti-psychologism. Besides, psychologist philosophers were only too ready to assign to judgement (as content

[10] H. Sluga, *Gottlob Frege*, London, Routledge, 1980, p. 60.

[11] W. Windelband, *Lehrbuch der Geschichte der Philosophie*, 1892, p. 539. (Repr. of the 13th ed. of 1935, Tübingen, J. C. B. Mohr, 1980).

and act, if they managed to keep them apart) the pride of place in their accounts of logic; Benno Erdmann's *Logik* is a case in point, and its debt to Kant's philosophy is quite obvious to the reader.[12]

In order to understand Frege's foundational project we have to bear in mind a distinction which psychologistic logicians did not or could not draw, namely the distinction between our entitlement for holding a sentence true and the ultimate grounds on which the justification for its truth rests. As I said, Frege's realistic conception of truth is an essential ingredient of his conception of logic. Sluga rightly perceives that the adoption of Platonism is irreconcilable with Kantian *Kritizismus* and sets out to show, unsuccessfully in my opinion, that Frege was not a realist in Dummett's sense. A line of argument similar to Sluga's is put forward by Philip Kitcher in his paper "Frege's Epistemology" of 1979. Kitcher's interpretation rests on his reading of a statement Frege makes in the Preface to GGA in the context of his discussion of Erdmann's doctrines. Frege says:

> The question why and with what right we acknowledge a law of logic to be true, logic can answer only by reducing it to another law of logic. Where this is not possible logic can give no answer.[13]

Kitcher reads this statement as evidence that Frege thought that the question with what right we acknowledge a law of logic had already been settled. "A philosopher whom he greatly admired, had set out a general scheme for human knowledge. ... The philosopher was Kant."[14] In Kitcher's opinion Kantian transcendentalism provides the key to the understanding of Frege's foundational programme. I myself don't see that this follows from the passage quoted above, nor from what Frege says elsewhere.

In a well-known passage of *Grundlagen*, where Frege declares that the Kantian distinctions between the *a priori* and the *a posteriori*, between the synthetic and the analytic do not concern the content of a judgement but our justification for making it, he qualifies his statement thus:

---

[12] Not very surprisingly, perhaps, if we remember that Benno Erdmann was not only an eminent Kant scholar, who edited several volumes of the Akademie-Ausgabe, but also the author of a book called *Kant's Kritizismus*, published in 1878.

[13] G. Frege, *Grundgesetze der Arithmetik*, vol. 1 1893 (=GGA I), vol. II 1903 (=GGA II, Jena, Pohle, repr. Hildesheim, Olms, 1962; Engl. part. transl. by M. Furth, *The Basic Laws of Arithmetic*, Los Angeles, Univ. of California Press, 1967, p. 15.

[14] P. Kitcher, "Frege's Epistemology", p. 241.

> When a proposition is called *a posteriori* or analytic in my sense, this is not a judgement about the conditions, psychological, physiological and physical, which have made it possible to form the content of the proposition in our consciousness; nor is it a judgement about the way in which some other man has come, perhaps erroneously, to believe it true; rather, it is a judgement about the ultimate ground upon which rests the justification for holding it to be true. (GLA, § 3)[15]

If a proof of the analytical (i. e. purely conceptual) nature of arithmetical propositions could be secured, Kant's appeal to an *a priori* intuition to justify their necessity could be dropped as redundant. Frege shares Kant's opinion that an appeal to intuition is indispensable in the case of geometrical propositions, on the ground that the denial of geometrical axioms (unlike that of analytical propositions) does not lead to inconsistency (GLA, § 14). Whereas the acceptance of specific geometrical axioms is forced on us by the peculiarity of our pure intuition of space, no appeal to intuition is required in the case of arithmetical propositions. Having discarded Kants's account of necessity,[16] Frege sets himself two formidable tasks: (a) to defend the thesis that pure conceptual thought, issuing in analytical propositions, can extend our knowledge and (b) to show that logic, far from being barren, can beget peculiar "objects", i. e. numbers. Having rejected Kant's thesis that without sensibility no objects are given to us, he squarely faces the question: "How then are numbers given to us, if we cannot have any idea or intuition of them?" (GLA, § 62). The surprising, indeed revolutionary, answer he gives to this question is that we should rather concentrate our attention on the assertions in the context of which number-words occur and investigate the grounds on which the justification of their truth rests. This

[15] G. Frege, *Die Grundlagen der Arithmetik* (=GLA) Breslau, Koebner, 1884, *Centenarausgabe* edited by C. Thiel, Meiner, Hamburg 1986; Engl. trans. by J. L. Austin, 2nd. rev. ed. Oxford, Blackwell 1967.

[16] Contrast the above-quoted statement from GLA with Kant's characterization of necessity in KrV (Einleitung, B17). It seems to me that Kant here wants to draw attention to the actual processes whereby we apprehend the content of synthetic (mathematical) judgements: the relevant question to be asked in this context is not, according to Kant, one about what we ought to add in thought to a given concept, but rather what we *really* think in thinking that thought, albeit obscurely, for only then does it become clear that the necessity of a certain predicate inhering in a given concept is brought about by an intuition which must be added to that concept. Frege's notion of objective justification, on the other hand, relates to the logical form of sentences expressing arithmetical propositions, and thus, in a sense, to that which we "ought to think". The further question of whether and how the content of such logically analysed sentences is related to that which we *really* think is discarded by him as *logically* irrelevant.

way of recasting epistemological issues in semantic terms (i. e. as questions concerning meaning, reference and conditions of justification) constitutes, in my opinion, the background against which Frege's remedy against the "corrupting incursion of psychology into logic"[17] has to be found.

To the three fundamental principles enunciated in the Preface to GLA Frege adhered all his life. And yet, after 1884, he never again made use of the terms "analytic" and "synthetic". Dummett surmises that perhaps he no longer thought that "arithmetical truths in the plural are analytic, but that arithmetic in the singular is a branch of logic and need not to draw upon experience or intuition for the basis of its proofs".[18] In his *Logik in der Mathematik* of 1914,[19] after having raised the question how we explain the difference in cognitive significance between two identity statements like "606 = 606" and "137 + 469 = 606", he does not answer it by appealing to the activity of counting, calculating, constructing, etc., as a Kantian would have done, nor does he mention his own previous account of analyticity, put forward in GLA. He now appeals exclusively to the argument he had given in "Über Sinn und Bedeutung". After having mentioned examples of numbers, planets, stars, mountains, all on the same footing, he pauses to remark on the marvellous potentialities of language, repeating the argument which we already found in the draft of the letter to Jourdain and which recurs at the beginning of the Third Logical Inquiry.

On the other hand, appealing to the manipulation of signs, to techniques of calculation, and to constructions would have been regarded by Frege as playing into the hands of the formalists. Whatever be may have meant in 1924 by a "logical source of knowledge", I am sure he did not mean the ability to manipulate signs and carrying out constructions. Benno Kerry, who concurred with Frege in mistrusting naive formalism, sees the latter as stemming from Kant's emphasis on construction. Hilbert himself – not a *naive* formalist, to be sure – acknowledges, in his essay *Über das Unendliche,* his debt to Kant's philosophy. "Contentual" meta-mathematics is, in Hilbert's eyes, an embodiment of Kan-

[17] G. Frege, GGA, I, p. XIV, Engl. transl. p. 12.

[18] M. Dummett, "Frege and the Paradox of Analysis", in *Frege and Other Philosophers,* Oxford, OUP, 1981, pp. 17–52, p. 28.

[19] G. Frege, *Nachgelassene Schriften* (=NS), ed. by H. Hermes, F. Kambartel, F. Kaulbach, 2nd revised and enlarged ed. Hamburg, Meiner, 1983; Engl. transl. of the 1st edition (1969) by P. Long and R. White, *Posthumous Writings* (=PW), Oxford, Blackwell, 1979.

tian insights, and as far as I understand Hilbert and Kant, Hilbert's claim does not seem far-fetched at all. The following statement is extremely instructive:

> Kant already taught – and indeed it is part and parcel of his doctrine – that mathematics has at its disposal a content secured independently of all logic and hence can never be provided with a foundation by means of logic alone; that is why the efforts of Frege and Dedekind were bound to fail. Rather, as a condition for the use of logical inferences and the performance of logical operations, something must already be given to our faculty of representation [*in der Vorstellung*], certain extralogical concrete objects that are intuitively [anschaulich] present as immediate experience prior to all thought. [...] in mathematics, in particular, what we consider is the concrete signs themselves, whose shape, according to the conception we have adopted, is immediately clear and recognizable.[20]

Even a superficial reader of GGA soon realizes that without the account of how a sentence of the formal language is determined as true – the sense expressed by an assertion of a sentence of GGA being the thought that its truth-conditions are fulfilled (GGA I, § 32) – Frege would have not a shade of an argument for opposing the doctrines both of psychologistic logicians and of formalist mathematicians. There would be no explanation of what it is for a sentence to be objectively true or false "independently of the judging subject" (GGA, I, p. XVII).

The psychologistic logicians add to a thought all sorts of irrelevant paraphernalia, since they lack the notion of objective truth and have access only to that of "true for us" ("us" meaning something like mankind at a given stage of its evolution). The formal mathematicians (e. g. Heine and Thomae) claim to be able to strip the signs of all content and they also employ a very crude notion of a sign and the sameness of a sign (GGA, II, § 90); no wonder that, as soon as within the theory of a game they want to say something significant about admissible and inadmissible configurations of signs they are forced to reintroduce through the back-door the contentual understanding of the rules governing the use of the signs concerned.

Frege thought he could prove that such an approach was misguided by showing how the thought expressed by a sentence of GGA is uniquely determined by the truth-conditions of the sentence, no matter whether *we* are or ever will be in a position to recognize whether or not

---

[20] D. Hilbert, "Über das Unendliche", Engl. trans. in *From Frege to Gödel*, ed. by J. van Heijenoort, Harvard Univ. Press, 1967, p. 376.

they obtain. The originality of Frege's Platonism does not reside in his belief that there is a mathematical reality external to our minds but in his view that the sense and reference of mathematical statements are determined once their truth-conditions have been laid down. The disrupting effect of Russell's paradox consisted, among other things, in showing that this kind of requirement was not fulfilled by certain sentences of GGA, thereby jeopardizing Frege's entire conception of semantics.[21]

I claimed that it is misleading to suppose, as Sluga and Kitcher do, that in Frege's opinion a generic appeal to transcendentalism would provide an answer to psychologism and a groundwork for objectivity. How, then, is the passage of GGA quoted by Kitcher to be interpreted? If we read the quoted statement in the context where it belongs (and how else should we read it?), it will ring a very different bell. For Frege continues:

> If we step away from logic, we may say: we are compelled to make judgements by our own nature and by external circumstances; and if we do so, we cannot reject the law – of Identity, for example; we must acknowledge it unless we wish to reduce our thought to confusion and finally renounce all judgement whatever. I shall neither dispute nor support this view: I shall merely remark that what we have here is not a logical consequence. What is given is not a reason for something's being true, but for our taking it to be true. Not only that: this impossibility of our rejecting the law in question hinders us not at all in supposing beings who do reject it; where it hinders us is in supposing that these beings are right in so doing, it hinders us in having doubts whether we or they are right.[22]

Far from being reticent, as Kitcher insinuates, Frege is here refusing to endorse the view that Kitcher attributes to him. Frege has no difficulty to concede that no *a priori* contradiction is involved in supposing that there may be beings who accept as sound inferential principles different from ours; a discrepancy between their laws of holding true and ours could not be discarded as spurious by appeal to the conditions of the possibility of thought in general. One need not think of exotic tribes to find examples of the situation here envisaged. It is enough to suppose that the beings in question suspend judgement on specific applications of certain principles which we accept; they need accept as true the negation of the law of identity or of a classical tautology. If we think of

---

[21] Or, at any rate, his conception of logical objects.

[22] GGA, I, p. XVIII; Engl. transl. p. 15.

mathematics, the situation is even more clear-cut: Kronecker and Cantor could be seen as representatives of such different tribes. What Frege says is that meeting a deviant tribe, though deeply puzzling, would not make us *doubt* whether *we* are wrong in accepting such laws, nor would it make us call into question our inferential practices. However, this unshakable commitment to certain inferential principles which we find in ourselves cannot be turned into a justification of the ultimate grounds on which the necessity of the logical laws themselves rests. If one tried to turn it into such a justification, then this would amount to acknowledging that *logical necessity stems from us*, from our human nature, from our mental and physical constitution. Must we then conclude with Wundt (or Kitcher) that logic needs the theory of knowledge for its foundation and methodology for its completion, for otherwise "logical laws would hold as given and unexplainable matters of fact"?[23] Frege held that logical laws are neither capable nor in need of justification: must we rest content with this statement?

Twenty years after the *Begriffschrift*[24] Frege comes back to his simile according to which the logical rules are "images" [*Abbilder*] of logical principles which because of their forming the basis of our thinking cannot be expressed in the concept-script, and he now says: "In the domain of objects there hold certain laws, and one can understand how they are mirrored in the rules that hold good for the corresponding symbols".[25] The possibility of a logical justification of an arithmetical sentence ultimately depends on there being objects standing in certain logical relations. But is there a way of forging a link between the grounds on which the laws of truth rest and our conscious striving to heed "the rules which hold for the corresponding signs"? In GGA Frege himself gives us a clue as to where a possible link can be found when he sums up the discussion by remarking that the source of the dispute between the psychologistic logicians and himself lies in their different conceptions of the True. I suggest that we interpret Frege's clue as follows: it is the notion of truth which informs our judgements and shapes our understanding of their contents which forces us to ac-

---

23 W. Wundt, *Logik*, vol. I, Stuttgart, Enke, 2nd ed. 1893, p. 2.

24 G. Frege, *Begriffsschrift: Eine der arithmetischen nachgebildete Formelsprache des reinen Denkens*, Halle, Nebert, § 13; repr. in *Begriffsschrift und andere Aufsätze*, ed. by I. Angelelli, Hildesheim, Olms. Engl. transl. and ed. by T. W. Bynum, Oxford, OUP, 1972.

25 G. Frege, *Über die Zahlen des Herrn Schubert*, Jena, Pohle, 1899, repr. in *Kleine Schriften*, ed. by I. Angelelli, Hildesheim, Olms, 2nd. rev. edition 1990, pp. 240–261, p. 253; cf. also GGA II, §§ 138–155.

cept those propositions which we consider neither capable nor in need of justification and choose as axioms in our logical system. The reason why we acknowledge as valid the rules which we have, as it were, extracted from the axioms is that they can be used as inferential principles which preserve truth: this is why "anyone who has once acknowledged a law of truth has by the same token acknowledged a law that prescribes the way in which we ought to judge, no matter where, or when, or by whom the judgement is made".[26] We cannot transcend our notion of *truth*, and, for this reason we cannot transcend our logic either. Frege's notion of truth is realistic in Dummett's sense: truth is not only objective but "independent of the judging subject".

Anti-psychologism and the third realm

In the previous section I have explored the path which leads to a defeat of psychologism through the theory of meaning. However, in many of his writings Frege alludes to a different path to securing the objectivity of content, which does not go via language but is straightforwardly metaphysical in its nature. So far we have been dealing with thoughts conceived as the senses of declarative sentences; however, in 1918 thoughts are characterized as self-subsistent entities, independent in the strongest possible sense of our apprehension of them.[27] All thoughts collectively are said to form a third realm. In the First Logical Inquiry the answer to the threat of psychologism seems to rest ultimately on a metaphysical postulation. One might go even further and claim that, since the whole theory of sense and reference is absent from the *Logische Untersuchungen*, Frege has here called into question his previous conception of semantics. To be sure, on account of Russell's contradiction, certain features of Frege's semantics had to undergo modification, but it is hard to believe that by 1918 Frege had abandoned also his entire conception of the sense and reference of linguistic expressions. Such a surmise – of which I do not know whether it has ever been put forward – would, I think, be untenable. In the précis of his logical views which Frege sent to Darmstaedter in July 1919 the theory of sense and reference still occupies the place of honour. Besides,

[26] GGA I, Preface, p. XVII; Engl. transl. p. 15.
[27] Cf. M. Dummett, "Frege's Myth of the Third Realm" (1986), in *Frege and Other Philosophers*, op. cit., pp. 249–262.

in his *Logische Untersuchungen* Frege offers very detailed discussions of the way language functions (indexicals, negation, and truth-functional connectives) including an analysis of assertoric force.

Now, independently of how we feel about third realms, as interpreters of Frege's philosophy we are faced with the question whether by 1918 Frege had reached the conclusion that a flight into a third realm was the only route to secure the desired objectivity of thoughts and thus the only defense against psychologism. Could it be that in the meantime Frege had became skeptical as regards the justification he had given in his previous writings of the role of the notion of sense in his account of logic and language? Could the way thought is exiled into a third realm in "Der Gedanke" be seen as an admission of defeat? What evidence is there in Frege's writing?

Here is a suggestion, which I myself consider extremely feeble and which I explore mainly in order to dramatize certain difficulties that Frege's doctrine of sense encounters on its own ground. Frege might have come to the conclusion that the sense of a proper name is something that belongs to psychology; perhaps he found himself unable to rebut Russell's challenge, perhaps he could find no argument for explaining why a word – and a proper name in particular – bears in the language the "objective" sense which Frege ascribes to it.[28] As a consequence he may then have assigned the function of securing objectivity to non-actual, a-temporal, a-spacial, self-subsistent entities.

These entities are intended to work like the gold that backs the paper currency of sentences, to borrow Geach's apt metaphor.[29] However, as Frege realizes, sheer postulation is useless unless accompanied by an account of how the market works, i. e. how the gold is related to the paper currency. At the close of "Der Gedanke" he offers a streamlined account of judgements of perception, which perhaps could be generalized. It goes roughly as follows: a non-sensible element, belonging presumably to the third realm and therefore itself thought-like in character, has to be acknowledged, for it is its presence that discloses the outer world and turns sense-impressions into perception and judgement. Perhaps the mutual understanding which we seem to reach through language is made possible by the presence of this non-sensible

[28] Cf. M. Dummett, "Frege's Distinction between Sense and Reference" (1975), repr. in *Truth and Other Enigmas*, London, Duckworth, 1978, pp. 116–144, p. 130.

[29] P. Geach, "The Identity of Propositions" (1967), repr. in *Logic Matters*, Oxford, Blackwell, 1972, pp. 166–74, p. 171.

element: it turns our apprehension of the sentences of natural language (conceived as physical entities) into a grasp of the thoughts they help to convey. The thoughts thereby grasped transcend the world of inner experience and the outer world of sense-impression. Perhaps following the same train of thought Frege conjectures in 1924 that there may be beings who, unlike us, can grasp these thoughts without the sensible medium of language and without any accompanying idea or representation:

> The connection of a thought with one particular sentence is not a necessary one; but that a thought of which we are conscious is connected in our mind with some sentence or other is for us men necessary.[30]

In 1924 the relative independence of thinking and speaking is seen by Frege as the only way of explaining how logical paradoxes are possible in the first place. This is not a matter of a struggle in which thinking is at war with itself, but the paradoxes are due to the imperfection of the linguistic means of expression which even creeps into the formal language. As a diagnosis of the source of the contradiction Frege's analysis must strike us as unconvincing; I merely mention it as testimony to a development of the view put forward in the First Logical Investigation. The exile of thoughts in the third realm could then be seen as issuing from Frege's failure to see how the objectivity of thoughts and their constituent senses could be secured in the absence of a viable semantics.

A symptom of a very different kind of dissatisfaction could be detected in the way Frege handles the example of Gustav Lauben in "Der Gedanke". He there comes to the conclusion that Leo Peter and Herbert Garner, who attach different definite descriptions to that name, "speak a different language", for although they use the name "Gustav Lauben" to refer to the same man they are unaware of so doing. Perceiving this as a defect, Frege modifies the example and decrees that "Dr. Gustav Lauben" and "Gustav Lauben" should be treated as different names. Now, the circumstance that two speakers might attach different definite descriptions to the same proper name does not turn its sense into something psychological. The different definite descriptions can be spelled out and become "possession of many". The name "Gustav Lauben" in the mouth of Leo Peter and Herbert Garner still differs in status from an interjection like "ouch!" – Frege's example of something psychological par excellence (GGA, I, p. XIX).

---

[30] NS, p. 288; PW, p. 269.

It is undeniable, however, that the account of proper names given by Frege in "Der Gedanke" bears a certain similarity to Russell's. In 1918 Russell takes the whole point of the proper names of a natural language to be that of calling up ideas to the mind of speaker and hearer. In his lectures on Logical Atomism of 1918 Russell candidly states:

> When one person uses a word, he does not mean by it the same thing as another person means by it. I have often heard it said that that is a misfortune. That is a mistake. If would be absolutely fatal if people meant the same things by their words. [...] We should have to talk only about logic – a not wholly undesirable result.[31]

Frege would of course have heartily disagreed with this statement. And yet, Frege's failure to draw a line between modes of presentation of an object grounded in the specific information of individual speakers and modes of presentation immanent in language makes the contrast between his account of names and Russell's less sharp than it may look at first sight.

In footnote 2 of "Über Sinn und Bedeutung" Frege had pointed out that different speakers may attach different descriptions to the name "Aristotle" and that therefore the same statement would have a different significance for different speakers: "Aristotle was born in Stagira" carries no news for those who attach to the name "Aritstotle" the definite description "The greatest philosopher of antiquity born in Stagira", whereas it may be news for someone who uses that name to refer to Aristotle while unaware that Aristotle is the unique satisfier of that description. In 1892 Frege says that no harm ensues as long as these speakers attach the same reference to the name; in an ideal language, however, these oscillations of sense should be avoided. But in 1918 he says that people who attach the same reference to a name without being aware of so doing speak a different language.

Now, obviously each of us has a different set of proper names and attaches different information to the same proper names; and each of us knows it and cannot possibly intend his words to carry for the hearer the very same detailed information which they carry at a certain time for himself. However, we should not describe the situation by saying that each of us speaks a different language, but rather that each of us has different beliefs concerning the bearer of the name. At least we should say so from a Fregean stand-point: communication rests on that

[31] B. Russell, *The Philosophy of Logical Atomism*, repr. in D. Pears (ed.) *Russell's Logical Atomism*, London, Fontana, 1972, Lecture II, p. 50.

ingredient of sense which is shareable, communicable and can be the common possession of many. The lack of a distinction between the information of a single speaker and community-wide information is a defect of Frege's doctrine but I believe that it is a defect which can be corrected without depriving the notion of sense of its cognitive significance. I think that Frege was not aware of the tension involved in this and that one should discard any conjecture according to which a perplexity as to the wholly psychological character of the senses of proper names might have shaken his faith in the non-psychological character of the senses of declarative sentences themselves in which those names occur.

But could an intervention of thoughts from the third realm help to disentangle the objective and the subjective ingredients in the thoughts conceived as senses of declarative sentences? The answer must, in my opinion, be negative. Not only have we no conception of how these entities are related to the thoughts expressed by the sentences of natural language, but we should not even know where to look for an answer to the question whether two people who utter the same sentence attach the same a-temporal thought to it. How can we know whether it is the same self-subsistent thought which two people get hold of when uttering the same sentence (without indexicals)? We cannot. Here an act of faith would be required. Frege should have remembered that at an analogous point he himself had detected a decisive weakness of psychologism, against which he had fought throughout his philosophical career. If the sense of a sentence consisted in the representation which each of us attaches to it we would be at a loss what to make of the disagreement among different speakers of the language: for what looks like a disagreement could then turn on a hidden difference of representational content, while what looks like agreement could depend on ignoring such a difference. If what the speaker takes the sentences to mean is equated with the complex of representations associated with the sentence in the speaker's mind, then it is unintelligible how anybody can ever hope to get his meaning across. Of course, when thinking or speaking there may occur in our minds all sorts of images, feelings and recollections, but they do not constitute the semantic contents which we intend to communicate.

A first step to take in order to get out of this tangle is to revert to the conception of sense as the sense of a linguistic expression. A second step, which Frege only hinted at, is to try to understand how, in detail, the notion of truth enters into the account of the practice of assertion.

As I said, Frege's conviction was that only a concept of truth realistically understood could feature in an adequate account of assertion – quite independently of the threat of psychologism. This, however, is a very large claim which I do not propose to take up here.

Appendix

After having distinguished between the three realms – of thoughts, of things in the external world, and of ideas – Frege sets out to forestall a possible objection which could be raised against his neat tripartition.

The section of "Der Gedanke" devoted to rebutting the objection to the three realms bears many similarities to a section of his *Logik* of 1897,[32] where he complains about the sliding into idealism of physiological psychology and remarks on the strange coincidence between Idealism and Realism. The editors of the *Nachlaß* append a footnote in which they refer to Wundt's work *Grundzüge der physiologischen Psychologie* of 1874. For reasons I will not enter here I doubt very much that Wundt was Frege's target in 1897, and I am sure that Wundt cannot be the thinker censured in 1918. In a similar vein Frege refers in "Der Gedanke" to the "Sinnespsychologie" and notices the odd transformation of realism into idealism. Who is the philosopher whose "odd objection" he seems to hear? My surmise is that the philosopher in question is Ernst Mach.

It seems very plausible that what Frege had in mind was the author of the *Analyse der Empfindungen,* whose first edition was published in 1886. In Germany Wilhelm Schuppe and Richard Avenarius, the editor of the *Vierteljahrsschrift für wissenschaftliche Philosophie*, the journal in which Frege published his "Über Begriff und Gegenstand", put forward views very similar to those of Mach. Mach is the thinker from whom derives the notion of monism (that is, the coincidence of idealism and realism) and the thesis that we start with sensations of external things only to discover that external things are posits useful in the economy of thinking. Much the same applies to the "I" conceived as a substance. Hence the well-known slogan "Das Ich ist unrettbar"- "The I is irrecoverable". Apart from these analogies with Frege's objector, what really clinches the case is the following statement from "Der Gedanke":

---

[32] G. Frege, NS, pp. 156–7, PW, p. 144 f.

> Does everything need an owner without which it can have no existence? I have considered myself as the owner of my ideas but am I not myself an idea? It seems to me as if I were lying in a deckchair, as if I could see the toes of a pair of polished boots, the front part of a pair of trousers, the waistcoat, buttons, some hair of a beard, the blurred outline of a nose. Am I myself this entire complex of visual impressions, this aggregate idea? It also seems to me as if I saw a chair over there. That is an idea. I am not actually much different from the chair myself, for am I not myself just a complex of sense-impressions, an idea? Why need this chosen owner of these ideas be the idea I like to call "I"? [...] The dependence which I found myself induced to ascribe to the sensations as contrasted with the sentient being disappears if there is no longer any owner. What I called ideas are then independent objects. No reason remains for granting an exceptional position to that object which I call "I".[33]

I am sure that Frege's contemporaries had no difficulty to read between the lines and to recall the famous drawing in ch. 1 of Mach's *Analyse der Empfindungen* (see the illustration at the end of the present article).

Gottfried Gabriel has pointed out to me that this suggestion is not new, but has already been put forward by Gregory Currie in the last chapter of his book on Frege.[34] Currie, however, does not greatly elaborate his own surmise, and this is somewhat surprising, if we remember that the declared aim of his book is to show that Frege conceived of his own work in philosophy as a contribution to (Kantian) epistemology. Currie correctly observes that the author of the First Logical Inquiry "was capable of deviating markedly from Kant's position",[35] and this too reinforces the impression that a comparison between Mach's and Frege's respective contributions to epistemology would have been highly appropriate.

In "Der Gedanke" Frege tries to meet the challenge of the skeptic who questions the way in which the distinction between ideas and thoughts is drawn. Having a thought differs dramatically, in Frege's opinion, from our having a *Vorstellung*: an idea or a sensation. Ideas, unlike thoughts, require a bearer, actually, in Frege's opinion, *one* bearer, an individual self. The skeptic questions not only our naive belief in an external world, independent of the inner world of conscious-

---

[33] G. Frege, *Der Gedanke. Eine logische Untersuchung*, repr. in *Kleine Schriften*, cit. pp. 342–362, p. 356; Engl. trans. Blackwell 1984, p. 365.

[34] The surmise is formulated at p. 182 of Currie's book *Frege. An Introduction to His Philosophy*, The Harvester Press, 1982, and is spelled out in some detail in the last footnote of the last chapter of the book.

[35] Ibidem, p. 185.

ness and ideas. He also questions the belief that there is such a thing as a self which may count as the bearer of representations and is clearly different from them. The skeptic whom Frege alludes to between the lines by putting forward a no-ownership thesis for feelings, sensations, and ideas, undermines the distinction between what is subjective (and necessarily private) and what is objective (and potentially accessible to all). If ideas need no bearer, then they are not different from thoughts. But is the skeptic right in suggesting that the contents of my consciousness alone can be objects of my thought? If at least one object could be found which does not belong to the contents of my consciousness, then the skeptic's reduction of the world to *my* world of ideas, and of the latter to nobody's world in particular (on account of the no-ownership thesis), would be blocked. In his reply Frege does not distinguish sharply between the thesis that ideas need no bearer and the thesis that ideas need no one (single) bearer. He argues that there is at least one *object* (namely myself) which can be the object of my thinking without being an idea. It may very well be that I may think of myself under this or that idea, but I am not myself the content of that idea.

Frege correctly points out that whenever I reflect upon myself it is myself and not an idea of myself which is the object at which my reflection is aimed. Yet one may grant the point without also granting that the self at which I seem to be directing my thought is itself an object – i. e. a self-subsistent substance. The gist of Kant's criticism of rational psychology was precisely that the latter turns the self of transcendental apperception into a substance, conceived as a simple, enduring and immutable object. But – Kant contends – there is no such object for us to know, no object which is to be found by inspecting the world of inner experience or the outer world. There is no trace of Kant's argument in Frege's purported demonstration of having found at least one thing which can be made the object of one's thought without being turned into an idea under which one thinks of such an object. Be that as it may, the self does not stand a good chance of fulfilling Frege's requirement for what to count as an object, i. e. something which subsists independently of its ever being thought of by any thinking being.

Frege's answer is so brief, that it would be rash to classify it under this or that well-known label. It is however an answer which is distinctly unsympathetic to a naturalistic dismissal of the self. Frege also hints that certain thoughts, on account of their intrinsic nature, are not open to skeptical doubts: Thus I may be in doubt whether there is a green leaf over there, but not whether I am having a visual impression of something green. It would be hasty to assimilate the kind of certainty Frege alludes to here to the certainty Russell claimed for sense-data. In fact, as we saw above, Frege's central claim is that a non-sensible element is necessary to transform sensation into perception. This applies, I think, to entities and events in our inner world, as well as to those of the external world. It is the presence of a non-sensible element which opens up the outer world to us, "for without this non-sensible something, everyone would remain shut up in his inner world".[36] The distinction between what is true and what is true for me would be lost, and the idea, so dear to Frege, that mankind has a common treasure of (true) thoughts would be unintelligible. Whether this non-sensible element is to be construed as a thought-like ingredient, or as the faculty of reason *tout court*, is a question which may here remain unanswered.

In conclusion, Frege's so-called epistemology does not provide him with any good argument for deflecting the skeptical challenge. Here, too, a more promising line of attack on skeptical doubts can be found by turning our attention to the role language plays in shaping our idea of objectivity: I doubt, however, that such a line of argument could persuade a realist of Frege's calibre.

[36] G. Frege, *Der Gedanke*, p. 360; Engl. trans., p. 369.

Fig. 1.

Frege's 'Epistemology in Disguise'[1]

GOTTFRIED GABRIEL

Abstract

The following article addresses questions concerning the place of epistemology in Frege's philosophy. It first treats Frege's limiting remarks on the traditional modalities of judgment in the "Conceptual Notation". Using a proof-theoretical interpretation or reconstruction, I argue for the claim that there can be no modal logic for Frege, although modal distinctions still have a legitimate place in his epistemology. A central issue is thus the relationship between the concept of necessity (the apodictic) and the epistemological idea of the a priori. It is shown that Frege does not confine the range of apodictic judgments to the a priori. Apodictic judgments are only necessary in so far as they are logically provable from other judgments. Their being necessary is thus relative to certain premises and leaves the kind of truth open, from logically true to a posteriori true. Frege's concept of the a priori excludes such possibilities and thus proves to be a "stronger" concept. These results occasion – beginning with the distinction between proof and justification – a detailed analysis of the difference between logic and epistemology. In the course of this analysis it becomes clear that it is not the proving (deductive) method of logic but the justifying (argumentative) method of epistemology which constitutes the method of philosophy.

Philosophy of Language

Today, after a period in which prevailing philosophical views interpreted Frege primarily as a philosopher of language, or at least a seman-

---

[1] Quotes are taken from the following English translations: Conceptual Notation and related Articles, ed. T. W. Bynum, Oxford 1972; The Foundations of Arithmetic, transl. J. L. Austin, 2. rev. ed. Oxford 1953, repr. 1959; The Basic Laws of Arithmetic. Exposition of the System, ed. M. Furth, 2. ed. Berkeley 1966, repr. 1982; Collected Papers on Mathematics, Logic, and Philosophy, ed. B. McGuinness, Oxford 1984; Posthumous Writings, ed. H. Hermes, F. Kambartel, F. Kaulbach, Oxford 1979; Philosophical and Mathematical Correspondence, ed. G. Gabriel, H. Hermes, F. Kambartel, C. Thiel, A. Veraart, Oxford 1980. Page numbers in square brackets refer to the German original. I am grateful to Michael Wrigley, Matthias Schirn, Peter McLaughlin and Finn Summerell for helping me with the translation.

ticist[2], more and more philosophers want to read him as an epistemologist.[3] This interpretation, I admit, faces the following problems: (1) the expression "epistemology" occurs only infrequently in Frege's writings; and (2) when the expression does occur, one might get the impression that the epistemological distinctions refer only to the subjective aspect of knowledge.

In what follows, I shall attempt to show in what sense the impression I just mentioned turns out to be wrong.[4] With respect to problem (1), my goal will be to create a certain symmetry between the two readings. It is also worth noting that the expression "philosophy of language" does not occur at all in Frege's writings. This is not, of course, an argument to be taken seriously: in our view of the discipline today, one could have been a philosopher of language without ever having used the expression at all. (For example, the fact that the expression "aesthetics" first occurs in the writings of A. G. Baumgarten during the 18th century does not hinder us from discussing the aesthetics of the Middle Ages.) We can better approach the subject by discussing not the disciplines but the themes in Frege's writings. The entire program of logicism in the "Conceptual Notation", "Foundations of Arithmetic" and "Basic Laws of Arithmetic", Frege's major works, can be seen as an attempt to clarify the "epistemological nature" of arithmetic. Frege says this explicitly.[5] The fact that knowledge-extending ("erkenntniserweiternde") judgments in arithmetic can be obtained on a purely logical basis is then used to support the claim that logic has a cognitive value. This view is used to counter "the legend of the sterility of pure logic".[6] It is but a slight exaggeration to say that Frege's main works are dedicated to the epistemological aim of obtaining a new understanding of analytical judgments which differs from Kant's views. Even Frege's classic paper on semantics exhibits epistemological elements, for throughout "On Sense and Meaning" Frege attempts to come to grips with the cognitive value of identity statements. This is especially significant in arithmetic, where Frege emphasized how numerical equations can be analytical, on one hand, and extend our knowledge, on the other.

[2] This classical reading is due to M. Dummett.

[3] Cf. H. D. Sluga (Gottlob Frege, London 1980) and G. Currie (Frege. An Introduction to His Philosophy, Sussex 1982). See my Introduction in R. H. Lotze, Logik. Drittes Buch. Vom Erkennen, Hamburg 1989, pp. XIII–XVII.

[4] Here, I am rejecting such well-intended subjectivist reconstructions as that of P. Kitcher, Frege's Epistemology. The Philosophical Review 88 (1979), pp. 235–262.

[5] Basic Laws of Arithmetic, p. 3 [vol. I, VII], p. 29 [1].

[6] Foundations of Arithmetic, § 17.

My thesis is that Frege had an immediate interest in logical and epistemological questions, but only a mediate interest in questions in the philosophy of language. This is clear in Frege's attitude to language. As the sensible cloak ("das sinnliche Kleid") of thought, it is simply the means to an end: the end itself is the thinking of thoughts – thoughts that become knowledge only in judgment. Of course, language is not just any means, one that could be exchanged for any other. For human beings, language is an unavoidable condition of mutual understanding. Frege's attitude towards language is determined by his insight that it is transcendental in character for *anthropological* reasons. For him this view seems to be more of an unwanted but unavoidable concession. The emphatic stress put on language by later linguistic transcendentalists – as found in the followers of Wittgenstein – is foreign to Frege. Reductionist remarks like Wittgenstein's "The thought *is* the meaningful sentence" (Der Gedanke ist der sinnvolle Satz)[7] are not to be found in Frege's works. In this respect, we must read him more as a follower of Leibniz and less as a forerunner of Wittgenstein. Thus, Frege can emphasize that essential matters are thinkable that do not need language in order to be thought. And it is clear at the same time that the Platonist in Frege regrets that human beings are not privy to this divine perspective.[8] In fact, Frege describes his preoccupation with language as a struggle with an opponent who denies him pure access to thought:

> "So one fights against language, and I am compelled to occupy myself with language although it is not my proper concern here."[9]

Frege thus becomes a philosopher of language out of necessity and not of his own free will. In this respect, the philosophy-of-language reading tries to make a virtue of Frege's necessity.

The struggle *with* language, which, for the reasons mentioned above, can only be fought *within* language itself, allows us to see the peculiarity of language as an antagonist. The actual goal is the construction of a logical language to serve a particular purpose: carrying out the epistemologically motivated logicist program. The additional philosophical benefit of this effort is not to develop a theory of language but to "break the power of the word over the human mind [Geist]".[10] The fact that Frege, despite his critical attitude towards language, deserves

[7] Tractatus 4, Emphasis G. G.
[8] Posthumous Writings, p. 269 [288].
[9] Collected Papers, p. 360, note 6 [Der Gedanke, 66].
[10] Conceptual Notation, p. 106 [VI].

to be regarded not only as the father of modern logic but also as the father of modern philosophy of language[11] has little to do with his own intentions. It results simply from his reception by subsequent authors. One cannot deny that Frege's investigations have important results which can be considered part of the philosophy of language. One should not, however, overlook the intention of the author's investigations, especially when these help to refute certain criticisms. To settle the controversy between the philosophy-of-language and the epistemological readings, one might offer the compromise of saying that Frege is a linguistic philosopher who methodically pursued aims within epistemology by using means drawn from logic and semantics.

The interpretation of the concept "sense" has a special "exemplary" significance in the controversy mentioned above. If we examine Frege's treatment of indexicals and *proper names,* we see that "sense" is rather an epistemological than a semantic category. Frege is concerned with assuring the identity or univocality of a thought. This alone makes it possible for thoughts to be the bearers of truth-values. Let us look at the following Fregean argument:

> "One could scarcely falsify the sense of the word "true" more mischievously than by including in it a reference to the subjects who judge. Someone will now no doubt object that the sentence "I am hungry" can be true for one person and false for another. The sentence, certainly – but not the thought; for the word "I" in the mouth of the other person denotes a different man, and hence the sentence uttered by the other person expresses a different thought."[12]

If "sense" were to be understood as "lexical meaning", as denoting a category in language-philosophical semantics, Frege could not argue in this way, for the lexical meaning of the utterance "I" is the same in all cases. What differs is the informational content: it changes with every speaker.[13] The epistemological primacy of (historical) proper names is even clearer. It is generally agreed that such names cannot be entered in a *semantic* lexicon at all.[14] Their entry in an *encyclopedic* lexicon serves only to record information about the name-bearer, not to indicate lexi-

---

11 See M. Dummett, Frege's Distinction between Sense and Reference; in: Dummett, Truth and Other Enigmas, London 1978, pp. 116–144, p. 119.

12 Basic Laws of Arithmetic, p. 14 [vol. I, XVI f.].

13 One can assert that the usage of the term "sense" fluctuates in Frege (when he speaks of the "Gegebenheitsweise" of objects) only if one takes "sense" from the very beginning to be merely a language-philosophical category.

14 Here one finds only general remarks such as "John: male proper name".

cal synonymity (between the name and the description assigned to it). It seems to me that in the discussion initiated by Saul Kripke the question of whether proper names have a "sense" or not turns on conflicting views of "sense" as lexical meaning on one hand and propositional or informational content on the other. The second view – and only this view, for it alone holds that names have propositional content – is suitable for securing Frege against Kripke's objections. The "price" is giving up a strong language-theoretical interpretation of "sense" in favor of an epistemological one. One cannot have it both ways: it is impossible simultaneously to hold that proper names have a "sense" and to maintain a language-theoretical interpretation of this concept.

Having made these basic points clear, I would now like to discuss the relationship between epistemology and logic. This relationship is especially important since Frege, according to his own logicist program, also considers his foundations of arithmetic to be a part of logic.[15] In any case, logic is the basis for deciding questions about the "epistemological nature" of the sciences.[16]

Today, such questions would be considered to belong to the philosophy of science. Frege's epistemology is not exhausted by such questions: in marginal comments and in the later work "Thoughts", he also treats topics like the reality of the external world.[17] In what follows, I will limit my remarks to his theory of scientific knowledge. I hope to show that it turns out to be a theory of proving. In this "proof theory" Frege is concerned with justified judgments about the provability of

---

[15] What about Frege's semantics? There are two components: (1) the distinction between sense and reference (meaning), which is motivated by epistemology; (2) the categorial semantics of such concepts as "negation", "generality", etc., which are part of the content of logic. We must remember that Frege's logic has content and thus cannnot be exhausted by the formal calculus.

[16] M. Schirn takes the following view: The goal of proving arithmetic truths from primitive logical laws and definitions alone is a matter for logicians and not epistemologists (Logik, Mathematik und Erkenntnistheorie. Archiv für Geschichte der Philosophie 69 (1987), S. 92–109, S. 92). I maintain on the other hand that proving arithmetical theorems is a matter for logicistic mathematicians, who coincide with logicians at this point. The aim of showing analyticity, on the other hand, is an epistemological one. In the same way, it seems incorrect to me to deduce from the role which the logical chains of conclusion play in proofs in the sciences, that the proofs themselves fall within the competence of the logician. Schirn claims this for geometry on p. 96. Here the formulation of logical laws is confused with the use of them to give proofs.

[17] See G. Prauss, Freges Beitrag zur Erkenntnistheorie. Überlegungen zu seinem Aufsatz 'Der Gedanke'. Allgemeine Zeitschrift für Philosophie 1 (1976), Heft 1, pp. 34–61. In the following section, I will attempt to show that the fluctuation between logical and psychological epistemology in Frege, as seen by Prauss, is only apparent.

scientific judgments.[18] The essential distinction in my discussion is that between justification and proof.

Logic and Epistemology

The relationship between logic and epistemology is shaped by Frege's limitation of logic to the "laws of truth". This results in a negative limitation in which no distinctions are made in the non-logical "remainder". Thus, theory of knowledge, psychology of knowledge and traditional modal logic are lumped together to await further distinctions. At this point, it is especially helpful to compare Frege's remarks on the modalities of judgment in the "Conceptual Notation" with his interpretations of "analytic", "synthetic", "a priori" and "a posteriori" in the "Foundations of Arithmetic".

That the distinctions of traditional modal logic are part of epistemology can be seen from the fact that apodictic and problematic judgments are held to make reference to a knowing subject. Frege follows Kant and Lotze in this respect.[19] He begins with Kant's view that modality "does not contribute to the content of the judgment"[20] but determines "the relationship of an entire judgment to the capacity for knowledge".[21] In Kant, the capacity for knowledge is understood as transcendental consciousness. It cannot be interpreted empirically or psychologically. Frege's view is not so univocal. The subject seems to be the particular judging subject. This is shown by the following remark:

> "If *I* call a proposition necessary, *I* thereby give a hint about *my* grounds for judgment."[22]

Such a formulation will not do for modal-logic distinctions that are to be independent of *individual* subjects and thus at least intersubjective. Frege's excluding modalities from logic in the narrow sense is not as problematic as the subjectifying of modality, for this would not even permit us to assign modality to an epistemology that differs from the psychology of knowledge.

---

[18] Proof theory is not meant here in the sense of D. Hilbert.

[19] For details see my Introduction in R. H. Lotze, Logik. Erstes Buch. Vom Denken, Hamburg 1989, pp. XXVI–XXXV.

[20] Kritik der reinen Vernunft B 100.

[21] Logik, § 30; Kritik der reinen Vernunft B 266.

[22] Conceptual Notation, § 4, emphasis G. G.

To keep our terminology straight, we must remember that in the 19th century logic and epistemology were not separate disciplines. Such a division was only intimated within logic in the distinction between the theory of elements ("Elementarlehre") and the theory of method ("Methodenlehre"). Frege is primarily interested in drawing a line between epistemology and logic in order to protect logic from intrusions by psychology. Epistemology is introduced to a certain extent as a buffer zone between logic and psychology. Distinguishing sharply between epistemology and the psychology of knowledge does not seem to have been particularly important to Frege. This is the only explanation for the fact that in other cases extremely different things are lumped together without distinction. Consider what falls under the collective concept "colouring". We must also remember that, in Frege's day, epistemology had not yet established itself as a non-psychological discipline. In this respect, Frege had occasion to emphasize his agreement with the neo-Kantian, H. Cohen, "that knowledge as a psychic process does not form [constitute] the object of the theory of knowledge, and hence, that psychology is to be sharply distinguished from the theory of knowledge".[23]

Despite some formulations which might give the impression that Frege wanted to leave distinctions of modal logic to the psychology of knowledge, one should not overlook the fact that he speaks of "*grounds* for judgment" in this context. Whoever makes an *apodictic judgment* by taking the judgment *J* to be necessary, "suggests the existence of general judgments" from which *J* "can be inferred".[24] In other words, the apodictic judgment makes a metajudgment about *J*. This metajudgment can only be considered justified when there are, in fact, general judgments from which *J* follows. Thus the justification for the necessity of *J* consists in a proof of *J* from general judgments.[25] When Frege says of the main variant of the problematic judgment that the speaker maintains "that *he knows* no laws from which the negation would follow",[26]

---

[23] Collected Papers, p. 111 [Rezension von H. Cohen, Das Prinzip der Infinitesimal-Methode und seine Geschichte, 329].

[24] Conceptual Notation, § 4.

[25] Frege characterises the apodictic metajudgment not as a judgment but as a "hint". Thus Frege cannot turn here to an expression like "metajudgment" because he limits the usage of "judgment" to cases where the judgment-stroke is in use. In Frege's formalism, a metajudgment cannot be represented as a judgment. Implicitly, this leads to a narrowing of the concept of truth (see below).

[26] Conceptual Notation, § 4, emphasis G. G. Here we have problematic judgment as subjectively coloured judgment. At this point, I will not discuss the interpretation of the judgment of possibility in the sense of the particular judgment. See my Introduction to R. H. Lotze, Logik. Vom Denken, Hamburg 1989, p. XXIX.

this metajudgment can only count as justified when there actually are no laws from which the negation of the judgment in question would follow.

It should not be controversial that, following Frege's line, the traditional apodictic and problematic judgments are to be interpreted as epistemic metajudgments. The question is whether the epistemic character refers merely to subjective *propositional attitudes* or to objective *proof-theoretical* contexts which refer not to subjective but to ideal knowledge.[27] The course Frege chose for his epistemology speaks for the latter reading. Although this proof-theoretical thought is first fully manifest in "Foundations", it is suggested in "Conceptual Notation". Thus (following Leibniz and Kant) Frege distinguishes clearly between the questions of proof (concerning validity) and psychology (concerning genesis). He continues:

> "The firmest method of proof is obviously the purely logical one, which, disregarding the particular characteristics of things, is based soley upon the laws on which all knowledge rests. Accordingly, we divide all truths which require a proof [die einer Begründung bedürfen] into two kinds: the proof [Beweis] of the first kind can proceed purely logically, while that of the second kind must be supported by empirical facts."[28]

This passage is not entirely easy to understand and has kept interpreters busy. It seems that Frege makes a distinction between logical and empirical truths, and that he advocates the Leibnizian dichotomy between a priori truths of reason and a posteriori truths of fact without regard for the Kantian distinction between the analytic and synthetic a priori. In particular, there seems to be no room for the synthetic a priori truths

---

[27] The use of personal pronouns does not necessarily imply a subjectification. In his Correspondence (p. 182 [118]), Frege writes: "Only after a thought has been recognized *by me* as true, can it be a premise *for me*" (emphasis G. G.). On page 78 f. [126 f.] we find the supplementary remark that an assertion is to be understood speaker-independently. This indicates that Frege does assign "making a judgment" as the act "of a knowing subject" to psychology (Posthumous Writings, p. 253 [273]) but not the judgment itself as an intersubjectively accessible representation of this act in conceptual notation. This is also the way to see modal-logical relationships between judgments. For this reason, I must reject the interpretation of P. Kitcher (Frege's Epistemology, pp. 242 f.) that, for Frege, knowledge is a psychological condition that is *generated* by proof. It is in fact the other way round: proof guarantees knowledge. The claim to knowledge is justified when a proof for what is known can be *redeemed*. The final sentence in the review of Cohen (quoted above) also speaks against the psychologized interpretation of epistemological distinctions.

[28] Conceptual Notation, Preface, p. 103 [III].

of geometry in this view. On the other hand, the "intuition a priori" is expressly mentioned at a later point in the text.[29] One might try to resolve this minor inconsistency by taking "purely logical" proofs to be those which use only logical rules, while those which make reference to experiential facts employ inductive reasoning. Geometry would now fall under Frege's distinction because the truths in need of proof are established logically from truths which are not themselves logical in nature. These are the "axioms" which (following the above citation) do not themselves "require a proof". Thus we would end up with the following epistemological distinctions between the academic disciplines: The truths of arithmetic are logical through and through, for they can be proved logically from logical laws; the truths of geometry are synthetic a priori because they can be proved logically from axioms of pure intuition; and the truths of natural science would be synthetic a posteriori (except for a conceivable synthetic a priori element of a pure law of nature), to the extent that they must be supported with the non-logical reasoning of induction. This interpretation assumes, however, that not only logical but also inductive reasoning can *prove* claims. A point against this interpretation is that Frege recognizes induction as a method for justifying truths.[30] This is not justification in the sense of proving, only of confirming ("Bewähren").[31]

We cannot escape the fact that, for Frege, proofs are exclusively deductions (or chains of deductions) conforming to logical rules. These deductions differ from one another only in the type of premises. A "purely logical" proof, with which we are concerned here, is distinguished by the fact that its premises are of a purely logical nature. In Frege's classification, our only choice is to regard geometrical truths as facts of experience. This is the interpretation favored by P. E. B. Jourdain. Frege makes the following clarifying remark: "the truths of geometry, in particular the axioms, are not facts of experience, at least if by that is meant that they are founded on sense-perceptions".[32] Even if we want to see this remark as an attempt to classify the geometrical

[29] Conceptual Notation, § 23.

[30] Foundations of Arithmetic, § 3, second note; cf. § 47.

[31] Foundations of Arithmetic, § 2: "But it is in the nature of mathematics always to prefer proof (Beweis), where proof is possible, to any confirmation (Bewährung) by induction." This formulation is terminologically incompatible with the recognition of induction as a method of *proof*.

[32] Correspondence, p. 183, note 10.

truths as *general* truths of experience, we will have to leave matters with the remark that he expressed himself very imprecisely.[33]

Our conclusions thus far show that Frege provides proof-theoretical criteria for decisions about the epistemic nature of truth and entire sciences. The common denominator of these criteria is deductive logic. By means of the gap-free chain of deductions ("Lückenlosigkeit der Schlußketten"), Frege hopes to find the exact conditions under which a proposition is true. These criteria are further differentiated in the "Foundations of Arithmetic". Here Frege emphasizes the epistemological significance of striving for proof in the following words:

> "The aim of proof is, in fact, not merely to place the truth of a proposition beyond all doubt, but also to afford us insight into the dependence of truths upon one another."[34]

"Rigorous proof", that is, logical deduction, enables us to become acquainted with the "conditions of validity" of a judgment. The expressions "a priori", "a posteriori", "synthetic" and "analytic" were defined by Frege as proof-theoretical meta-predicates for judgments and entire sciences. These should show "the ultimate ground upon which rests the justification [Berechtigung] for holding it to be true".

> "The problem becomes, in fact, that of finding the proof of the proposition, and of following it up right back to the primitive truths. If, in carrying out this process, we come only on general logical laws and on definitions, then the truth is an analytic one, bearing in mind that we must take account also of all propositions upon which the admissibility of any of the definitions depends. If however, it is impossible to give the proof without making use of truths which are not of a general logical nature, but belong to the sphere of some special science, then the proposition is a synthetic one. For a truth to be a

---

[33] Frege's view of geometry is not so unequivocal. Although he eliminates sense perception as a source of geometrical knowledge, he does not seem to be a decisive apriorist in the sense of Kant. His early remark in the doctoral dissertation, "that the whole geometry rests ultimately on axioms which derive their validity from the nature of *our* intuitive faculty" (Collected Papers, p. 1 [3], emphasis G. G.) does not eliminate the view that the axioms of geometry describe an anthropological fact and, in this sense, would merely have to be considered relative to the "facts of experience". Geometry thus remains a science of *general* facts and this circumstance would (in Frege's terminology) distinguish it from a posteriori sciences. The latter make reference to "truths which cannot be proved and are *not general*" (Foundations of Arithmetic, § 3, emphasis G. G.) In any case, "logical laws of inference" also find application in geometry (Posthumous Writings, p. 168 [183]. See Dummett on the difficulties with Frege's view: Frege and Kant on Geometry. Inquiry 25 (1982), pp. 233–254.

[34] Foundations of Arithmetic, § 2. Cf. Posthumous Writings, p. 204 [220 f.] concerning "proof".

> posteriori, it must be impossible to construct a proof of it without including an appeal to facts, i. e., to truths which cannot be proved and are not general, since they contain assertions about particular objects. But if, on the contrary, its proof can be derived exclusively from general laws, which themselves neither need nor admit of proof [die selber eines Beweises weder fähig noch bedürftig sind], then the truth is a priori."[35]

Especially the last sentence raises the question of the relationship between Frege's definition of "a priori" in the "Foundations" and "apodictic" in "Conceptual Notation". As proof-theoretical meta-predicates, they would seem to coincide. In apodictic judgments, too, the provability of a judgment from "general judgments" is claimed, and here we may understand general judgments as "laws", especially because Frege uses this expression himself in his analogous definition of the problematic judgment. Frege does not seem to want to distinguish between general judgments and laws (for example, natural laws).[36] In the formalism of his conceptual notation, he cannot express this difference, nor even the difference between factual generality and "causal connections".[37] Frege certainly underestimated this distinction. He seems to want to explicate the causal "because" by the logical "if-then".[38] A distinction between the various types or grades of generality can only be found outside of logic. It is based on precisely these proof-theoretical distinctions. In his definition of the term "a priori" (cf. the above citation), Frege not only speaks of "generality" and "laws" as in "Conceptual Notation", but also of "general laws" which "neither need nor admit of proof".

As a hermeneutic aid to understanding Frege's distinction between "apodictic" and "a priori" we might turn to a parallel passage in Lotze's "Logic". This alternative remains attractive despite the complaints of some interpreters that Frege should not "be hauled back into Lotzean fields".[39]

> "Every proof is a syllogism, or a chain of syllogisms, which completes the premises required for the given proposition *T*, so that they

---

[35] Foundations of Arithmetic, § 3.

[36] Frege simply glosses the "character of a law" as "generality of content" in other remarks (Correspondence, p. 69 [103]).

[37] Conceptual Notation, § 12.

[38] See his analysis of the sentence "because ice is less dense than water, it floats on water" in "On Sense and Meaning" (Collected Papers, p. 175 [48]).

[39] J. Schulte subscribes to this view in the afterword of his new edition of Frege's "Grundlagen der Arithmetik" (Stuttgart 1987, p. 147).

> fit into one another in such a way that *T* follows as their *necessary consequence* [emphasis G. G.]. But the validity of every conclusion depends upon the validity of its premises: these again might be established by fresh proofs, but this procedure would go on *ad infinitum* without any result were there not a number of universal propositions which we accept as immediate truths, which therefore neither need nor are capable of proof [eines Beweises weder bedürftig noch fähig sind], but are themselves the ultimate grounds by appeal to which we may decide in every case whether a conclusion is correctly or incorrectly drawn from its premises."[40]

In Frege's apodictic judgment, it is maintained that a "necessary consequence" (Lotze) of the specified premises obtains. Apodicticity only expresses a *relative* necessity, namely, necessity relative to general premises which might be purely factual. Frege speaks of a priori knowledge, as we have seen, only when the provability of the premises can be traced back to such general propositions "which themselves neither need nor admit of proof [die eines Beweises weder fähig noch bedürftig sind]".[41] Among these are the axioms of intuition, the basic laws of logic and the laws of nature.[42] The proof-theoretical difference between the meta-predicates "apodictic" and "a priori" is that the former expresses "relative" provability and the latter "absolute" provability.[43] But how "absolute" is this a priori? Can one penetrate beyond it in a foundational way? Lotze has a method for testing the certainty of evidence. Does something like this exist for Frege?

Proof ("Beweis") and Justification ("Rechtfertigung", "Begründung")

We need to investigate the relationship between proof and justification. We have seen from the example of induction that not every

---

[40] R. H. Lotze, Logik, Leipzig 1874, second ed. 1880, Engl. transl. ed. B. Bosanquet, 2 vol., Oxford 1888, repr. New York & London 1980, § 200.

[41] See our citation of Foundations of Arithmetic, § 3. Frege's formulation deviates from Lotze's (in German) only in the sequence of the expressions "fähig" and "bedürftig". The view itself is not original, dating back at least to Aristotle. A. B. Levison has already remarked on Frege's conservative view of proof: Frege on Proof. Philosophy and Phenomenological Research 22 (1961/62), pp. 40–49, p. 43.

[42] Frege indicates in the footnote to this passage (§ 3) that there is also an a priori element in natural science.

[43] See also L. Haaparanta, Frege and His German Contemporaries on Alethic Modalities; in: S. Knuuttila (ed.), Modern Modalities, Dordrecht 1988, pp. 239–274. She subscribes (p. 253) to the subjectivist reading ("private grounds of judgment") for modal-logical concepts while taking an objectivist line for the epistemological ones

justification of a judgment must take the form of a proof. In Frege's terminology, there are no inductive proofs, only inductive justifications.[44] Are there other types of proofs? Frege does not seem to want to acknowledge any. He views indirect proof as a direct proof that carries along the assumption of proof as a condition without actually assuming or claiming it. Such moves are forbidden in Frege's view, because proofs can only be found using true, acknowledged premises.[45]

Frege retains the view that proofs are limited to logical forms of deduction even after giving up his logicist program. Conceivable non-logical methods of proof in arithmetic, such as complete induction, are assigned the status of basic laws.[46] But – if there are no other types of proof – aren't there at least other types of justification? Frege expressly acknowledges the possibility of non-proving justification and assigns the task to epistemology.[47] The most general propositions outside of logic are just as incapable of logical *proof* as the basic laws of logic and the rules of deduction themselves.[48] At this point we must rely on conceivable, *non-logical,* that is, epistemological, efforts at justification. Nonetheless, Frege does not generally demand such a final justification. Frege concedes to epistemology that there must be judgments whose justification rests on something besides logical conclusions, if (and here he restricts his remark) these judgments "stand in need of justification at all".[49]

The distinction between logic and epistemology is marked in the later works by the terminological difference between a "reason for something's being true" and a "reason for our taking something to be true".[50] Frege's use of the personal pronoun in the second expression

("ideal and 'ultimate' grounds of judgment"). I defend an objectivist reading in both cases and see the crucial difference in how far back the justification is pursued. I should also like to take issue with the view that the distinction between "a priori" and "a posteriori" is "a genuinely logical distinction". Rather it (also) serves to *evaluate* logic, but with an epistemological intent.

44 Foundations of Arithmetic, § 3, second note.

45 Posthumous Writings, pp. 245 f. [264 ff.].

46 Posthumous Writings, pp. 203f. [219 f.].

47 Posthumous Writings, p. 3 [3]. Cf. p. 175 [190], key sentence 13.

48 The same is also true of singular facts characterized as "truths which cannot be proved and are not general" (Foundations of Arithmetic, § 3). For this reason, the implicit limitation of G. Currie, that epistemology tells us "how to ground our *first* premises in a noninferential way" is too narrow (Remarks on Frege's Conception of Inference. Notre Dame Journal of Formal Logic 28 (1987), pp. 55–68, p. 57, emphasis G. G.).

49 Posthumous Writings, p. 3 [3].

50 Basic Laws of Arithmetic, p. 15 [vol. I, XVII].

clearly indicates the reference to a knowing subject. Here, Frege is following Kant's epistemic interpretation.[51] Reasons for taking something to be true may be reasons for a knowing subject, but they are still reasons. One should not make the mistake of considering these different types of reasons – and these are types of argument or of justification – to be analogous to his laws of truth and laws of takings-to-be-true.[52] Laws of takings-to-be-true are, unlike the validity laws of truth, mere empirical and psychological laws about the genesis of truths. In this case it might appear as though reasons for our taking something to be true are mere causes of subjectively being convinced. These might affect one person differently from the next. But causally subjectifying the reasons for our taking things to be true is not at all what Frege is aiming at. His distinction of logical and non-logical forms of justification succeeds in separating epistemology from logic in the desired way without undermining epistemology's efforts at justification. When Frege writes, "it is not the holding something to be true [das Führwahrhalten] that concerns us but the laws of truth [die Gesetze des Wahrseins]", he is only rejecting references to "how we actually think or arrive at our convictions", to the psychological causes of our taking something to be true. He does not refrain from making references to the justifying grounds for taking something to be true.[53]

For Frege, logical justification coincides with deductive inference, understood as "proving" truths from truths. When he calls the reasons in a non-logical justification "reasons for taking something to be true", he states that here the judgment does not stand in a deductive relationship to the reason, i. e., he means that the judgment is not grounded in a deductive chain of reasoning back to a truth. Sometimes he even seems to go so far as not to accept non-logical reasons as truths *because* they are not capable of deductive order. When he tries to distinguish epistemology from logic by claiming that logic is concerned only with such grounds of judgment "which are truths", this only makes sense if epistemology is concerned with grounds which are not truths.[54] This

---

[51] Cf. Kritik der reinen Vernunft B 848.

[52] Basic Laws of Arithmetic, p. 13 [vol. I, XVI].

[53] Posthumous Writings, p. 145 [157] . On the distinction between causes which "actually produce" conviction and reasons which "justify" conviction, see p. 147 [159]. We can say with G. Currie that justification is "subject-involving but not subjective" (Remarks on Frege's Conception of Inference, p. 62).

[54] Posthumous Writings, p. 3 [3]. Cf. the distinction between "grounds of judgment" and "grounds of belief" in Wittgenstein's "On Certainty".

way of expressing things clearly seems less than sensible because Frege does not treat epistemological questions as pseudo-questions. And so in other passages we also find the distinction between true and false in cases of classical epistemological arguments:

> "Therefore the thesis that only what belongs to the content of my consciousness can be the object of my awareness, of my thought, is false."[55]

Frege persistently maintains the distinction between justification and proof. Thus he avoids speaking of a *proof* of the logical nature of arithmetic. In the "Foundations", he only hopes "to have made [this] probable".[56] This reserved judgment cannot be explained simply by noting that Frege did not fulfill his demands of "utmost rigour"[57] in keeping "the chain of reasoning [Schlußkette] free of gaps"[58] until he united "Conceptual Notation" and "Foundations" in the "Basic Laws of Arithmetic". He is not concerned with a "proof" here. The passages remain expressed in the tone of epistemology. While for Frege the sense of the word "true" cannot include "reference to the subjects who judge",[59] this relationship plays a role in his arguments about the relationship between logic and arithmetic. It is essential that here in the epistemic formulations (such as "can we be persuaded [kann man sich überzeugen] that the root of the matter is logic alone") a frame of justification is specified: "a basis upon which to judge the epistemological nature" of arithmetical laws.[60] Reasons are not grounds of proof or truth: they are grounds of judgment ("Beurteilung"), grounds for "our taking something to be true".[61] Provability cannot itself be proved, it can only be justified. The distinction between proof and justification is of interest for logic itself.

> "The question why and with what right we acknowledge a law of logic to be true, logic can answer only by reducing it to another law of logic. Where that is not possible, logic can give no answer. If we

[55] Collected Papers, p. 366 [Der Gedanke, 72]. Whether Frege would accept such statements as "judgments" seems to be questionable.

[56] See Foundations of Arithmetic, § 87 and Posthumous Writings, pp. 37 f. [42] on the probability of the completeness of the basic laws.

[57] Foundations of Arithmetic, § 4.

[58] Conceptual Notation, p. 104 [IV].

[59] Basic Laws of Arithmetic, p. 14 [vol. I, XVI].

[60] Basic Laws of Arithmetic, p. 3 [VII].

[61] See also the conclusion to the foreword of Basic Laws of Arithmetic (p. 25 [XXVI]) and its epistemic vocabulary ("convictions [Überzeugungen]", "improbable [unwahrscheinlich]").

step away from logic [i. e., if we move into epistemology, G. G.], we may say: we are compelled to make judgments by our own nature and by external circumstances; and if we do so, we cannot reject this law – of Identity, for example; we must acknowledge it unless we wish to reduce our thought to confusion and finally renounce all judgment whatever. I shall neither dispute nor support this view; I shall merely remark that what we have here is not a logical consequence [logische Folgerung]. What is given is not a reason for something's being true [Grund des Wahrseins], but for our taking it to be true [Führwahrhaltens]."[62]

Let me make a few concluding remarks: We have seen that Frege recognizes non-logical reasons as reasons, and thereby acknowledges epistemology as an argumentative basic discipline which is to be distinguished from the psychology of knowledge. Non-logical justifications are examined and applied in epistemology, for example, when answering questions about the proof-theoretical status of certain sciences. Distinctions are made between grounds of truth (logic), reasons for our taking something to be true (epistemology), and causes for our taking things to be true (psychology of knowledge). If we move from the type of justification to the subject matter of the disciplines, Frege seems content with a blurred distinction between logic and epistemology. This results from the fact that, as is commonly the case, he uses the term

---

[62] Basic Laws of Arithmetic, p. 15 [vol. I, XVII]. M. Dummett (Frege and Kant on Geometry, p. 237) interprets this passage as follows: "no justification whatever can be given for accepting those laws of logic which cannot be derived from other laws." The opposite is true: this passage is itself a justification, namely an *epistemological* one, based on reasons for taking things to be true, for a transcendental turn of mind. See G. Gabriel, Frege als Neukantianer. Kant-Studien 77 (1986), pp. 84–101, pp. 92 f.; G. de Pierris, who at least considers transcendentalism in Frege, comes to the same conclusion as Dummett: that an epistemological justification of logical basic laws is out of the question for Frege (Frege and Kant on A Priori Knowledge. Synthese 77 (1988), pp. 285–319, pp. 305, 310). However, she does not seem to correctly apprehend the difference between proof and justification. Frege does not consider basic logical laws to be provable. It is striking, on the other hand, that de Pierris speaks of "proof" where only "justification" is appropriate: "to prove that arithmetic is analytic" (p. 291), "be proved that it is a priori" (p. 308), "to give a proof that we are justified in believing [the laws of logic]" (p. 310). The distinction between proof and justification is also missing in E. Picardi, Assertion and Assertion Sign. Atti del Convegno Internazionale di Storia della Logica: Le teorie delle modalità (1987), Bologna 1989, pp. 142 f. By obtaining a proof using certain laws of logic I have a justification that the proven sentences are logical, a priori or synthetic a priori, but not a proof of it. One must first see that Frege's own concept of proof is so narrow that proofs only occur in his own works in the derivations in conceptual notation in "Conceptual Notation" and in "Basic Laws of Arithmetic". One has to wonder how to categorize the philosophical "remainder". In my view, the only way to view this remainder is as an effort at "argumentative" justification.

"logic" in two ways. One is the system of laws of truth, the other is the discipline in which these laws are worked out.[63] Logic in this second sense is part of philosophy, and here the distinction between logic and epistemology in the Fregean sense is not possible at all. For discussing logic (1) within logic (2) requires us to use arguments which are not proving arguments in the sense of logic (1). These are, according to Frege's own definition, epistemological arguments.[64] In logic (2), we are compelled to go so far beyond logic (1) that sometimes we must violate categorial distinctions in logic (1) in order to clarify them.[65] For philosophy, this means that there are no proofs within the discipline itself. It consists merely of "elucidations [Erläuterungen]" and "hints [Winke]" which do not thereby lose their argumentative character.[66] As the final result of my considerations here, I would like to point out that this insight is already nascent in Frege, although it only matures in the writings of Wittgenstein.[67]

---

63 We find both uses alongside one another, for example, in the Posthumous Writings, p. 6 [6], where it is clear that Frege writes a Logic (2) in order to free Logic (1) from language-related shortcomings. Cf. as well p. 252 [272].

64 In "Thoughts [Der Gedanke]" Frege is concerned with traditional epistemological problems, too, like the question of reality, even though the work is part of his series of papers "*Logical* Investigations".

65 For the consequences of this ineffability of Logic (1) for the style of Logic (2) see my Der Logiker als Metaphoriker. Freges philosophische Rhetorik; in: G. Gabriel, Zwischen Logik und Literatur, Stuttgart 1991, pp. 65–88.

66 A critical comment on Frege's treatment of the question whether elucidations (of the "logical simples") and explicative definitions ("zerlegende Definitionen") have cognitive value can be found in my book Definitionen und Interessen, Stuttgart-Bad Cannstatt 1972, chap. 2.4. This question arises because the two are not part of the deductive system, which includes only stipulative definitions ("aufbauende Definitionen").

67 The consequences stated here concerning the role of proofs and epistemological arguments are similar to some of the theses of J. Weiner's Frege-book (Frege in Perspective, Ithaca/London 1990). I especially agree with her main point that Frege's philosophical method is essentially elucidative (pp. 274 ff.), but I obviously disagree with the following view: "The only *justification* Frege would recognize for a logical law is a *proof* from another logical law" (p. 227, emphasis G. G.). My point is that there is indeed justification of logical laws which is not proof, namely, *epistemological* justification.

Frege on Knowing the Third Realm[1]

TYLER BURGE

Anyone who reads Frege with moderate care is struck by a puzzle about the central objective of his work. His main project is to explain the foundations of arithmetic in such a way as to enable us to understand the nature of our knowledge of arithmetic. But he says very little about our knowledge of the foundations. A full treatment of this and associated puzzles would require more room than I have here.[2] I want to give a short solution to the puzzle, and then discuss one aspect of it that I find interesting.

The short solution is that Frege accepted the traditional rationalist account of knowledge of the relevant primitive truths, truths of logic. This account, which he associated with the Euclidean tradition, maintained that basic truths of geometry and logic are self-evident. Frege says on several occasions that such primitive truths – as well as basic rules of inference and certain relevant definitions – are self-evident. He did not develop these remarks because he thought they admitted little development. The interesting problems for him were finding and understanding the primitive truths, and showing how they, together with inference rules and definitions, could be used to derive the truths of arithmetic.

This short solution seems to me correct – as far as it goes. It does, however, leave out a lot. Frege thought that knowledge of the axioms of geometry required intuition – an imaginative or broadly perceptual ca-

Reprinted with the kind permission of Oxford University Press from *Mind* 101 (1992), 633–50.

[1] I am indebted to Tom Ricketts for clarifying his views, discussed in note 16. I have also benefited from remarks from various participants at a conference on early analytical philosophy held at the University of Chicago in honour of Leonard Linsky.

[2] An auxiliary puzzle attends this primary one. Most of Frege's philosophical work is directed at correcting what he regards as the misunderstandings embedded in normal practice and language – misunderstandings that he thought had prevented a correct understanding of the fundamental notions present in his account of the foundations. But he has even less to say about the epistemology of his analysis and elucidation of the notions that interested him than he does about knowledge of the foundations.

pacity (1968, pp. 19–21). Knowledge of the basic truths of logic simply required reason. He regarded both types of basic truths as self-evident, but the differences between the two types of knowledge are significant. That is one complication. Another is that Frege uses a variety of terms that are translated "self-evident". His sophisticated understanding of the notion is neither psychologistic nor purely proof-theoretic. He does not mean by it what most contemporary philosophers would mean by it. His uses of it relate in interesting ways to his basic philosophical views. A third complication is that there are complex relations between Frege's appeals to self-evidence and an appeal he makes to pragmatic epistemological considerations. This appeal makes his rationalism original and gives it, I think, special relevance to modern problems. Although these points are worth developing, I will not discuss them here. Instead, I shall discuss an intensification of the puzzle in the light of the short solution that I have just given.

Frege assumes that only truths are self-evident. He also assumes that it is rational to believe what is self-evident, given that it is well understood. Frege believes in other types of purely mathematical justification for arithmetical judgments besides self-evidence and derivation from self-evident truths.[3] But these other types also involve only reason. The key idea in what follows is that Frege assumes that we can know arithmetic and its foundations purely through reason, and that individuals are reasonable and justified in believing basic foundational truths (e. g. 1979, p. 175; 1983, p. 190).

Frege held that both the thought contents that constitute the proof-structure of mathematics and the subject matter of these thought contents (extensions, functions) exist. He also thought that these entities are non-spatial, non-temporal, causally inert, and independent for their existence and natures from any person's thinking them or thinking about them. Frege proposed a picturesque metaphor of thought contents as existing in a "third realm". This "realm" counted as "third" because it was comparable to but different from the realm of physical objects and the realm of mental entities. I think that Frege held, in the main body of his career, that not only thought contents, but numbers and functions were members of this third realm.[4] (Cf. 1968, p. viii;

[3] I distinguish purely mathematical justifications from justifications of mathematics that derive from applications to the empirical world – which he also seems to have believed in, but which I lay aside.

[4] Frege's logic is not committed to thought contents, only to extensions and functions. But this is an artifact not of his views about logic, but of his interests in deriving

1967a, pp. 15–16; 1962, p. xvii). Entities in the other realms depended for determinate identities on functions (concepts) in the third realm. Since logic was committed to this realm, and since all sciences contained logic, all sciences were committed to and were partly about elements of this realm. Broadly speaking, Frege was a Platonist about logical objects (like numbers and truth values), functions, and thought contents. I shall say more about Frege's Platonism later, but I think that I have said enough to enable me to introduce the problem that I want to discuss.

The problem is that of understanding how reason alone could justify one in believing that a thought is true, when the thought has a subject matter that is as independent of anyone's thinking as Frege indicates it is. How could mere reasoning give one any ground for believing that a realm of entities is one way rather than another, when that realm is so independent of that reasoning? How could reasoning and understanding have any tendency to tell one how things in such a realm really are?

This problem is clearly kin to a problem about the relation between the knowledge and truth of mathematics that is commonly discussed today.[5] The contemporary problem is that of understanding how our beliefs about mathematics could have any tendency to be true, given that we do not appear to bear causal-perceptual relations to the subject matter of mathematics. This may be seen as a problem for Frege. But it is not one that he would have naturally formulated for himself. His attitude toward the point that numbers and thought contents are not causally effective ("wirklich") seems to have been "so what?".[6] He showed no special interest in the causal theory of knowledge, or in cashing out his occasional physical-contact metaphors of "grasping" thoughts. The idea that mathematical or logical knowledge should be judged by reference to the standard of empirical knowledge would have seemed foreign to him.

---

arithmetic from logic. For that, he did not need to refer to thought contents (*Gedanken*). But he clearly envisioned a logic which was committed to thought contents. In the correspondence with Russell, for example, he indicates the need for special names of senses to avoid the "ambiguity" of indirect discourse or propositional attitude attributions (cf. 1980, p. 153; 1976, p. 236).

[5] Benacerraf (1983).

[6] Actually, he does provide an argument: objective sense-perception requires perceptual belief; but perceptual belief requires grasp of thoughts in the third realm – a non-causal relation; so one cannot cite the element of causal interaction in sense perception as providing grounds for thinking that knowledge cannot involve non-causal cognitive relations to abstract entities (1984, pp. 369–370; 1967b, p. 360).

Like Frege I see no reason to think that mathematical or logical knowledge is questionable because it apparently lacks causal-perceptual relations to its subject matter. But I formulated a problem that made no reference to causal-perceptual relations. This formulation seems not to import assumptions foreign to Frege. A theory of knowledge should not make it puzzling how being reasonable could be conducive to having true beliefs. Frege's rationalist theory of knowledge combines with his Platonism to raise a question at just this point. Why did he not discuss the question?

Some recent interpretations of Frege suggest that it is a question that is somehow precluded by his philosophy, or that it rests on fundamental misreadings of his views. One might question the notion of "subject matter" that the formulation of the problem uses. Or one might claim that Frege's notion of truth or of logic blocks a "meta-standpoint" from which one could raise the question. Or one could doubt whether Frege's Platonism should be understood in the way that the "third realm" metaphor suggests, and maintain that in talking about numbers or thought contents, Frege was really talking about our language or our cognitive practices in such a way that no gap between our beliefs and the numbers was even formulable. I will not criticize in detail all such lines for short-circuiting our question for Frege, though I will remark on some of them in a general way. I think none provides good grounds for ignoring the question. In fact, Frege himself gives an answer to it. The reason why he did not discuss it in detail is similar to the reason why he did not discuss knowledge of the foundations in detail. He believed that he had little to add to a traditional answer. I think that his answer is worth understanding.

Let us back up a bit. I want to explain in more detail what I mean by saying that Frege was a Platonist about logical objects, functions, and thought contents.

First, some preliminary disclaimers. Although I think that Frege maintained a metaphysical view about numbers and other such entities, I do not believe that this view dominated his thinking. His is, for the most part, the relaxed Platonism of a mathematician who simply assumes that there are numbers, functions, and so on, and who regards these as an abstract subject matter which can be accepted without special philosophical explanation, which is clearly different from mental or physical subject matters, and which mathematics seeks to characterize correctly. One can see this attitude toward functions very prominently in "On Function and Concept". Frege highlighted the inter-subjective objectivity of scientific theorizing. He believed that standard mathematical practice told one most of what was true about mathematical entities, and he thought that one could know mathematical truths independently of any philosophy. Indeed, he assumes that ordinary mathematical practice yields "certain" knowledge even prior to the execution of his foundationalist program (1977, § 13; 1968, § 2).

Most of Frege's uses of his metaphysical view are defensive. His metaphysical remarks ward off idealist, physicalistic, psychologistic, reductive, or deflationary positions because he thinks that they prevent clear understanding of the fundamental notions of logic and arithmetic. As I shall later show, he does give his Platonism extra-mathematical work. But he does not think out this side of his philosophy as someone would who was concerned about certainty or who believed that logic and mathematics had no other cognitive underpinning than that provided by philosophy.

Another preliminary point about Frege's Platonism is that although he uses the Platonic metaphor of vision on occasion, when characterizing our knowledge, he shows no interest in developing the metaphor. He appeals to no faculty other than reason in his account of our mathematical knowledge. Moreover, as I have intimated earlier, his epistemological views are complex, and involve not only Platonic elements, but elements not at all associated with traditional Platonism. The discussion in what immediately follows will be concerned with the Platonic character of Frege's *ontology*. For now, I lay epistemology aside.

As is well-known, Frege thought that extensions – including numbers – functions – including concepts – and thought contents are imperceptible, non-spatial, atemporal, and causally inert.[7] He emphasizes that numbers (1968, p. 108), concepts (1968, p. vii), and thought contents (1967a, p. 23; 1962, p. xxiv) are discovered – not created. He sharply

---

[7] Numbers are counted imperceptible (1968, p. 85; 1979, p. 265; 1983 p. 284). Thoughts are termed imperceptible (1984, p. 369; 1967b, p. 360). Numbers are counted non-spatial (1968, pp. 58, 61, 85, 93). Thoughts are counted non-spatial (1984, pp. 369–370; 1967b, p. 360). Concepts or other functions are counted atemporal and by implication imperceptible, non-spatial, and causally inert (1968, p. vii, 1968, p. 37; 1984, p. 133; 1967b, p. 122). He also suggests these points about concepts indirectly (1967a, p. 23; 1962, p. xxiv; 1984, p. 198; 1967b, pp. 181–182). Numbers are counted atemporal (1984, p. 230; 1967b, p. 212). Thoughts are counted atemporal (1984, pp. 369–370; 1967b, p. 360). Numbers are counted causally inert (1968, p. 85; 1967a, pp. 15–16; 1962, p. xviii). Thoughts are said to be causally inert (1967a, p. 23; 1962, p. xxiv; 1979, pp. 137–138; 1983, pp. 149–150; 1984, pp. 230, 371; 1967b, pp. 212, 361–362).

distinguishes the act of thinking, which does occur in time, from the thought contents that we "grasp" or think, which are timeless. So in coming to know thought contents that denote numbers, concepts, and the like, one discovers objects, concepts, and relations that are what they are timelessly, independently of any causal influence. One comes to "stand in relation", as Frege says, with non-spatial, atemporal entities. (1984, pp. 363, 369; 1967b, pp. 353–4,360; 1967a, p. 23; 1962, p. xxiv.)

Frege calls numbers, concepts, and thought contents "objective". By this he means, partly, that they are not intrinsically borne by a mind, as a pain or an after image is. He says that they are subject to laws. They are common property to different rational beings (1968, p. 26; 1984, pp. 363 ff.; 1967b, pp. 355 ff.). Much of Frege's discussion of atemporal entities centres on their objectivity. For many of his purposes, the inter-subjectivity and lawfulness of logic are its key properties.

Many of these things might be maintained by someone who was not a Platonist. One might make the remarks about imperceptibility, non-spatiality, atemporality, and causal inertness, if one glossed them as part of a practical recommendation or stipulation for a theoretical framework, having no cognitive import – or as otherwise not being theoretical claims or claims of reason. Carnap might have said at least some of those things, though only given certain background qualifications. Or one might have some other basis for qualifying these remarks, reading them as "non-metaphysical" or as lacking their apparent ontological import. Moreover, certain idealists might say these things. Kant might have said them, given certain background qualifications. He could have seen numbers as just as genuinely existent and discoverable as physical objects are. And he could see their objective status in terms of the possibility of inter-subjective agreement on laws governing them. Platonism has no monopoly on claims to lawlike or inter-subjective objectivity about non-spatial, atemporal entities. So we need to say more in order to distinguish Frege's view from alternatives.

I would not take very seriously a reading of Frege as a Carnapian. Discussing my attitude would require going more into his methodology and epistemology than I plan to. I think it clear, however, that Frege was trying to provide a rational foundation for mathematics – in a way that Carnap would have regarded as misguided. Frege saw reason, not practical recommendation, as giving logical objects to us (e. g. 1968, p. 105). There is nothing remotely akin to Carnap's Principle of Tolerance either in Frege's philosophical pronouncements, or even more emphatically, in his temperament.

What interests me more is the distinction between Frege's Platonism, on the one hand, and certain idealisms or certain vaguer "practice" oriented anti-Platonisms on the other. Platonism, as I understand the doctrine, regards some entities (for Frege, some objects and all functions) as existing non-spatially and atemporally. Further, it avoids commenting on them as having special status, including being dependent for their existence or nature (as opposed to their discovery) on practice or mental activity. They are in *no way* derivative, instrumental, fictional, or otherwise second-class. The relevant entities are fundamental. It would be incompatible with Platonism to regard them as essentially part of an appearance or perspective for a thinker – as Kant would have – though they may impose constitutive conditions on such appearances or perspectives. Platonism rejects any deeper philosophical commentary that would indicate that the nature or existence of these atemporal entities is to be regarded as in any way dependent on something mental, linguistic, communal, or on anything like a practice or activity that occurs in time. In Kant, we find a non-Platonic explanation of mathematical structures in terms of a mental activity, "synthesis", that underlies the categories and the forms of spatial and temporal intuition. And in Hegel abstract structures are held to be abstractions from spirit in history. Recently, some philosophers have sought to avoid being "metaphysical", contenting themselves with generalized remarks that mathematical objects are grounded in some unspecified way in linguistic or mathematical practice. Such views can admit non-spatio-temporal entities and can grant them objective status. But they are not Platonic in my sense. They regard atemporal entities as derivative from human practices – such as linguistic activity. I see such views as covertly idealist. Idealism regards actual activity or practice as implicated in the nature and existence of non-spatio-temporal structures. Platonism holds that structure is more fundamental than actual activity.

Frege's Platonism shows itself in two ways. One is that he never enters the commentary that an idealist (or a deflationist) would enter on his claims about non-spatio-temporal entities, or about their objectivity or their discoverability. He takes them to be fundamental. The other is that he claims, more than once, that the assumption of the relevant entities explains the inter-subjective objectivity of science and communication. I will discuss these points briefly, in turn.

There is, as far as I can see, no evidence that Frege thought that the existence or nature of these non-spatio-temporal entities was to be explained in terms of human language, human inference, human practices (including the *activity* of judgment), or other patterns of human activity in time. Frege thought of extensions, functions, and thought contents as genuinely existing entities.[8] He opposed thinking of such entities as having some derivative status. He inveighs against any suggestion that they are products of the mind, mere symbols, or otherwise dependent on events in time.[9] Had he maintained that extensions, functions, or thought contents were dependent on human conceptualization or human language, judgment, or inference (actual or possible), he would have said so, and thereby qualified the numerous remarks that have traditionally invited the Platonic interpretation of his work. He never does say so. His claims that atemporal entities are independent of us are unqualified.

On several occasions, Frege compares the objectivity and existence of numbers, concepts, or thought contents with the existence and objectivity of physical objects. He compares numbers to the North Sea as regards objectivity (1968, p. 34). In doing so, he very explicitly indicates that the entity that we call "the North Sea" is what it is completely independently of our imposing the boundaries or making a map that we use to associate that entity with the name "the North Sea". He elaborates this comparison elsewhere:

> Just as the geographer does not create a sea when he draws boundary lines and says: the part of the ocean's surface bounded by these lines I am going to call the Yellow Sea, so too the mathematician cannot really create anything by his defining. Nor can one by pure definition magically conjure into a thing a property that in fact it does not possess – save that of now being called by the name with which one has named it. (1967a, p. 11; 1962, p. xiii)

He compares a mathematician's relation to numbers with the astronomer's relation to the sun (1979, p. 7; 1983, p. 7) and to the planets (1968, p. 37). He says that like geographers, mathematicians cannot create, but can only discover "what is there and give it a name" (1968, p. 108; cf. also 1967a, pp. 23–24; 1962, p. xxiv; 1979, p. 137; 1983,

---

[8] He quantified over them with quantifiers of different types. He used first-order quantifiers for the objects, second-order quantifiers for the functions. The quantifiers are appropriately read as involving existential commitments.

[9] For extensions and numbers, cf. 1968, passim, 1967a, pp. 10, 12; 1962, pp. xiii, xiv; 1967a, pp. 15–16; 1962, p. xviii; 1984, p. 230; 1967b, p. 212. For concepts or functions, cf. 1968, p. viii; 1984, p. 133; 1967b, p. 122. For thoughts, cf. 1984, pp. 363, 370; 1967b, pp. 353–354, 360–361.

p. 149). He compares our epistemic relation to numbers and concepts (and probably thought contents) to our grasping a pencil:

> The picture of grasping is very well suited to elucidate the matter. If I grasp a pencil, many different events take place in my body... but the pencil exists independently of them. And it is essential for grasping that something be there which is grasped... In the same way, that which we grasp with the mind also exists independently of this activity... and it is neither identical with the totality of these events nor created by it as a part of our own mental life. (1967a, pp. 23–24; 1962, p. xxiv; cf. 1979, p. 137; 1983, p. 149)

Thought contents exist independently of thinking "*in the same way*", he says, that a pencil exists independently of grasping it. (The artifactual character of pencils plays no role in his understanding of the analogy, as other examples indicate.) He says that thought contents are true and bear their relations to one another (and presumably to what they are about) independently of anyone's thinking these thought contents – "just as a planet, even before anyone saw it, was in interaction with other planets" (1984, p. 363; 1967b, p. 354). And he compares a thought's independence of our grasping it to the star Algol's independence of anyone's being aware of it (1984, p. 369; 1967b, p. 359).

All these comparisons suggest (and those of 1967a, pp. 23–24; 1962, p. xxiv; 1984, pp. 363, 369; 1967b, p. 354, 359 explicitly state) that numbers, functions and thought contents are independent of thinkers "in the same way" that physical objects are. Frege nowhere asserts or clearly implies that he maintains any sort of idealism – Kantian or otherwise – about the physical objects studied by the physical sciences. He nowhere qualifies the ontological status of physical objects. It is dubious historical methodology to attribute to a philosopher with writings that stretch over decades, a large, controversial doctrine, if he nowhere clearly states it in his writings. If Frege had believed in any such idealism about physical objects (or any doctrine qualifying their ontological status), he would have surely said he did.[10] Doing so would have

[10] Such passages as 1979, p. 137; 1983, p. 149, or any of the various passages about independence of mind that I discuss below, would require strong qualification, which Frege nowhere makes, to be compatible with any sort of idealism or deflationary reading. For an interpretation of Frege as a Kantian idealist, see Sluga, 1980 e. g. pp. 59–60, 115–116. Sluga cites mainly considerations that are external to Frege's texts. He also writes, "the central role of the Fregean belief in the primacy of judgments over concepts would seem to be explicable only in the context of a Kantian point of view". Sluga does not explain this remark. I think it misleading. Judgments and inferences are a source of discovery. But logical theory is about the

been necessary for a philosopher to balance the flat-out statements about mind-independence that Frege makes.[11]

Frege thought that to know the physical world, one has to grasp thoughts (which bore for him eternal denotational relations to concepts and extensions) that are eternal and eternally true. Logic is embedded in the content of any knowledge. Since logic is about (denotes) concepts and other functions, relations, and logical objects, all knowledge is at least partly about non-spatio-temporal entities. Moreover, logic concerns the forms of correct judgment and inference; and logical structure is discovered by reflecting on patterns of correct judgment and inference. But Frege does not give the slightest indication that he thought that either the physical world or the non-spatio-temporal entities inevitably appealed to in knowing it depend in any way on any activities of judgment, inference, or linguistic practice.[12]

---

forms of correct judgment and inference – not about judgments and inferences. Frege regards judgment as a form. (Cf. 1984, pp. 383–385; 1967b, pp. 372–374.) I know of no evidence that he saw this form as dependent for its nature or existence on actual activities of judgment, or on anything like Kantian synthesis; there is substantial evidence that he did not.

[11] Some philosophers have suggested that Frege's use of the context principle somehow suggests a qualification on his Platonism. Issues surrounding Frege's context principle(s) are, of course, extremely subtle and complex. But it seems to me that the suggestion must involve some confusion. The context principles govern relations between linguistic expressions and their senses or referents. They do not bear directly on the nature of the senses or referents themselves at all. At most one of the principles might be coherently thought to rule out certain naive forms of epistemological Platonism (those that require that we have perception-like intuition of mathematical objects). There are many complex issues here, and some of them are not completely independent of ontology. But I think that any simple appeal to the context principles to motivate opposition to my interpretation will confuse language and epistemology with ontology.

[12] An interpretation of Frege similar to Sluga's is proposed in Weiner (1990). In interpreting the North Sea comparison (1968, p. 34), Weiner notes Frege's remark that if we should happen to draw the boundaries of what we call "the North Sea" differently, what we now call "the North Sea" would still exist, though perhaps unrecognized. But she continues, "It is important to realize, however, that the claim that such unrecognized objects exist need not be a substantive metaphysical claim. For... to claim that unrecognized objects exist is simply to claim that it is possible to formulate (heretofore unformulated) concepts under which exactly one object falls" (p. 171). Weiner cites no texts to support this reading. I see no reason to think that existence claims for Frege are "simply" claims about possibility or about formulations; he gives every indication that they are not about possibility, language, or activity at all. Later she correctly claims that Frege believed our knowledge requires language or drawing boundaries. But she moves without argument from this remark about knowledge to one about the world: "Frege's view is that the physical world is not articulated – that we impose structure on it" (p. 184). The language of imposition is not present or implied in Frege. That concepts mark boundaries of the ocean is

Frege not only compares non-spatio-temporal entities to physical objects in their independence of us; he makes unqualified statements about the independence of such entities from anything about us. He repeatedly claims that both the truth of thought contents and thought contents themselves are independent of individuals' and groups' thinking the thoughts or recognizing them to be true (1967a, pp. 15, 23; 1962, pp. xvii, xxvi; 1968, p. 60; 1984, p. 363; 1967b, p. 354). He writes: "What we want to assert in using that proposition [that the number three is prime] is something that always was and always will be objectively true, quite independently of our waking or sleeping, life or death, and irrespective of whether there were or will be other beings who recognize or fail to recognize this truth." (1984, p. 134; 1967b, p. 123)

The lack of qualification in his claims of independence is especially striking in two passages: one where he writes that someone's thinking a thought has "nothing to do" either with its truth or with the thought content itself (1984, p. 368; 1967b, p. 359); and another where he writes that thought contents are not only true independently of our recognizing them to be true, but they, the thought contents themselves, are "absolutely independent of our thinking" (1979, p. 133; 1983, p. 145). Independence is independence. Frege's repeated remarks about mind-independence of non-spatio-temporal entities would not have been literally true, if they had been backed by a set of unstated qualifications of the sort that an idealist (or deflationary) interpretation of them would require. Ultimately the idealist asserts dependence of the thought-contents and timeless objects on some underlying practice or activity that makes possible the framework in which attributions of objectivity are

---

nowhere said to depend in any way on anything about language or people. (Similarly, with concepts demarcating possible numerations in such cases as packs of cards.) Frege writes: "To bring an object under a concept is merely to recognize a relation that already existed beforehand" (1984, p. 198; 1967b, pp. 181-182; cf. 1979, p. 137; 1983, p. 149). Weiner glosses Frege's claims that mathematical truths are independent of us by *excepting* an alleged presupposed need to impose structure and formulate boundaries linguistically (p. 201 ff.). She further writes, "discovering what is 'there' in the 'realm of the abstract' amounts to discovering what meets the descriptions that interest us" (p. 203). Weiner cites no texts to support either of these claims. Frege makes no exceptions or qualifications on his claims of independence; he notes no such presupposition. And it is at best deeply misleading to say that for Frege discovering mathematical structures "amounts to" discovering something associated with words, our interests, or ourselves. When our interests and descriptions happen to accord with mathematical truth, we do, of course, discover things that "meet" those interests and descriptions. But Frege explicitly says that our relation to logical truths and mathematical structures is "inessential" to their nature and existence (1984, p. 371; 1967b, p. 361).

made. No idealist – and no deflationist who thought that non-spatio-temporal entities were dependent on our language, practices, or judgments, or who thought that general philosophical assertions about them were "non-factual" – would have made such statements without careful, explicit qualifications. Frege enters no such qualifications.

Frege repeatedly inveighs against seeing logic (or mathematics) as embedded in language in the way that grammar is.[13] He thought that thought contents, logical objects, and logical functions bore no such essential dependence relation to the actual practice of thinking or language use. For Frege, the subject matter of logic is not the nature of human thinking or practice (1967a, p. 13; 1962, p. xvi), *even when that practice accords with the laws of truth*: "But above all we should be wary of the view that it is the task of logic to investigate actual thinking and judging, insofar as it is in agreement with the laws of truth" (1979, p. 146; 1983, p. 158 – the published translation is ambiguous in a way that does not match the German). This independence insures, for Frege, no scope for variation in the laws of logic between one group of thinkers and another (1979, pp. 7, 146; 1967b, pp. 7, 158; 1967a, p. 13; 1962, pp. xv–xvi).

Frege criticizes one Achelis who writes,

> ... the norms which hold in general for thinking and acting cannot be arrived at by the one-sided exercise of pure deductive abstraction alone; what is required is an empirico-critical determination of the objective principles of our psycho-physical organization which are valid at all times for the great consciousness of mankind.

Frege replies:

> [It appears that according to Achelis] the laws in accordance with which judgments are made are set up as a norm for how judgments are to be made. But why do we need to do this?... Now what is our

[13] Some interpreters of Frege have taken his views to be redescriptions of features of our practices of judgment or of linguistic use. Although Frege does describe logical structures that inform linguistic and cognitive practice, and does think that by reflecting on and reforming such practice we can discover these structures, I know of no evidence that Frege thought that the theory of judgment is really fundamentally about the activity or practice of judgment, much less linguistic practice. It is important to distinguish Frege's method of discovery (which does focus on language and activities of judgment) from Frege's views about the nature of thought contents and of judgment. It is also crucially important to realize that Frege was interested in judgment as a norm-yielding form, not in judgment as a human activity. Frege thought that thought contents and the form of judgment bore no essential relation to either language or activities (practices) of judgment, potential or actual, of human beings.

> justification for isolating a part of the entire corpus of laws and setting it up as a norm?... Are [the laws of logic] like the grammar of a language which may, of course, change with the passage of time? This is a possibility we really have to face up to if we hold that the laws of logic derive their authority from a source similar to that of the laws of grammar... if it is normal to judge in accordance with our laws of logic as it is normal to walk upright (1979, p. 147; 1983, p. 159)

Frege thought one can discover logic by reflecting on linguistic and mathematical practice. But he makes it very clear that his logical theory is not about practices, and does not take its authority from such practices. They are not what ground the normative structures that logic articulates.[14]

A second way Frege's Platonism shows itself lies in his attempts to explicate the success of science, the fact of intersubjectively objective cognitive practices, and indeed the authority of logic, in terms of the timelessness of the truths and structures of logic. In *The Basic Laws of Arithmetic* he states that the laws of logic (which he also calls the laws of truth) are authoritative *because* of their timelessness: "[the laws of truth] are boundary stones set in an eternal foundation, which our thought can overflow, but never displace. It is because of this that they have authority for our thought if it would attain to truth." (1967a, p. 13; 1962, p. xvi).

Frege frequently claims that only because individuals are not the bearers of thought contents is scientific communication possible (1967a, p. 17; 1962, p. xix; 1979, pp. 133, 137–138; 1983, pp. 144, 149–150; 1984, pp. 363, 368; 1967b, pp. 354, 359). But he sometimes goes further. In the Logic manuscript of 1897 he indicates that it is the timelessness of the subject matter of logic – the laws' not containing conditions that might be satisfied at some times and not at others – that enables logic to provide completely universal laws of truth (1979, p. 148; 1983, p. 160). The idea seems to be that all true thoughts are

[14] Frege writes that there is no contradiction between something's being true and everyone's taking it to be false (1967a, p. 13; 1962, pp. xv–xvi), making it clear that he does not believe in some general connection between thought contents (or intentional contents, or what are expressed in language) and actual judgments and practice, that would close any possible gap between mind and subject matter. There is more evidence for this fact in his discussion of scepticism in "The Thought". The example he gives in 1967a, p. 13 and 1962, pp. xv–xvi concerns an empirical truth. As I discuss below, he indicates that some truths – simple truths of arithmetic and basic logical truths – can be denied only through madness, and that any attempt to deny them in a thoroughgoing way will undermine judgment itself.

eternally true if they are true at all; but some have temporal subject matters. Some true laws even contain conditions that might be satisfied at certain times but not at others. But the laws of logic cannot be about temporal subject matters and cannot contain such conditions. For if the truth of some thought follows from the truth of others, then it must always follow. So to account for the universal aspect of entailment, one must assume that the subject matter of logic is eternal. (The conclusion of this argument, though not the argument itself is stated in 1967a, p. 13; 1962, p. xvi.)

In "Thoughts" Frege gives two more arguments that scientific objectivity (of communication and of knowledge of the physical world, respectively) is explicable only on the view that thought contents belong to a "third realm" that is neither mental nor physical. In the first argument (1984, p. 363; 1967b, p. 354) he holds that scientific communication cannot be understood on the assumption that thought contents are ideas in particular people's minds. He had previously maintained that thought contents are clearly not perceptible or knowable on the basis of perceptions (1984, pp. 354, 360; 1967b, pp. 345, 351). He concludes (1984, p. 363; 1967b, p. 354) that in order to understand the objectivity of the communal scientific enterprise, one "must recognize" the third realm. The timelessness of the truths of this realm and the fact that their truth is independent of whether anyone takes them to be true are clearly seen as part of an account of how a "science common to many on which many could work" is possible.[15]

In the second argument (1984, p. 368; 1967b, p. 359), Frege indicates that a "firm foundation of science" must be facts that are independent of men's varying states of consciousness. Facts are, he maintains, true thoughts. True thoughts have the requisite independence: not only are they not part of anyone's "inner" mental world; their truth "has nothing to do with" someone's thinking them. The work of science consists in the discovery of true thoughts (which provide a "firm foundation" for the science). Moreover, Frege argues that the applicability of mathematical truths to investigations at any time (he cites application of mathematics in an astronomical investigation into events in the distant past) is possible *because* a mathematical thought's truth, and the thought content itself, are timeless. So, he concludes, ex-

[15] Although this argument is not explicit in his 1968, the attitude behind it is not hard to discern in the introduction (1968, pp. vii–viii). I think that the argument is the least interesting of the three arguments I am discussing.

plicating the objectivity of science and the temporally neutral applicability of mathematics requires that both the thoughts and their truth be timeless.[16]

These arguments take for granted the existence of the objectivity manifested in intersubjectively accepted norms for communication and the checking and confirming of scientific results. Frege thinks that we need no reassurance about its solidity. He is not concerned with scepticism. He regards ordinary certainties as certain (1977 § 13; 1968, p. 2). He does not seek foundations, nor does he appeal to his Platonism, to

[16] Thomas G. Ricketts (1986) opposes reading Frege as "the archetypical metaphysical platonist" (p. 65), according to which "the mind-independent existence of things is for Frege a presupposition of the representational operation of language: it explains how our statements are determinately true or false apart from our ability to make or understand them". This description of Platonism does not fit the Platonism I attribute. Frege was clearly not trying to give a general explanation of linguistic representation or even of intentionality in judgment. But – in contrast to Ricketts – I think that Frege thought, as the previously cited passages indicate, that assuming the mind-independence of all thought contents, concepts, and logical objects, was necessary to understanding the objectivity of scientific practice and the universal applicability of logic and mathematics. I do not think that he thought that such objectivity would somehow be in jeopardy in the absence of such an explanation. Logic was for him epistemically prior to philosophy of logic. It is rather that such an explanation accounts for what is involved in judgment, logical inference, and logical truth. Ricketts elaborates: "The crucial feature of this [Platonist] line of interpretation is its taking ontological notions, especially that of an independently existing thing, as prior to and available apart from logical ones, from notions of judgment, assertion, inference, and truth" (p. 66). Ricketts also thinks that Frege's claims about the objectivity associated with judgment are not meant to be factual claims, and that there is therefore no possible explanation for Frege of objectivity. As I indicate in the text, I see no evidence for a relevantly applicable distinction in Frege between factual and non-factual claims. Moreover, Frege's Platonism does not involve any claim about the priority of ontological notions over logical notions. (I do not see even initial plausibility to attributing this assumption to Frege.) Logic and ontology are mutually entangled in Frege. Logic is about what is, as Frege says (1984, p. 351; 1967b, p. 342); it has an ontology. But logic is the most general science. So no thought about being could be independent of its notions. Moreover, Frege's most fundamental ontological categories (function and object) are logical categories. Nor does Frege's appeal to ontology in his account of the objectivity of science and the universality of logic imply that he thought that ontological notions were prior to logical ones, much less available apart from them. The explanation is not a definition, derivation, or reduction. All the key ontological notions he uses both presuppose and include logical notions. Rather he thought that a full understanding of logic involved appeal to notions like logical object, function, thought content, mind-independence, timelessness, causal inertness, imperceptibility, non-spatiality. Frege thought of himself as describing the ontological features that logic must have. Logical and ontological notions are interrelated for Frege; and all the relevant logical objects and functions are timelessly related to the relevant notions (thought components). Frege sees the whole logical structure, not just objects, in a Platonic fashion.

bolster confidence in an otherwise doubtful scientific enterprise. He does not view philosophy in the grand manner, as protecting science against otherwise dangerous philosophical worries. He articulates his Platonism because he finds a refusal to qualify the timelessness of mathematical structures, or to explain them in terms of something more familiar and temporal – such as our minds or practices – provides the best understanding of scientific inter-subjective objectivity. He thinks his view shows why practices that have been found to be firm are in fact firm (1968, p. 2).

Let us turn to Frege's views about how we know this third realm of entities. As I indicated earlier, I am prescinding from complexities in Frege's epistemology. What is important for our purposes is that Frege thought that our knowledge of the primitive logical truths and inference rules depended on a logical faculty – reason (1968, p. 21; 1980, pp. 37, 57; 1976, 37, 89; 1962 II, § 74; 1984, p. 405; 1967b, p. 393; 1979, pp. 267–273; 1983, pp. 286–292.) The question is: how could Frege believe that reason alone could give one knowledge of an atemporal realm of entities that are completely independent for their existence, nature, and relations to one another, of anyone's reasoning?

Frege is aware that foundational questions about our knowledge of mathematical structures ultimately come down to questions about knowledge of the primitive truths and inference rules. He is admirably clear that logic does not answer these questions: "The question why and with what right we acknowledge a law of logic to be true, logic can answer only by reducing it to another law of logic. Where that is not possible, logic can give no answer" (1967a, p. 15; 1962, p. xvii); "We justify a judgment either by going back to truths that have been recognized already or without having recourse to other judgments. Only the first case, inference, is the concern of Logic" (1979, p. 175; 1983, p. 190). Frege lays fundamental epistemological questions aside in much of his work, especially in *Basic Laws of Arithmetic.* But it would be a serious misunderstanding to think that he thought that the questions were off limits.[17] For he expresses a consistent interest in them.

Of course, he thought that one could not and need not argue for the basic logical truths. But he did see them as a source for the justification

[17] Cf. also 1979, pp. 3, 175; 1983, pp. 3, 190. Contrast Ricketts (1986, p. 81) "There is, as far as Frege is concerned, nothing to be said about the justification for our recognition of those basic laws of logic to be truths"; and Weiner (1990, pp. 71–72). Frege says a good bit about the epistemology of belief in the basic laws, scattered through his writings. I shall not discuss these passages in this paper, however.

of the belief in them by a person who understood them. He thought that they were self-evident.

We justify our judgment of the basic truths, as he said, without having recourse to other judgments (1979, p. 175; 1983, p. 190).

One needs to bear in mind here a three-fold distinction that Frege often carries along in his writings (it is very explicit in "The Thought" 1984, p. 352; 1967b, pp. 342–343): (a) psychological explanation of belief or judgment, including an account of its acquisition, (b) justification of our belief or judgment, and (c) grounding for logical truth. Frege always lays aside psychological explanation. But he repeatedly discusses the justification of "our" belief or judgment in logical truths as well as the grounding of logical truth. Understanding grounding of truth is a matter of understanding the natural order of truths, which is independent of thinking or practice, and the "same for all men" (1968, p. ii, 17; 1967a, pp. 13, 15; 1962, pp. xvi, xvii). One understands the grounding of truth when one understands the natural order of logical and mathematical proofs, and the primitive truths on which such proofs rest. What grounds logical truths are the primitive logical truths.

One of Frege's primary motivations for understanding logical truth and the proof structure of logic was to understand the nature of justification for our mathematical judgments. In *Foundations of Arithmetic* Frege begins in §§ 1 and 2 by announcing an interest in the proof structure of mathematics. But he immediately associates this structure with the question of the justification of belief. In § 2 he says that the aim of proof is, partly, to place a proposition beyond doubt. In § 3 he says that "philosophical motives" underly his inquiry into the foundations of mathematics: The motives turn out to centre on answering the question "Whence do we derive the justification of our assertion [of mathematical truths]"? The question of whether arithmetic is analytic turns out to concern the justification for making a judgment. He refines this to read, it concerns "the ultimate ground upon which rests the justification for holding [a proposition] to be true".

What is important about this passage is not only Frege's concentration on justification for judgment, but also his belief that the justification of an arithmetical judgment derives from the mathematical foundation (*Grund*) – from the primitive truths. The problem [of finding the justification for assertion or judgment], he says, is to be solved by "finding the proof of the proposition and following it back to the primitive truths." (1968, p. 4). One might ask, how can a problem of the justification of our beliefs or judgments be solved by citing primitive truths? How can such truths be primary in an account of justification?

Frege's line is made clearer in "The Thought" where he characterizes laws of truth as general laws which concern not "what happens" but "what is". Speaking of these laws about "what is" in the third realm, Frege says that "from the laws of truth there *follow* ["ergibt sich" – a non-technical term] prescriptions about asserting, thinking, judging, inferring" (1984, p. 351; 1967b, p. 342). Frege then calls these prescriptive epistemic laws "laws of thought" (1984, p. 351; 1967b, p. 342). This is a paradigmatic Platonic direction of explanation: from what *is* in an abstract realm to what is reasonable. What could be the nature of this derivation from general laws of truth – which concern logical objects and functions – to prescriptive laws about judgment? Frege writes: "[the laws of truth] are boundary stones set in an eternal foundation, which our thought can overflow, but never displace. It is because of this that they have authority for our thought *if it would attain to truth*" (1967a, p. 13; 1962, p. xvi).

Is it contingent that a judging subject "would attain to truth"? Frege is certainly insistent that the laws of truth are independent of their being taken to be true by anyone (1967a, pp. 13, 15; 1962, pp. xvi, xvii). Moreover, he thinks it not contradictory to suppose something's being true which everyone takes to be false (1967a, p. 13; 1962, pp. xv–xvi). On the other hand, Frege sees judgment as an advance from thought content to truth value. The function or aim of judgment is to reach truth. So to be a judging subject, one must have this aim or function insofar as one makes judgments. In this sense, the prescriptions of the laws of truth must apply to the judgments of judging subjects.

There is a second way in which Frege thinks that there is a deep, non-contingent relation between the laws of truth and prescriptive laws about judgment. To be rational, he thinks, one must be disposed to acknowledge the simplest logical truths. Judgments in contradiction with the laws of logic would constitute a kind of madness (1967a, p. 14; 1962, p. xvi). In fact, Frege appears to believe that failure to acknowledge primitive logical laws, like the principle that every object is identical with itself, and even certain truths of arithmetic, would throw thought into confusion and undermine the possibility of genuine judgment and thought (1968, p. 21). This suggests that he was inclined to believe that a disposition to acknowledge basic logical truths and inferences – and a disposition not to deny non-basic but relatively simple truths of arithmetic – was a condition not only for being rational but for being a judge or thinker at all.

Here it is worth looking very carefully at a famous passage in *Basic Laws*. Frege considers a supposed possibility in which beings had laws of thought (prescriptions for judgment) that contradicted ours. He claims that such beings would exhibit a "hitherto unknown type of madness", and indicates that such beings' procedures for taking things to be true would be in radical disaccord with the laws of truth (1967a, p. 14; 1962, p. xvii). Shortly afterwards, he writes:

> If we step away from logic, we may say: we are compelled to make judgments by our own nature and by external circumstances; and if we do so, we cannot reject this law – of Identity, for example; we must acknowledge it unless we wish to reduce our thought to confusion and finally do without all judgment whatever. I shall neither dispute nor support this view; I shall merely remark that what we have here is not a logical consequence. What is given is not a ground (*Grund*) for something's being true, but for our taking it to be true. Not only that: this impossibility of our rejecting the law in question hinders us not at all in supposing beings who do reject it; where it hinders us in supposing that these beings are right in so doing... Anyone who has once acknowledged a law of truth has by the same token acknowledged a law that prescribes the way in which one ought to judge, no matter where, or when, or by whom the judgment is made. (1967a, p. 15; 1962, p. xvii)

Frege is taking a hands-off attitude toward the epistemological issues for the purpose of his mathematical treatise. But, given his own beliefs, what does he neither dispute nor support in the view he states? Some have thought that in citing the limits of logic, he is prescinding from any judgment about grounds for our taking something to be true, as opposed to the ground for its being true. Some have even held that grounds for our taking something to be true are thought by Frege to be psychologistic, and of no interest to him. These are serious misreadings of the passage.

Understanding grounds for our *taking* something to be true had long been what motivated his inquiry into the foundations of arithmetic. (Cf. especially 1968, p. 3, where he uses exactly the same German terms as he does in the above-cited passage: grounds for taking [holding] something to be true.) One page earlier in *Basic Laws* Frege characterized the laws of logic in the double way I have described: not only as laws of truth but as laws that "prescribe the way in which one ought to think" (1967a, p. 14; 1962, p. xvi) – as laws of thought. What Frege takes no position on is whether we are compelled to acknowledge the laws by our own nature and by external circumstances.

This is indeed a psychological matter. He thinks that any such psychological law would admit of conceivable exceptions – mad beings that do reject the law. But where he writes, "we must acknowledge it unless we wish to reduce our thought to confusion and finally do without all judgment whatever", he is speaking in his own voice. For he had already indicated that he believes that renouncing the laws of arithmetic (which are less basic for him than the basic laws of logic) would be to reduce thought to confusion and make thinking impossible (1968, p. 21). Frege thinks that acknowledging these laws, at least implicitly in one's actual thinking, is necessary for having reason and for being a non-degenerate thinking and judging subject. (He apparently believed that although a mad person could reject a law, abiding by such rejection would reduce thought to confusion, and by degrees undermine judgment altogether.) These are normative not psychological judgments. Although they are not logical consequences, they are part of Frege's epistemic view.

So let us summarize the view that Frege maintains. He holds that justification for holding logical laws to be true rests on and follows from primitive laws of truth. He spells out this dependence of epistemic justification on the laws of truth in two ways. He thinks that laws of truth indicate how one ought to think "if one would attain to truth". But a judging subject necessarily would attain to truth, insofar as it engages in judgment. So any judgment by a particular person necessarily is subject to the prescriptive laws set out by the primitive laws of logic. One is justified in acknowledging them because doing so is necessary to fulfilling one's aim and function as a judging subject.

Frege's second way is: acknowledgement of certain laws of truth is necessary for having reason and for engaging in non-degenerate thinking and judging. One is rationally entitled to judge the primitive laws of logic to be true because the nature of reason – and even non-degenerate judgment – is partly constituted by the prescription that one acknowledge at least the simple and basic laws of truth. To put it crudely, reason and judgment – indeed mind – are partly defined in terms of acknowledging the basic laws of truth.[18]

[18] One can see this view alluded to in the passage where Frege claims that logic can, with anti-idealist and anti-psychologistic qualifications, be seen as the study of not minds but Mind (*Geist*) (1984, p. 369; 1967b, p. 359). One can also see it in his claim:

Our problem was to explain how, for Frege, mere reason could give grounds to believe that a subject matter is any particular way, given that the subject matter is atemporal, causally inert, and independent of thinking about it. Most current approaches to the substantive problem look for some analog of causal interrelation in our mathematical knowledge. More traditional views – both Platonic and idealist – see the relation as individuative or constitutive.

An idealist line is to make the subject matter constitutively dependent on thinking, synthesis, or practice. Frege's line is to hold that, although the laws of truth are independent of judging subjects, judging subjects are in two ways not independent of the laws of truth. First, to be a judging subject is to be subject to the prescriptions of reason, which in turn are provided by the laws of truth (logic). For judgment has the function of attaining truth; and the laws of logic – which are constituted by atemporal thoughts and atemporal subject matter – provide universal prescriptions of how one ought to think, given that one's thinking has the function of attaining truth. Second, being a judging subject is to have or have had some degree of reason. Having or having had some degree of reason requires acknowledging, at least implicitly in one's thinking, the simplest, most basic logical truths and inferences; and doing so commits one to an atemporal subject matter. Questions of "access" to the third realm are on reflection seen to be misconceived. For, to reverse somewhat Gertrude Stein's dictum about Oakland, there is no there there.

Why was this line not more prominent in Frege's philosophy? He thought that his primary contribution lay in identifying primitive truths and inference rules, and in deriving arithmetic from them. He accepted the traditional rationalist-Platonist line about the relation between reason and primitive truths. He did not think it needed substantial elaboration. Like Frege, I think that this neglected line is not to be dismissed. Unlike Frege, I think it may be worth developing.

"We might with alteration of a well-known proposition say: the proper object of reason is reason. In arithmetic we are not concerned with objects which we come to know as something alien from without through the medium of the senses, but with objects given directly to reason, which as her most proper objects are completely transparent to her" (1968, p. 115). Cf. also 1977 § 23. and 1968, § 26. These quotes are not idealist, as they have sometimes been taken. They are expressions of the view that the basic forms and objects of logic constitutively inform minds, and help define what it is to be mind or reason.

References

Benacerraf, Paul 1983: "Mathematical Truth" in *Philosophy of Mathematics: Selected Readings*, ed. Benacerraf and Putnam, Cambridge: Cambridge University Press.
Frege, G. 1962: *Grundgesetze der Arithmetik*. Hildesheim: Georg Olms.
Frege, G. 1967a: *The Basic Laws of Arithmetic*, trans. and ed. M. Furth, Los Angeles: University of California Press.
Frege, G. 1967b: *Kleine Schriften*. Hildesheim: Georg Olms.
Frege, G. 1968: *Foundations of Arithmetic*, trans. J. L. Austin. Evanston: Northwestern University Press.
Frege, G. 1977: *Begriffsschrift*. Hildesheim: Georg Olms.
Frege, G. 1976: *Wissenschaftlicher Briefwechsel*. Hamburg: Felix Meiner Verlag.
Frege, G. 1979: *Posthumous Writings*. Chicago: University of Chicago Press.
Frege, G. 1980: *Philosophical and Mathematical Correspondence*. Chicago: University of Chicago Press.
Frege, G. 1983: *Nachgelassene Schriften*. Hamburg: Felix Meiner Verlag.
Frege, G. 1984: *Collected Papers*. Oxford: Basil Blackwell.
Ricketts, Thomas G. 1986: "Objectivity and Objecthood; Frege's Metaphysics of Judgement", in *Frege Synthesized*, Dordrecht: D. Reidel.
Sluga, Hans 1980: *Gottlob Frege*. London: Routlege and Kegan Paul.
Weiner, Joan 1990: *Frege in Perspective*. Ithaca: Cornell University Press.

Part III:
Philosophy of Language

Fregean Theories of Truth and Meaning

Terence Parsons

Until a few years ago, most philosophers assumed that Frege's semantics provides an appropriate framework for the study of natural language. Then, various criticisms convinced people that Frege was on the wrong track. This is unfortunate. One criticism (in Kripke 1980) attacks the view that the sense of a name is constituted by the descriptions a speaker associates with the name. This view, however, is not Frege's. Its attribution to "the popular Frege" is based on an extrapolation from a footnote in "On Sense and Reference" (Frege 1892), and it ignores his rejection of the view in other parts of his writings (such as Frege 1918). Even if Frege had endorsed this view, it conflicts with so much of his semantical theory that it should have been seen as a mistake on his part, and should never have been thought to affect the semantical theory as such.

Some philosophers have refuted other views of "the popular Frege" by construing senses as constituted entirely by the "general" aspects of our beliefs about the world, aspects that our twin earth counterparts would share. This is also inconsistent with things that Frege said about senses in Frege 1918, and again the conflict is between a view about senses that Frege never held and the rest of his system.

Frege's semantical views form a framework that encapsulates much of what everybody agrees on in the semantics of simple English. In addition to avoiding mistaken inferences, it can yield ones that we take as unexceptional, such as:

> Kim believes everything that Samantha believes.
> Samantha believes that the pope is infallible.
> ∴ Kim believes that the pope is infallible.

or:

> Mary believes what John said.
> John said that snow is white.
> ∴ Mary believes that snow is white.

The theory thus holds out some promise of providing a general framework for semantics within which more specific theories may be articulated.

It is surprising how little work has been done in this area. When discussing Fregean semantics, the work of Alonzo Church comes naturally to most people's minds. However, Church's systematic work (1951, 1973–74, 1993) focusses not on Frege's semantics but on his ontology of senses and non-senses; it is an axiomatic theory about the entities in Frege's "third realm." There is no development of the relations of *referring* and *expressing* that link language to these "semantical" entities. Further, the semantical development that guides Church's theorizing is not the one that Frege attributed to natural language; it is the one he urged for a scientifically useful notation, a notation that would function quite different than natural language. My goal is to focus instead on Fregean accounts of the semantics of natural language. The principal work in print on this topic is Linsky 1967, which contains a substantial chapter on these issues, along with the beginnings of a formal treatment. (See also Linsky 1983.) I hope to expand on his pioneering work.

My aim is to develop a precise and rigorous account of certain semantical theories inspired by Frege: to produce theories of truth and meaning, to see what assumptions are needed for what results, and to explore some of the options that are left open. One of the major surprising results is that theories of truth and meaning turn out to be independent of Frege's doctrine that sense determines reference.

The principal complication in Frege's theory is the issue of whether there is a hierarchy of senses and references associated with each meaningful piece of language. I think that the hierarchy is ill-motivated and is not necessary in an adequate semantics for natural language, and I develop versions of the theory that dispense with it. But the issues are varied and subtle, and a complete resolution of the issue goes beyond this essay, so I investigate what the theory looks like with the hierarchy as well as without it.

In order to avoid taking too much space I presuppose various Fregean views without justification. These include the view that, *in a technical sense of 'refer'*, sentences refer to truth values and predicates refer to functions mapping objects to truth values, and the view that the senses of functional expressions are themselves functions.

1. The Semantics of Sense and Reference

1.1 Referring and Expressing

Frege takes for granted that words refer to things in the world. Thus, the parts of the simple sentence '*John is uneasy*' refer:

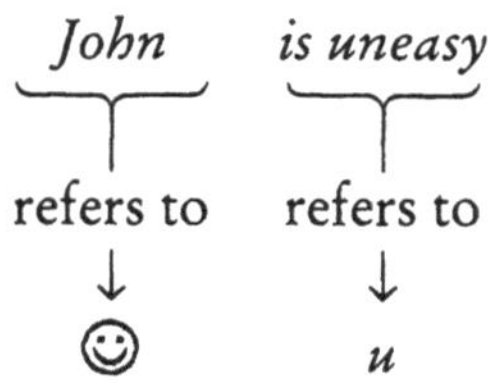

The name 'John' refers to John himself, the person. The predicate, 'is uneasy', refers to a function, the function that maps uneasy things to truth and all other things to falsehood.

Actual things, such as the person John, cannot be communicated. Frege assumes that literal linguistic communication takes place as follows: the speaker has a certain cognitive content in mind, and utters a sentence which expresses *that* cognitive content. Both language and mind, then, have common contents. The hearer who knows the language knows which cognitive content is expressed by the sentence, and thus knows what is being communicated. The cognitive content expressed by the sentence is itself a product of the meanings associated with the words. So individual words express "cognitive content-parts":

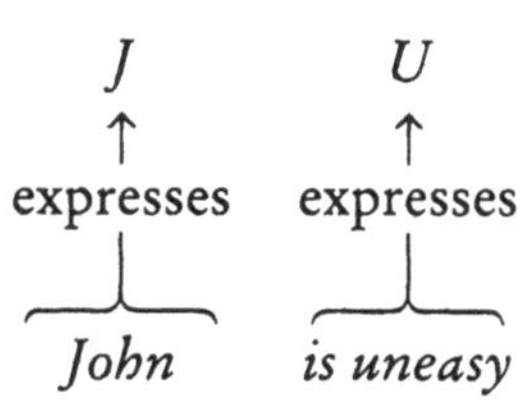

Words thus have a dual role; they refer to things, and they also express cognitively relevant meanings, which Frege calls *senses*.

It is not just simple words and phrases that refer to things and express meanings. Whole sentences also have this dual role; they express "thoughts," which are their meanings, and they are also true or false. Truth and falsehood are analyzed in terms of referring:

John is uneasy

refers to

↓

t ← truth if John is uneasy; otherwise falsehood

And expressing thoughts is a species of expressing:

T ← the thought that John is uneasy

↑

expresses

John is uneasy

These, too, will require discussion.

1.2 Indirect (Shifted) Reference

Probably the most distinctive aspect of Frege's semantics is his treatment of propositional attitude contexts. If we wish to discuss cognitive contents, saying, for example, which things a speaker believes, we typically do so by means of that-clauses. For example, we say that Agatha believes *that snow is white*. Frege assumes that in saying this we *refer* to what she believes (if what we say is true). We do this by using a that-clause which contains the very sentence that, in independent uses, expresses what she believes. In the context of a that-clause, the sentence comes to refer to what it normally expresses. The parts of the sentence are similarly affected:

Mary believes that John is uneasy

refers to refers to

↓ ↓

J *U* ← the senses that are usually expressed

Mary believes that John is uneasy

refers to

↓

T ← the proposition that John is uneasy

Although not proposed for this purpose, Frege's account solves certain famous logical puzzles, such as the apparent failure of substitutivity of identicals in belief contexts: if Mary believes that the morning star is a planet it does not follow that she believes the evening star is a planet, even if they are the same thing. For, on Frege's analysis, when you substitute the phrase '*the evening star*' for '*the morning star*' in an indirect context like '*Mary believes that* __ *is a planet*', you are substituting phrases that, in *that* context, refer to their customary senses, which are different from each other. The identity of their *customary* referents does not permit their intersubstitutivity in contexts in which reference is shifted, and so the terms refer to different things.

Frege himself viewed the shift of reference just discussed as a defect of natural language, and he proposed that it be banned from a language constructed for serious scientific purposes. He says: "To avoid ambiguity, we ought really to have special signs in indirect speech, though their connection with the corresponding signs in direct speech should be easy to recognize." (Frege 1902). For example, we could use $John_0$ for the name 'John' as it occurs in ordinary contexts, and $John_1$ for its occurrence in indirect contexts, writing:

$$John_0 \ is \ uneasy_0$$

for '*John is uneasy*' when it occurs alone or in complex ordinary contexts, and:

$$John_1 \ is \ uneasy_1$$

when it occurs in indirect contexts. So we would symbolize '*Mary believes that John is uneasy*' as

$$Mary_0 \ believes_0 \ John_1 \ is \ uneasy_1.$$

Here we assume that symbols never shift reference at all; a reference shift in ordinary language is captured by a symbol shift in the respectable scientific idiom. In case of reembeddings of indirect contexts the symbols shift anew; for example, in '*Agatha believes that Mary believes that John is uneasy*' we have doubly new symbols for '*John*' and '*is uneasy*':

$$Agatha_0 \ believes_0 \ Mary_1 \ believes_1 \ John_2 \ is \ uneasy_2.$$

Eventually, every word of English would be associated with an infinite number of different formal language symbols in this way. This is *not* the view that I am investigating. The view I am investigating is Frege's view about how natural language does work: we do not have an infinite number of unambiguous symbols, we have instead a small number of words that shift their sense and reference with context.

1.3 Predicates and Sentences as Referring Expressions

I follow Frege in treating predicates and sentences as referring expressions; predicates refer to functions, and sentences refer to truth values. This suggests that predicates and names may appear in language anywhere that ordinary names may appear. If this were so, the following would be meaningful:

Snow is white is uneasy.
Is clever is uneasy.

Frege endorsed the first of these, and rejected the second. But these were his views about how language *ought* to work, not about how it does work. Obviously, neither of the displayed strings of words are grammatical English. I assume that sentences, names, and predicates occupy distinct places in language, so that they may not be meaningfully interchanged. Applying a technical notion of reference to all of them does not entail that they are grammatically interchangeable.

Alongside Frege's technical notion of reference is a technical notion of expressing, which relates each semantically significant piece of language to a sense. For example, Frege takes literally the maxim that a well-formed indicative sentence *expresses a complete thought*; expressing is a relation that relates complete sentences to a certain sort of senses that he calls thoughts and that I call propositions.

Senses of predicates are functions, making poignant the principle that what a complex phrase expresses "is a function of" what its parts express. For example, the sense expressed by a predicate maps the sense expressed by a singular term to the sense expressed by their combination:

J *U*
↑ ↑
expresses expresses
John *is uneasy*

$U(J)$ ← what U maps J to, namely, the thought that John is uneasy
↑
expresses

John is uneasy

Likewise, the reference of the predicate maps the reference of the singular term to the truth-value of their combination:

John *is uneasy*

refers to refers to
↓ ↓
☺ *u*

John is uneasy

refers to
↓
u(☺) ← what *u* maps John to: truth if John is uneasy, falsehood if he isn't

1.4 The Syntax of the Object Language Without Reference Shifting

I introduce here the first portion of the syntax of the object language whose semantics will be discussed throughout this paper. I postpone discussion of "shifted" contexts until the next section.

I adopt Frege's view that symbols of a language sometimes refer to themselves. In order to avoid extensive use of quotes, I use bold-face to indicate when this happens. Italic '*X*' and '*Y*' are variables ranging over terms (simple or complex, function or object) of the object language. 'Z' is a schematic letter.

In this initial simplified part of the language, there are three types of terms: names, sentences, and function terms. They can be simple or complex. The simple ones are given by:

Simple Terms:	Basic Individual Terms:	**a, b, c, ...**
	Basic Function Terms:	**F, G, H, ...**

There are no basic sentential terms, since sentences are complex. I assume, without giving details, that function terms are divided into various types depending on the number of arguments they can take, on the kinds of arguments that are allowed in each place, and on the type of output that they yield. The resulting type structure can be quite complex; I avoid giving details here.

Complex Terms:	$X(Y)$	when X is functional and Y is not functional.
	$X(Y,U)$	when X is functional and Y and U are not functional.

Complex terms are classified as individual or sentential, depending on the types of their major function terms. Complex terms can themselves appear as arguments of functional terms whenever simple terms of the same type can.

Illustrations:
Function terms include both predicates and connectives. Predicates are function terms that take individual terms as arguments and yield sentential terms as outputs:

Snow is white	**W(s)**
John loves Mary	**L(j,m)**
Fred is Mr. Jones	**=(f, j)**

Connectives are functional terms that take sentential terms as arguments and yield sentential terms as outputs:

¬S	**¬(S)**
S & R	**&(S,R)**
S ≡ R	**=(S,R)**

Note that in symbolizing the biconditional, I have used the ordinary identity sign. This is because when sentences are viewed as referring to truth values, a biconditional sentence is true if and only if each side refers to the same thing (to the same truth value).

1.5 The Syntax of Reference Shifting

According to Frege, in non-extensional contexts there is a shift of reference from the ordinary reference of words to what they normally express. No mechanism for this shift is built into our object language yet. How shall we accomplish this?

To address this issue I begin by examining the syntax of indirect contexts in ordinary language. Some familiar examples are:

1. *Necessarily, snow is white*
2. *It is necessary that snow is white*
3. *Mary believes that snow is white*
4. *Mary believes everything that John believes*

Normally, examples such as these are analyzed as containing non-truth functional operators; typically, modal logics, deontic logics, and logics of knowledge and belief all assume this. Using '□' for the modal operator for necessity, (1) or (2) are usually symbolized as

□ *(Snow is white)*

and (3) as:

Mary believes-that (Snow is white).

On this type of analysis, the *'that'* is fused with the adjective or verb, and reference-shifting is automatic when the adjective or verb is present.

An alternative analysis, and one that I think is preferable, is to see the *'that'* as going with the embedded sentence itself to form a "that-clause," and to treat the adjective *'necessary'* as a property of propositions, and the verb *'believes'* as a relation between persons and propositions. This suggests the forms:

Necessary (that snow is white)
Believes (Mary, that snow is white),

in which necessity is a property of propositions, and belief is a relation between people and propositions. To see why this is important, consider an argument employing (4):

Mary believes everything that John believes
John believes that snow is white
∴ *Mary believes that snow is white*

This is certainly valid, and its validity ought to be explainable. The explanation assumes that belief is a two-place relation, and that *'that snow is white'* functions as a singular term. Using 't' for *'that snow is white'*, the form of the argument is:

$(x)[B(j,x) \rightarrow B(m,x)]$
$B(j,t)$
$\therefore B(m,t)$

This is valid in the ordinary predicate calculus. If instead we insist on representing belief by means of a non-truthfunctional operator, the argument cannot be symbolized in this familiar way.

For reasons such as this I will assume that the adjectives and verbs of ordinary English that are involved in the phenomena of indirect reference are so involved solely because they are predicates of propositions, and not because they have a different logical form than other verbs and adjectives. That is, they can be construed as ordinary predicates. Opacity is due to other phenomena, such as occurrence in that-clauses. That-clauses themselves are terms that in ordinary contexts refer to propositions.

I accommodate shifted contexts in the object language by using a reference-shifting symbol '^'. This symbol plays the role of 'that' in English as the standard way of introducing subordinate clauses. It needs to be accompanied by brackets for punctuation:

Indirect Complex Terms:	If X is a term then so is ^[*X*].

If X is a functional term then so is ^[*X*], and if *X* is nonfunctional then so is ^[*X*]. When *X* is a sentence, the '^' plays the role of '*that*' in English. As an illustration, the sentence '*Mary believes that snow is white*' is represented by

B(m,^[W(s)]),

where '^[W(s)]' represents the English phrase '*that snow is white*'.

1.6 The Metalanguage

If Frege is right about the nature of language, the metalanguage should look just like the object language. That is, its syntax will be exactly the same as the object language, and its semantics is meant to follow the same principles as those of the object language. It will contain some additional vocabulary, such as the terms 'express' and 'refer', but this will be all. Thus it will contain names, predicates, and sentences, indirect contexts, connectives, and so on. Eventually, both metalanguage and object language need to be expanded to contain quantifiers. Except for *de re* quantification, the addition of quantifiers raises no new issues of principle, and they will not be discussed in this paper.

I assume that we have available in the meta-language the ordinary principles of the predicate calculus. In addition, the following two principles are worthy of note. First, the metalanguage obeys the principle of substitutivity of identicals outside of indirect contexts, so that this is valid:

$$a = b \rightarrow [\ldots a \ldots = \ldots b \ldots]$$

whenever the context '... x ...' is not one in which 'x' is within a shifted context (i. e. within square brackets preceded by '^'). This will hold true for all uses of identity, including identities expressed with sentences and with predicates.

Second, we need a special axiom schema in the metalanguage (or a rule of inference) indicating that if a whole context is shifted, this is the same as shifting its parts:

ML	^[X(Y)] = ^[X](^[Y]).

For example, the result of shifting '*snow is white*' is the same as shifting both '*snow*' and '*white*' and combining them. The importance of this principle is not so much its truth as its intelligibility, for it assumes that symbols can be shifted even when not within sentential that-clauses. An instance of the principle is:

^[*White*(*snow*)] = ^[*White*](^[*snow*])].

The lefthand side of this represents the that-clause '*that snow is white*', but its righthand side has no ready reading in English. This is awkward, since our goal is not just to find some interesting formal results, but to produce a theory that is understandable to us all, building on our ability to understand our own native language. I think that we have to see this as an extension of our language, analogous to the introduction of mathematical notation in arithmetic. The extension permits us to formulate some principles which yield a semantical account. The final output of that account, a theory of meaning and truth, contains an infinitude of results that are perfectly meaningful and unproblematic, results such as

> '*Mary believes that snow is white*' is true iff Mary believes that snow is white.

and:

'Snow is white' means that snow is white.

These do not raise the problem discussed above; the problem occurs only in the theoretical machinery used in arriving at the results, in which we need to appeal to assumptions such as "*'white'* expresses ^[white]." Whether this is justifiable cannot be determined until we are familiar with the results and how they are arrived at.

1.7 The Semantics

The semantics proceeds by way of an account of Frege's two semantical relations of referring and expressing. If it were not for reference shifting, the theory would be quite simple; we would invoke principles that identify what the simple parts of language refer to and express:

ref{**Z**} = Z for each simple expression Z.
expr{**Z**} = ^[Z] for each simple expression Z.

and principles that identify what the complex pieces of language refer to and express, given what their parts refer to and express:

ref{*X*(*Y*)} = ref{*X*}(ref{*Y*})
expr{*X*(*Y*)} = expr{*X*}(expr{*Y*}).

But the account needs to be a bit more complicated because in shifted contexts a phrase changes what it refers to and what it expresses. This change is systematic with context. For example, in

Agatha thinks that Bill believes that snow is white,

'Agatha' is not shifted at all, *'Bill'* is shifted once, and *'snow'* is shifted and then shifted anew:

shifted (spanning ^[Believes(Bill, ^[White(snow)])])
shifted (spanning ^[White(snow)])

Thinks(Agatha, ^[Believes(Bill, ^[White(snow)])]).

↑ Agatha: shifted 0 times
↑ Bill: shifted 1 time
↑ snow: shifted 2 times

To encode this, our referring and expressing relations will be relativized to a number indicating the kind of context in which the symbol occurs. The reference of a phrase such as '*Agatha*' which is not embedded in an indirect context will be $\text{ref}_0\{Agatha\}$, the reference of the same phrase embedded in a single indirect context will be $\text{ref}_1\{Agatha\}$, and so on. I assume at the outset that it is a matter for investigation what happens when a phrase occurs in an indirect context which is itself embedded in another indirect context. So I make no assumption as to whether or not $\text{ref}_1\{Agatha\} = \text{ref}_2\{Agatha\}$. Similarly for whether or not $\text{expr}_1\{Agatha\} = \text{expr}_2\{Agatha\}$.

I assume that function signs refer to functions at any level of embedding, and that they also express functions at any level of embedding. Sentential signs and names refer to and express things other than functions at any level of embedding.

Now for the official semantic theory. First, we adopt the natural analogue of Frege's principle of functionality at every level of shifting. The following holds for every n≥0, for every *X* and *Y* when *Y* is not of the form ^[Z]:

FUNC	$\text{ref}_n\{X(Y)\} = \text{ref}_n\{X\}(\text{ref}_n\{Y\})$.
	$\text{expr}_n\{X(Y)\} = \text{expr}_n\{X\}(\text{expr}_n\{Y\})$.

We also have to know what happens when *X*'s argument is of the form ^[Z]. The answer is given by the following, for n≥0:

INDIR	$\text{ref}_n\{X(\wedge[Z])\} = \text{ref}_n\{X\}(\text{ref}_{n+1}\{Z\})$.
	$\text{expr}_n\{X(\wedge[Z])\} = \text{expr}_n\{X\}(\text{expr}_{n+1}\{Z\})$.

The above principles tell us what happens when we make bigger phrases out of smaller ones. For the simple terms of the language, every instance of the following is an axiom, where '**Z**' is a basic name of the object language and 'Z' is the corresponding symbol of the metalanguage:

AT.0	$\text{ref}_0\{\mathbf{Z}\} = Z$
AT.1	$\text{ref}_1\{\mathbf{Z}\} = \wedge[Z]$
AT.2	$\text{ref}_2\{\mathbf{Z}\} = \wedge[\wedge[Z]]$
AT.3	$\text{ref}_3\{\mathbf{Z}\} = \wedge[\wedge[\wedge[Z]]]$
...	...
...	...

Lastly, we state Frege's hypothesis that tells us that a phrase in a single that-clause refers to what it expresses when it is not embedded at all. This hypothesis links the reference of a phrase in a single level of embedding with the sense of its occurrence in isolation:

LINK $\mathrm{expr}_0\{X\} = \mathrm{ref}_1\{X\}$, whenever X is not of the form ^[Y].

These are all of the principles that we need for a theory of truth and meaning.

Notice that none of the semantical principles assign senses or references to phrases of the form '^[*X*]'. The idea is that the reference-shifting '^[...]' shifts the reference of its contents, but it does not itself form a larger phrase with sense and reference of its own. This is not the only way to analyze it, and other options are of interest; I skip discussion of them here for want of space.

2. *Some Fruits of the Semantics*

2.1 The Principles of Extensionality (Compositionality)

Frege takes for granted that the reference of a whole remains unchanged when one of its parts is replaced by another that has, in that context, the same reference, and he also assumes that what is expressed by a whole remains unchanged when a part is changed to another that, in that context, expresses the same sense. These are sometimes called principles of extensionality, or of compositionality. He is also customarily represented as holding that the reference of a whole *is a function of* the references of its parts, and that what a whole expresses *is a function of* what its parts express. These principles are satisfied by the present account. Consider, for example, the following application of FUNC to the formal analogue of 'Snow is white':

$$\mathrm{ref}_0\{\mathbf{W}(\mathbf{s})\} = \mathrm{ref}_0\{\mathbf{W}\}(\mathrm{ref}_0\{\mathbf{s}\}) \qquad \text{FUNC}$$

This illustrates that the unembedded reference of **W(s)** is a function of the unembedded reference of its parts, that is, it is identical to what you get by taking the reference of one part, **W**, which is a function, and applying that function to the reference of the other part, s. It is also apparent, because of the substitutivity of identicals in the metalanguage, that substituting something else for **W** or **s** that has the same unembedded reference would yield no change of reference in the righthand side, and thus no change in the lefthand side either.

In case the sentence in question contains embedded contexts, the principles still hold. Consider the formal analogue of '*Mary believes that snow is white*':

$$\mathbf{B(m,\wedge[W(s)])}.$$

Applications of INDIR and FUNC yield the following:

$$\mathrm{ref}_0\{\mathbf{B(m,\wedge[W(s)])}\} = \mathrm{ref}_0\{\mathbf{B}\}(\mathrm{ref}_0\{\mathbf{m}\},\mathrm{ref}_1\{\mathbf{W(s)}\}) \quad \text{INDIR}$$
$$= \mathrm{ref}_0\{\mathbf{B}\}(\mathrm{ref}_0\{\mathbf{m}\},\mathrm{ref}_1\{\mathbf{W}\}(\mathrm{ref}_1\{\mathbf{s}\})) \quad \text{FUNC}$$

The last line is entirely in function-argument form, illustrating that the unembedded reference of **B(m,^[W(s)])** is a function of the references that its parts have *in this context*, keeping in mind that the parts of **B(m,^[W(s)])** are **B**, **m**, **W**, and **s**, and that the first two of these occur here in contexts of level 0, and the latter two occur in contexts of level 1.

This sort of example also illustrates the Fregean solution to the non-interchangeability of '*Hesperus*' and '*Phosphorus*' within belief contexts. Suppose that Mary believes that Hesperus is a star:

$$\mathrm{B(m,\wedge[S(h)])}.$$

This is subject to the principles:

$$\mathrm{ref}_0\{\mathbf{B(m,\wedge[S(h)])}\} = \mathrm{ref}_0\{\mathbf{B}\}(\mathrm{ref}_0\{\mathbf{m}\},\mathrm{ref}_1\{\mathbf{S(h)}\}) \quad \text{INDIR}$$
$$= \mathrm{ref}_0\{\mathbf{B}\}(\mathrm{ref}_0\{\mathbf{m}\},\mathrm{ref}_1\{\mathbf{S}\}(\mathrm{ref}_1\{\mathbf{h}\})) \quad \text{FUNC}$$

Now suppose that Hesperus is Phosphorus:

$$\mathrm{h = p}.$$

By AT.0 this yields:

$$\mathrm{ref}_0\{\mathbf{h}\} = \mathrm{ref}_0\{\mathbf{p}\}.$$

This, however, does not permit intersubstitutivity in the sentence about Mary's belief, since this does not guarantee that

$$\mathrm{ref}_1\{\mathbf{h}\} = \mathrm{ref}_1\{\mathbf{p}\}.$$

Other sorts of substitutions, however, are sanctioned. Consider the relation between '*Mary believes that snow is white*' and '*Mary believes what John said*', supposing that what John said was that snow is white. The first sentence differs from the second by having one part changed to another that *in that context* refers to the same thing. *'What John said'* occurs here in an unembedded context. This is a complex phrase, and we do not yet have the means to analyze its parts, but for present purposes we only need the fact that it occurs in an unembedded context and that it refers to the proposition that snow is white. Let me use '**J**' for 'what John said'. Then the formal symbolization of *'Mary believes what John said'* enters into this equation:

$$\text{ref}_0\{\mathbf{B(m,J)}\} = \text{ref}_0\{\mathbf{B}\}(\text{ref}_0\{\mathbf{m}\},\text{ref}_0\{\mathbf{J}\}) \qquad \text{FUNC.}$$

If what John said is that snow is white, then $\text{ref}_0\{\mathbf{J}\} = \text{ref}_1\{\mathbf{W(s)}\}$, and so we can substitute *'that snow is white'* for *'what John said'*:

$$\begin{aligned} \text{ref}_0\{\mathbf{B(m,J)}\} &= \text{ref}_0\{\mathbf{B}\}(\text{ref}_0\{\mathbf{m}\},\text{ref}_0\{\mathbf{J}\}) && \text{FUNC.} \\ &= \text{ref}_0\{\mathbf{b}\}(\text{ref}_0\{\mathbf{m}\},\text{ref}_1\{\mathbf{w(s)}\}) && \text{Subst Iden} \end{aligned}$$

One application of INDIR then completes the demonstration that *'Mary believes that snow is white'* and *'Mary believes what John said'* have the same reference – and thus the same truth value:

$$\begin{aligned} \text{ref}_0\{\mathbf{B(m,J)}\} &= \text{ref}_0\{\mathbf{B}\}(\text{ref}_0\{\mathbf{m}\},\text{ref}_0\{\mathbf{J}\}) && \text{FUNC.} \\ &= \text{ref}_0\{\mathbf{B}\}(\text{ref}_0\{\mathbf{m}\},\text{ref}_1\{\mathbf{W(s)}\}) && \text{Subst Iden} \\ &= \text{ref}_0\{\mathbf{B(m,{}^\wedge[W(s)])}\} && \text{INDIR} \end{aligned}$$

So our theory validates the inference:

Mary believes that snow is white
What John said is that snow is white
∴ *Mary believes what John said.*

This is the sort of result that should be a normal part of our semantical theory of belief; it is seen here as a consequence of the semantics of indirect contexts in general. Notice, however, that the premise

What John said is that snow is white

does not justify the claim that 'what John said' expresses the same sense as 'that snow is white', or that it has the same indirect reference. So we cannot infer that if Agatha knows that Mary believes that snow is white then Agatha knows that Mary believes what John said, given that what he said is that snow is white.

2.2 A Theory of Truth

We mentioned earlier that truth for a sentence S is the same as $ref_0\{S\}$. But how do we know this? The purpose of this section is to show how we can know this, and to show that it is so.

The key to a successful account of truth is to satisfy *Tarski's Material Adequacy Condition for Truth*. Suppose that we have defined a predicate, 'Tr'. Tarski's material adequacy condition is a test for how to tell whether the extension of 'Tr' actually contains all of the true sentences and none of the false ones. It is a test for whether the "matter" (the *extension*) of the predicate is right. Tarski's proposal is that we know this if our account of truth entails each instance of

$$\mathrm{Tr}(s) \equiv p,$$

where 's' names a sentence of the object language and 'p' is the metalinguistic version of 's'. The reason that this test shows 'Tr' to have the right extension is this. Suppose, first, that p is true. Then this fact, together with our theory, entails Tr(s), provided that our theory entails the relevant instance of the biconditional displayed above. That is, this fact, together with our "adequate" theory, entails that the sentence in question is in the extension of 'Tr'. Now suppose instead that p isn't true. Then this fact, together with the theory, entails $\neg\mathrm{Tr}(s)$; that is, it entails that s is not in the extension of 'Tr'.

Tarski's material adequacy condition is not itself a theory of truth; it is a test of correctness for a theory of truth. Within our semantic theory we already have a materially adequate theory of truth. We have this because we can define for any sentence X:

$$\mathrm{True}\{X\} =_{df} \mathrm{ref}_0\{X\}.$$

Recall our *intent* that truth correspond to our technical notion of customary reference for sentences. If X is a sentence, then '$\mathrm{ref}_0\{X\}$' should refer to truth iff X is true, so this account should work correctly. But does it? The test is to see if this definition satisfies Tarski's material adequacy condition for truth. Can we prove, for arbitrary S

$$\mathrm{True}\{S\} \equiv S\ ?$$

Substituting our proposed definition of truth, the question is whether we can prove, for each sentence S

$$\mathrm{ref}_0\{S\} \equiv S.$$

Using identity for the biconditional, this has the form:

$$\text{ref}_0\{\mathbf{S}\} = \text{S}.$$

I refer to this as our formal version of Tarski's material adequacy condition:

Tarski's Material Adequacy Condition for Truth (Formal Version):
We should be able to prove the following for any sentence **S** of the object language:

$$\text{ref}_0\{\mathbf{S}\} = \text{S}.$$

To show that Tarski's material adequacy condition applies to 'snow is white', we need to show that our theory yields

True{**snow is white**} ≡ snow is white

or, in our notation

$$\text{ref}_0\{\mathbf{W(s)}\} = \text{W(s)}.$$

This is proved as follows:

$$\begin{aligned} \text{ref}_0\{\mathbf{W(s)}\} &= \text{ref}_0\{\mathbf{W}\}\,(\text{ref}_0\{\mathbf{s}\}) && \textsc{Func} \\ &= \text{W(s)} && \text{AT.0} \end{aligned}$$

Cases involving indirect contexts are more interesting. To show that '*Mary believes that snow is white*' is true iff Mary believes that snow is white, we need to show:

ref$_0${*Mary believes that snow is white*} = Mary believes that snow is white.

In formal notation this is:

$$\text{ref}_0\{\mathbf{B(m,\wedge[W(s)])}\} = \text{B(m,}\wedge\text{[W(s)])}.$$

Here is the proof:

$$\begin{aligned} \text{ref}_0\{\mathbf{B(m,\wedge[W(s)])}\} &= \text{ref}_0\{\mathbf{B}\}(\text{ref}_0\{\mathbf{m}\},\ \text{ref}_1\{\mathbf{W(s)}\}) && \textsc{Indir} \\ &= \text{B(m,ref}_1\{\mathbf{W(s)}\}) && \text{AT.0} \\ &= \text{B(m,ref}_1\{\mathbf{W}\}(\text{ref}_1\{\mathbf{s}\})) && \textsc{Func} \\ &= \text{B(m,}\wedge[\text{W}](\wedge[\text{s}])) && \text{AT.1} \\ &= \text{B(m,}\wedge[\text{W(s)}]) && \text{ML} \end{aligned}$$

When we iterate indirect contexts, the same principles apply. For example, for *'Agatha believes that Mary believes that snow is white'* we need:

$$\mathrm{ref}_0\{\mathbf{B(a,{}^\wedge[B(m,{}^\wedge[W(s)])])}\} = \mathrm{B(a,{}^\wedge[B(m,{}^\wedge[W(s)])])}.$$

The proof of the truth-equivalence is just like that above, though because of the iterated indirect context it also appeals to AT.2.

$$\begin{aligned}
&\mathrm{ref}_0\{\mathbf{B(a,{}^\wedge[B(m,{}^\wedge[W(s)])])}\} && \\
&\quad = \mathrm{Ref}_0\{\mathbf{B}\}(\mathrm{ref}_0\{\mathbf{a}\},\mathrm{ref}_1\{\mathbf{B(m,{}^\wedge[W(s)])}\}) && \textsc{Indir} \\
&\quad = \mathrm{B(a,ref_1}\{\mathbf{B(m,{}^\wedge[W(s)])}\}) && \mathrm{AT.0} \\
&\quad = \mathrm{B(a,ref_1}\{\mathbf{B}\}(\mathrm{ref}_1\{\mathbf{m}\},\mathrm{ref}_2\{\mathbf{W(s)}\})) && \textsc{Indir} \\
&\quad = \mathrm{B(a,{}^\wedge[B]({}^\wedge[m],ref_2}\{\mathbf{W(s)}\})) && \mathrm{AT.1} \\
&\quad = \mathrm{B(a,{}^\wedge[B]({}^\wedge[m],ref_2}\{\mathbf{W}\}(\mathrm{ref}_2\{\mathbf{s}\}))) && \textsc{Func} \\
&\quad = \mathrm{B(a,{}^\wedge[B]({}^\wedge[m],{}^\wedge[{}^\wedge[W]]({}^\wedge[{}^\wedge[s]])))} && \mathrm{AT.2} \\
&\quad = \mathrm{B(a,{}^\wedge[B]({}^\wedge[m],{}^\wedge[{}^\wedge[W]({}^\wedge[s])]))} && \mathrm{ML} \\
&\quad = \mathrm{B(a,{}^\wedge[B]({}^\wedge[m],{}^\wedge[{}^\wedge[W(s)]]))} && \mathrm{ML} \\
&\quad = \mathrm{B(a,{}^\wedge[B(m,{}^\wedge[W(s)])])} && \mathrm{ML}
\end{aligned}$$

2.3 An Application to a Problem concerning Modal Logic

We can now illustrate one of the conveniences of treating indirect contexts in terms of verbs plus that-clauses instead of in terms of operators. There is a long-standing problem for the theory of truth for the language of modalities. In conventional ways of doing modal logic, the recursion clause for the modal operator looks like this:

'□(S)' is true in world w *iff* 'S' is true in every alternative to w.

This is what you have if you are giving a recursive definition of *true in possible world w*. This notion of truth relative to a world is needed in the logic in order to define logical notions such as validity. But what if we are interested instead in the (non-relative) notion of truth? When we do the clause for negation we have something like this:

'¬S' is true *iff* Not('S' is true).

However the corresponding clause for necessity is problematic:

'□S' is true *iff* Necessarily ('S' is true).

This seems to invoke an implausible condition; the left-hand side might be true when the right-hand side is false, since it is not necessary that words mean what they do (see Gupta 1980, chapter 5 §§ 2–3). The solution is immediate if we abandon the treatment of necessity as an operator and construe it instead as a predicate of propositions. Then '*Necessarily*, S' has the form '*Necessary*(^[S])'. With this form, and with principles that we already have at our disposal, the truth-clause for necessity is:

True{**Nec(^[S])**}	= ref_0{**Nec(^[S])**}	Definition of 'true'
	= ref_0{**Nec**}(ref_1{**S**})	INDIR
	= Nec(ref_1{**S**})	AT.0
	= Nec(^[S])	AT.1

(The last line can use AT. 1 only if 'S' is atomic;'otherwise a longer argument is needed of the sort discussed in the following section.) This automatically avoids the difficulty, since it nowhere claims that it is necessary that a certain piece of language be true. The next to last line claims that **S** has an indirect reference that is a necessary proposition; it does not claim that it is necessary that **S** have such an indirect reference.

2.4 A Theory of Meaning

When we are doing semantics we are just as interested in a theory of meaning as in a theory of truth. Fortunately, in the framework developed so far we can produce a theory of meaning as easily as a theory of truth. That is, we can explicitly define a locution which satisfies the following Tarski-like material adequacy condition for *meaning*:

> **A Tarski-Like Material Adequacy Condition for Meaning:**
> A theory of meaning should entail every instance of:
>
> **S** means that S

The key to accomplishing this goal is to clarify the grammar of '**S** means that S'. I suggest that the grammatical form of

S means that S

be construed as

the meaning of **S** is that S,

that is, as

the meaning of **S** = that-S.

In Frege's semantics, the meaning of a symbol is the sense that it expresses, and so in technical terms we are discussing the condition:

$$\text{expr}_0\{\mathbf{S}\} = {}^\wedge[S].$$

Recall that by LINK, $\text{expr}_0\{X\} = \text{ref}_1\{X\}$, for any X. Thus the following is an equivalent formulation of this condition:

$$\text{ref}_1\{\mathbf{S}\} = {}^\wedge[S].$$

It simplifies the proofs to use this form.

> **The Material Adequacy Condition for Meaning:**
> A theory of meaning should entail every instance of:
> $\text{ref}_1\{\mathbf{S}\} = {}^\wedge[S]$

Our theory meets this condition. To illustrate this, consider the following instance of the condition:

'Snow is white' means that snow is white.

Formally, this is:

$$\text{ref}_1\{\mathbf{W(s)}\} = {}^\wedge[W(s)].$$

Its proof is:

$$\begin{aligned} \text{ref}_1\{\mathbf{W(s)}\} &= \text{ref}_1\{\mathbf{W}\}(\text{ref}_1\{\mathbf{s}\}) && \text{FUNC} \\ &= {}^\wedge[W]({}^\wedge[s]) && \text{AT.1} \\ &= {}^\wedge[W(s)] && \text{ML} \end{aligned}$$

With examples containing indirect contexts, the proof is equally easy. Consider:

'Mary believes that snow is white' means that Mary believes that snow is white.

In formal notation this is:

$$\text{ref}_1\{\mathbf{B(m,{}^\wedge[W(s)])}\} = {}^\wedge[B(m,{}^\wedge[W(s)])].$$

The proof is:

$$\begin{aligned} \text{ref}_1\{\mathbf{B(m,{}^\wedge[W(s)])}\} &= \text{ref}_1\{\mathbf{B}\}(\text{ref}_1\{\mathbf{m}\},\text{ref}_2\{\mathbf{W(s)}\}) && \text{INDIR} \\ &= {}^\wedge[B]({}^\wedge[m],\text{ref}_2\{\mathbf{W(s)}\}) && \text{AT.1} \\ &= {}^\wedge[B]({}^\wedge[m],\text{ref}_2\{\mathbf{W}\}(\text{ref}_2\{\mathbf{S}\})) && \text{FUNC} \\ &= {}^\wedge[B]({}^\wedge[m],{}^\wedge[{}^\wedge[W]]({}^\wedge[{}^\wedge[s]])) && \text{AT.2} \end{aligned}$$

$$
\begin{array}{ll}
= {}^\wedge[\mathrm{B}]({}^\wedge[\mathrm{m}],{}^\wedge[{}^\wedge[\mathrm{W(s)}]]) & \text{ML} \\
= {}^\wedge[\mathrm{B(m,{}^\wedge[W(s)])}] & \text{ML}
\end{array}
$$

When we iterate indirect contexts, the same principles apply. For example, to show:

> '*Agatha believes that Mary believes that snow is white*' means that Agatha believes that Mary believes that snow is white

we need to show:

$$\mathrm{ref}_1\{\mathbf{B(a,{}^\wedge[B(m,{}^\wedge[W(s)])])}\} = {}^\wedge[\mathrm{B(a,{}^\wedge[B(m,{}^\wedge[W(s)])])}].$$

The proof of the meaning-equivalence is just one degree more complex than the truth-equivalence from the previous section:

$$
\begin{array}{ll}
\mathrm{ref}_1\{\mathbf{B(a,{}^\wedge[B(m,{}^\wedge[W(s)])])}\} & \\
\quad = \mathrm{ref}_1\{\mathbf{B}\}(\mathrm{ref}_1\{\mathbf{a}\},\mathrm{ref}_2\{\mathbf{B(m,{}^\wedge[W(s)])}\}) & \text{INDIR} \\
\quad = {}^\wedge[\mathrm{B}]({}^\wedge[\mathrm{a}],\mathrm{ref}_2\{\mathbf{B(m,{}^\wedge[W(s)])}\}) & \text{AT.1} \\
\quad = {}^\wedge[\mathrm{B}]({}^\wedge[\mathrm{a}],\mathrm{ref}_2\{\mathbf{B}\}(\mathrm{ref}_2\{\mathbf{m}\},\mathrm{ref}_3\{\mathbf{W(s)}\})) & \text{INDIR} \\
\quad = {}^\wedge[\mathrm{B}]({}^\wedge[\mathrm{a}],{}^\wedge[{}^\wedge[\mathrm{B}]]({}^\wedge[{}^\wedge[\mathrm{m}]],\mathrm{ref}_3\{\mathbf{W(s)}\})) & \text{AT.2} \\
\quad = {}^\wedge[\mathrm{B}]({}^\wedge[\mathrm{a}],{}^\wedge[{}^\wedge[\mathrm{B}]]({}^\wedge[{}^\wedge[\mathrm{m}]],\mathrm{ref}_3\{\mathbf{W}\}(\mathrm{ref}_3\{\mathbf{s}\}))) & \text{FUNC} \\
\quad = {}^\wedge[\mathrm{B}]({}^\wedge[\mathrm{a}],{}^\wedge[{}^\wedge[\mathrm{B}]]({}^\wedge[{}^\wedge[\mathrm{m}]],{}^\wedge[{}^\wedge[{}^\wedge[\mathrm{W}]]]({}^\wedge[{}^\wedge[{}^\wedge[\mathrm{s}]]]))) & \text{AT.3} \\
\quad = {}^\wedge[\mathrm{B}]({}^\wedge[\mathrm{a}],{}^\wedge[{}^\wedge[\mathrm{B}]]({}^\wedge[{}^\wedge[\mathrm{m}]],{}^\wedge[{}^\wedge[{}^\wedge[\mathrm{W(s)}]]])) & \text{ML} \\
\quad = {}^\wedge[\mathrm{B}]({}^\wedge[\mathrm{a}],{}^\wedge[{}^\wedge[\mathrm{B(m,{}^\wedge[W(s)])}]]) & \text{ML} \\
\quad = {}^\wedge[\mathrm{B(a,{}^\wedge[B(m,{}^\wedge[W(s)])])}] & \text{ML}
\end{array}
$$

2.5 Definitions of Truth and Meaning

Although we have an adequate theory of truth (and of meaning), we do not yet have a *definition* of truth (or of meaning). This is because although we have defined truth and meaning, we have used semantical terminology ('ref' and 'expr') to do so, and this technical semantical terminology has not itself been defined. So we have what Tarski called a theory of truth, but not a definition of truth, and what he would have called a theory of meaning, but not a definition of meaning.

Can we define the notions of referring and expressing? Given the terminology that we have so far, we cannot, at least not without further assumptions. The problem is that we have an infinite number of axioms in our theory. The 'n' in '$\mathrm{ref}_n\{X\}$' is a quantifiable variable, and so we have only a finite number of axioms in FUNC, INDIR, and LINK. But we have an infinite number of axioms in AT.0, AT.1, AT.2, ... If there were a finite number of these, we could use them as the basis steps in recursive definitions of reference and expressing, but not with an infinite number of them. This is a technicality, but an important one if a definition of truth is sought; we have all of the ingredients of a recursive definition except for the fact that there are too many atomic cases (all of AT.n) to put into a single finite clause that would constitute the definition.

The impediment to formulating theories of truth and meaning is Frege's linguistic hierarchy.

3. *Introduction to Frege's Linguistic Hierarchy*

3.1 The Fregean Hierarchy and Definitions of Truth and Meaning

Words shift reference within the context of a that-clause. Does embedding a that-clause in another that-clause have an additional effect on words that are already shifted? For example, does '*snow*' refer to something different in '*Agatha believes that snow is white*' and '*Mary believes that Agatha believes that snow is white*'? This is the question of the Fregean Hierarchy: is each word in a language associated semantically with an infinite number of different senses and references?

The theory is already committed to the view that each word is associated with at least two different references, since any single embedding in an indirect context shifts reference to the customary sense. This is embodied in our principle:

LINK: $ref_1\{X\} = expr_0\{X\}$, where X is not of the form ^[Y].

The fact that this *is* a shift presupposes that the sense expressed by any symbol of the language is different from the reference of that symbol. This background assumption should be made explicit; it is:

S ≠ R: $ref_0\{X\} \neq expr_0\{X\}$, for any X.

This is obviously true in most cases, since most phrases do not refer to senses at all. This principle, though useful in the exposition, is not used in producing theories or definitions of truth or meaning, so its universal validity need not be worried about. In assessing theories, I will assume that this principle is true in all normal cases.

3.2 Single Shift: Reembeddings have no effect

Although embedding a word in an indirect context shifts its reference to what it normally expresses, there is no obvious reason to assume that there are any additional shifts upon further embedding. So one thesis to consider is that reembedding phrases in new indirect contexts has no effect at all on what they refer to or express. That assumption is embodied in the principles:

SINGLE-SHIFT:	$ref_n\{X\} = ref_1\{X\}$,	for n>1.
	$expr_n\{X\} = expr_1\{X\}$,	for n>1.

(See Parsons 1981 for the relation between the doctrine expressed in Frege 1892 and SINGLE-SHIFT.) The main advantage of this proposal is its simplicity together with its theoretical fruitfulness. For if SINGLE-SHIFT is true, we have no need of principles AT.N for n>1, and this solves the problem of producing definitions of reference and expressing, and thus of truth and meaning. We need only replace the infinite set AT.0, AT.1, ... by:

AT.0 $ref_0\{\mathbf{Z}\} = Z$

AT.+ $ref_n\{\mathbf{Z}\} = \wedge[Z]$ for $n \geq 1$.

All earlier appeals to AT.n for n>1 are now replaceable by uses of AT.+ together with SINGLE-SHIFT.

3.3 Infinite-Level Theories

The obvious competitor to SINGLE-SHIFT is that there are new things referred to and expressed with each new embedding:

ω-SHIFT:	$ref_n\{X\} \neq ref_m\{X\}$,	for all $n \neq m$.
	$expr_n\{X\} \neq expr_m\{X\}$,	for all $n \neq m$.

This is Frege's own view, as expressed in correspondence late in his life (Frege 1902). It is difficult to find any direct evidence for or against it (Frege offered none).

If there are new senses and references introduced at each level, it is neat to assume that there is some relation between the two resulting hierarchies; in particular, it is natural to generalize our LINK principle to all levels:

ω-LINK: $\text{ref}_{n+1}\{X\} = \text{expr}_n\{X\}$	for every n.

The main *a priori* argument in favor of Ω-LINK is that without it, or something like it, we are introducing theoretical structures with so little constraints on them that it will be impossible to find any evidence that bears on them. I will assume that Ω-LINK is part of any theory containing Ω-SHIFT. I suspect that the possibility of evidence for or against Ω-SHIFT may depend on making even more assumptions beyond this.

3.4 Rigid Theories

Suppose that we have an infinite level theory. The consequences of Ω-SHIFT depend on the relations among the various things that get referred to when a term is continually reembedded. One possibility is that although reembedding a phrase in a that-clause gives it a new sense, a word's higher level senses are all determined by its customary sense (and consequently, because of Ω-LINK, that its higher level references are all determined by its level-1 references). I call this *Upward Determination*:

$$\text{If } \text{expr}_0\{X\} = \text{expr}_0\{Y\} \text{ then } \text{expr}_n\{X\} = \text{expr}_n\{Y\} \text{ for all n.}$$

A theory that satisfies this constraint I call a *rigid* theory.

In a rigid theory, there is a "projection" function, ⇑, which allows us to "project up" the n+1 level sense of a word from its n level sense:

UP: $\text{expr}_{n+1}\{X\} = \Uparrow(\text{expr}_n\{X\})$	for any X and any n>0

By Ω-LINK, we also have a version of UP for reference:

UP: $\text{ref}_{n+1}\{X\} = \Uparrow(\text{ref}_n\{X\})$	for any X and any $n \geq 1$

The corresponding principle must, of course, hold for the metalanguage as well, so we have the schema:

$$\text{PROJ} \quad \wedge[\wedge[X]] = \Uparrow(\wedge[X])$$

This entails the analogue of UP:

$$\wedge[X] = \wedge[Y] \rightarrow \wedge[\wedge[X]] = \wedge[\wedge[Y]].$$

I say that there is such a function as $\Uparrow$, for this can be proved from the assumption of upward determination. But this is no guarantee that we have a term in our language for it. I make this apparently pointless remark because it is crucial for the question from the last section, whether we can produce *definitions* of referring and expressing:

> *Definitions of Truth and Meaning in Infinite-Level Rigid Theories:* If we have a term for $\Uparrow$ in the metalanguage, then we easily achieve definitions of truth and meaning by adopting UP, and by eliminating all of AT.2 and above, leaving only AT.0 and AT. 1. If we have no term for $\Uparrow$ then there is no apparent way to produce the definitions.

Here is an illustration of how the use of '$\Uparrow$' can replace uses of AT.2 in a proof of the material adequacy condition for truth for '*Agatha believes that Mary believes that snow is white*':

$$\begin{array}{ll}
\mathrm{ref}_0\{\mathbf{B(a,\wedge[B(m,\wedge[W(s)])])}\} & \\
\quad = \mathrm{ref}_0\{\mathbf{B}\}(\mathrm{ref}_0\{\mathbf{a}\}, \mathrm{ref}_1\{\mathbf{B(m,\wedge[W(s)])}\}) & \text{INDIR} \\
\quad = \mathrm{B}(a,\mathrm{ref}_1\{\mathbf{B(m,\wedge[W(s)])}\}) & \text{AT.0} \\
\quad = \mathrm{B}(a,\mathrm{ref}_1\{\mathbf{B}\}(\mathrm{ref}_1\{\mathbf{m}\},\mathrm{ref}_2\{\mathbf{W(s)}\})) & \text{INDIR} \\
\quad = \mathrm{B}(a,\wedge\mathrm{B}(\wedge m,\mathrm{ref}_2\{\mathbf{W(s)}\})) & \text{AT.1} \\
\quad = \mathrm{B}(a,\wedge\mathrm{B}(\wedge m,\mathrm{ref}_2\{\mathbf{W}\}(\mathrm{ref}_2\{\mathbf{s}\}))) & \text{FUNC} \\
\quad = \mathrm{B}(a,\wedge\mathrm{B}(\wedge m,\Uparrow\mathrm{ref}_1\{\mathbf{W}\}(\Uparrow\mathrm{ref}_1\{\mathbf{s}\}))) & \text{UP} \\
\quad = \mathrm{B}(a,\wedge\mathrm{B}(\wedge m,\Uparrow\wedge[W](\Uparrow\wedge[s]))) & \text{AT.1} \\
\quad = \mathrm{B}(a,\wedge\mathrm{B}(\wedge m,\wedge[\wedge[W]](\wedge[\wedge[s]]))) & \text{PROJ} \\
\quad = \mathrm{B}(a,\wedge\mathrm{B}(\wedge m,\wedge[\wedge[W](\wedge[s])])) & \text{ML} \\
\quad = \mathrm{B}(a,\wedge\mathrm{B}(\wedge m,\wedge[\wedge[W(s)]])) & \text{ML} \\
\quad = \mathrm{B}(a,\wedge[\mathrm{B}(m,\wedge[W(s)])]) & \text{ML}
\end{array}$$

Davidson 1965 suggests a link between the possibility of a definition of truth and the possibility of learnability of a language. The situation above illustrates that there is a difference between the questions of whether a truth definition is possible and whether the language is learnable. On the one hand, the metalanguage may already contain a term for the projection function; if it does, a truth definition is obtainable. On the other hand, there may not be in the language a term to express the projection function. In this case we may not be able to produce a truth definition, even though there are only a finite number of primitive terms that need to be independently learned; speakers can project higher level meanings from lower level ones, even though there is no symbol in their language to express a procedure for this. So our ability to produce a truth definition (and a meaning definition) may turn on technical or fortuitous issues, and may be not very well related to the question of whether or not the language is learnable.

3.5 Non-Rigid Theories

The remaining option is that there is no systematic way to predict the meaning or reference of a reembedded phrase from its meaning or reference when singly embedded. In Parsons 1981 I argue against the view that natural language is actually like this, but the arguments are not conclusive, and the opposite view is worth considering. I have no idea how to give definitions of truth and/or meaning within such a theory. Further, a language that works this way *would* seem to suffer from learnability problems; you apparently would have to learn an infinite number of unrelated meanings for each word of the language.

As far as I can see, the options surveyed in sections 3.2–3.5 are all self-consistent, and the only issues regarding their adequacy have to do with their elegance and their application to particular pieces of language. Unfortunately, they are so abstract that it is difficult to get any purchase on what would count as evidence for or against any of them. The situation improves somewhat when we turn to a powerful principle that has not yet been broached: the principle that sense determines reference. This is the subject matter of the next section, where we also return to the hierarchy and to the question of choices among the options.

4. *Does Sense Determine Reference?*

4.1 Frege's View

A slogan often associated with Fregean approaches to semantics is "sense determines reference." Frege himself never said this, so there is no canonical version of the doctrine. And indeed, it is subject to different interpretations, to be explored.

Frege 1892 held that each sense *presents* a unique thing (or perhaps nothing), and an unembedded sign refers to whatever it is (if anything) that its sense presents. Using '$\Downarrow$' for the notion of presentation, Frege's principle about presentation relates expressing and referring for unembedded contexts as follows:

$$\text{PRESENTATION:}\quad \text{ref}_0\{X\} = \Downarrow(\text{expr}_0\{X\}).$$

This principle governs both the parts of language and complex wholes made up of these parts, so the theory prescribes more relationships, and it is thus more powerful. For example, for the simplest of sentences, we now have the following relationships, where the short up arrows represent expressing, the short down arrows represent referring, and the longer ones represent the new presentation function:

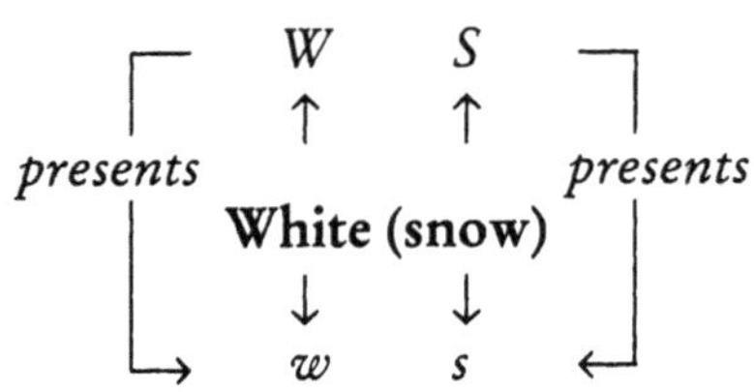

We also have:

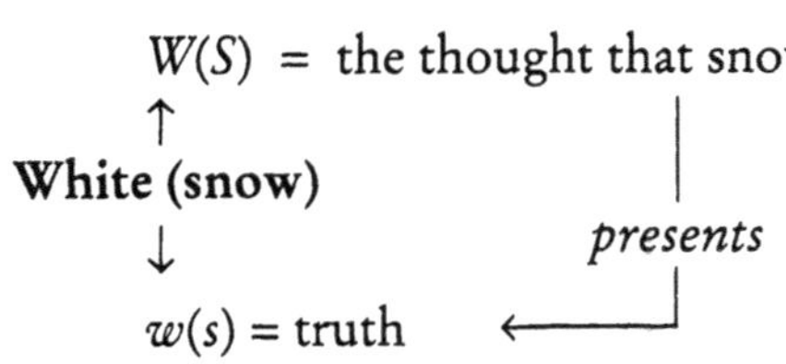

These various relationships interact to overdetermine certain semantic relations. For example, the reference of 'Snow is white' is what its sense presents, and it is also what the reference of its predicate maps the reference of its subject to. So we have these new equivalences:

$$\begin{aligned}\text{ref}_0\{\mathbf{W(s)}\} &= \Downarrow(\text{expr}_0\{\mathbf{W(s)}\})\\ &= \text{ref}_0\{\mathbf{W}\}(\text{ref}_0\{\mathbf{s}\})\\ &= \Downarrow(\text{expr}_0\{\mathbf{W}\})(\Downarrow(\text{expr}_0\{\mathbf{s}\})).\end{aligned}$$

The moral is that the theory now has far-reaching consequences that are not easily anticipated, and simplifications must be scrutinized with care.

Interestingly, no version of the claim that sense determines reference has been involved in formulations of the semantics given in previous chapters. We have been able to produce materially adequate theories of meaning and truth without even raising the issue of the connection between the two, except for Frege's claim that the reference of a singly embedded phrase is the sense of that phrase when unembedded. But that claim does not entail that any word or phrase ever has its reference in a given context determined by the sense it expresses in *that* context. This may be a natural assumption, but it is not needed for the results discussed above. So there is a new claim here to be evaluated by new considerations.

A consequence of claiming that sense determines reference is that no two expressions may (when unembedded) express the same sense without having the same reference:

> DETERMINATION:
> If $expr_0\{X\} = expr_0\{Y\}$ then $ref_0\{X\} = ref_0\{Y\}$.

A related corollary is the validity of this principle, for both the object language and the metalanguage:

If $\wedge[X] = \wedge[Y]$ then $X = Y$.

4.2 Propositional Truth

Once we have decided that sense determines reference, it follows that the senses of sentences determine truth values. So propositions "have" unique truth values. Frege 1918 argues that the notion of truth as applied to propositions is basic, and the notion of truth applied to sentences is derivative. In earlier chapters we have developed the derivative notion of sentential truth. What would we need for a theory of truth as applied to propositions?

Suppose that we have a term in our object language that stands for propositional truth. With it we can write the formal analogues of claims such as:

It is true that snow is white.
That snow is white is true.

These have the form:

$$T(^\wedge[\mathbf{W}(s)]).$$

The reference of 'T' is then a function which maps propositions to their truth values, to the truth values that they present. But the presentation function is, trivially, a function that maps propositions to what they present. In fact, the reference of the propositional truth predicate just *is* the presentation function as applied to propositions. This means that we can define the predicate 'T' as '$\Downarrow$'.

Is this correct? Well, a test of adequacy for propositional truth would be the analogue of the Tarski material adequacy conditions for sentential truth. For example, our theory should let us prove claims such as 'That snow is white is true iff snow is white':

$$T(^\wedge[\mathbf{W}(s)]) \equiv \mathbf{W}(s).$$

And we can do just this. Supposing that 'T' has been defined as '$\Downarrow$', what we need to prove is, e. g.

$$\mathrm{ref}_0\{\Downarrow(^\wedge[\mathbf{W}(s)])\} = \mathrm{ref}_0\{\mathbf{W}(s)\}.$$

The proof goes:

$$\begin{aligned} \mathrm{ref}_0\{\Downarrow(^\wedge[\mathbf{W}(s)])\} &= \mathrm{ref}_0\{\Downarrow\}(\mathrm{ref}_1\{\mathbf{W}(s)\}) && \textsc{Indir} \\ &= \Downarrow(\mathrm{ref}_1\{\mathbf{W}(s)\}) && \text{AT.0} \\ &= \Downarrow(\mathrm{expr}_0\{\mathbf{W}(s)\}) && \textsc{Link} \\ &= \mathrm{ref}_0\{\mathbf{W}(s)\} && \textsc{Pres} \end{aligned}$$

Notice that this proof did not exploit the internal structure of '$\mathbf{W}(s)$', so it can automatically be duplicated for any sentence of the object language.

The notion of propositional truth that meets the Tarski material adequacy condition is useful in analyzing certain inferences, such as this:

Mary believes that snow is white
Everything that Mary believes is true
∴ Snow is white

Although we do not yet have quantifiers in the language, it is clear how the argument will go:

$\mathbf{B}(\mathbf{m},^\wedge[\mathbf{W}(\mathbf{s})])$	Premise
$(\mathbf{x})(\mathbf{B}(\mathbf{m},\mathbf{x}) \rightarrow \mathbf{T}(\mathbf{x}))$	Premise
$\mathbf{T}(^\wedge[\mathbf{W}(\mathbf{s})])$	Predicate logic
$\mathbf{W}(\mathbf{s})$	The material adequacy condition

4.3 Taking Account of Embeddings

A general statement of the principle that sense determines reference needs to take account of the fact that sense and reference vary with context, and here variations abound. The generalized principle seems to demand that given any particular level of embedding, the reference of a word embedded that many times is predictable from the level of embedding together with the sense that it expresses when it is embedded that many times. A consequence of this assumption is the generalization of the principle called DETERMINATION in the previous section: if two phrases express the same sense in a context, then they have the same reference in that context:

> ω-DETERMINATION:
> If $expr_n\{X\} = expr_n\{Y\}$ then $ref_n\{X\} = ref_n\{Y\}$, for any n.

The purpose of this section is to discuss the consequences of the assumption that sense determines reference, in the light of the phenomenon of reference shifting due to embedding. This will involve us in considering a variety of somewhat technical assumptions. Many of these are so abstract that it is difficult to argue directly for or against them. So I will confine myself here to the task of exploring the consequences of adopting them, with particular attention paid to the question of the Fregean hierarchy of senses: does each word of the language have associated with it an infinite number of distinct senses? It turns out that apparently quite different theories end up being equivalent, in a certain sense, to one which assumes that there is no hierarchy at all. This sense of equivalence will be important in what follows, and so it needs careful formation. Simply put, I say that two accounts are *outwardly equivalent* if they assign the same sense and reference to isolated sentences:

> Theory A is *outwardly equivalent* to theory B if they both agree with regard to $ref_0\{S\}$ and $expr_0\{S\}$ for all sentences S.

If the theories assign the same customary senses and the same truth values to sentences, then they are outwardly equivalent, even if they differ regarding internal details, such as the senses of predicates, and what happens in embedded contexts. The rationale for this focus is that internal details are generally unavailable for direct testing; they are only testable indirectly in terms of how they affect the senses and references of the whole (unembedded) sentences in which they occur. Since the way in which internal parts affect the whole is highly theory specific, it is hard to find theory-neutral ways to address these directly. Some quite different theories are outwardly equivalent to one another.

4.4 Presentation in Indirect Contexts

When Frege turns to the semantics of reported speech, he claims that each word has an indirect sense that is different from its customary sense. Why should we think this? Frege's reasoning is this (1960: 59). He begins by noting the principle that is above called "LINK," that

> In reported speech one talks about the sense, e. g., of another person's remarks. It is quite clear that in this way of speaking words do not have their customary reference but designate what is usually their sense.

Although this might be challenged (Carnap 1947 did so), I am accepting it as a hypothesis to be explored. He then continues:

> In order to have a short expression, we will say: In reported speech, words are used *indirectly* or have their *indirect* reference. We distinguish accordingly the *customary* from the *indirect* reference of a word; and its *customary* sense from its *indirect* sense. The indirect reference of a word is accordingly its customary sense. Such exceptions must always be borne in mind if the mode of connexion between sign, sense, and reference in particular cases is to be correctly understood.

Frege does not say here that the indirect sense of a word is different from its customary sense, but context makes it clear that he assumed this, and ten years later (Frege 1902) he explicitly held this. But why?

One possible line of reasoning is that if the customary and indirect senses were the same, so would be the customary and indirect references. Since the indirect reference of a word is its customary sense, this would wrongly identify the word's customary sense with its customary reference. This would be universally true, and would obliterate the sense-reference distinction. Formally, the argument would go:

Suppose:	$\mathrm{expr}_1\{X\} = \mathrm{expr}_0\{X\}$	
Then:	$\mathrm{ref}_1\{X\} = \mathrm{ref}_0\{X\}$	Since sense determines reference
But:	$\mathrm{ref}_1\{X\} = \mathrm{expr}_0\{X\}$	By LINK
So:	$\mathrm{expr}_0\{X\} = \mathrm{ref}_0\{X\}$	Contradicting $S \neq R$.

This is a surprising result. How can the simple intuition that sense determines reference get us so quickly to the conclusion that words have multiple senses? The answer is that the argument does not employ the "simple" intuition that sense determines reference, it employs a highly theoretical extension of that intuition. The theoretical extension is found in the rationale for the inference of the second line from the first. This inference *does not follow* by the simple intuitions articulated so far in the principles DETERMINATION, PRESENTATION, or even ω-DETERMINATION. Frege is here taking some principles that are intuitively natural for unembedded contexts and generalizing them to contexts that it is unlikely we had in mind when judging them plausible. His argument seems to rest on the assumption that the link between what a word expresses and what it refers to is independent of context. He is assuming something like this:

> CONTEXT FREE DETERMINATION:
> If $expr_n\{X\} = expr_m\{Y\}$, then $ref_n\{X\} = ref_m\{Y\}$, for any n and m.

This, I suggest, is not nearly as apparent as DETERMINATION or even ω-DETERMINATION. Suppose we find it plausible that sense determines reference, based on considering typical examples of simple, unembedded contexts. Now we turn to embedded contexts, and we find that here, as well, sense should determine reference. There is still no rationale for Frege's conclusions. What Frege needs is this: after deciding that which sense is expressed may vary with context, he still holds that the connection *between* what is expressed and what is referred to may *not* rely on context. This is a much stronger principle.

It is likely that Frege adopted, in addition to PRESENTATION, a generalization of that principle to all contexts:

> ω-PRESENTATION: $ref_n\{X\} = \Downarrow(expr_n\{X\})$, for any n.

This is the principle that Dummett (1973, 267–68) attributes to Frege, under a different title (Dummett calls *this* the principle that sense determines reference). He criticizes Frege for adopting it. The principle yields Frege's conclusion about indirect senses, since it entails CONTEXT FREE DETERMINATION, as follows:

Suppose: $\text{expr}_n\{X\} = \text{expr}_m\{Y\}$.
Substituting these as arguments of $\Downarrow$ yields:
$\Downarrow(\text{expr}_n\{X\}) = \Downarrow(\text{expr}_m\{Y\})$.
Then, by Ω-PRES: $\text{ref}_n\{X\} = \text{ref}_m\{Y\}$.

The principle ω-PRESENTATION entails all of the principles discussed so far in this section. It also forces each word to express different senses for every level of embedding in indirect contexts, since Frege's argument for indirect sense generalizes recursively to every level. The resulting hierarchy of senses is sufficiently complex as to raise questions about its necessity.

5. *Simplifying the Hierarchy*

5.1 Truncating the Hierarchy

We have seen that if one wishes to reject the infinite hierarchy of senses and references in Frege's theory, one must reject ω-PRESENTATION. But this is easy, since there appears to be no rationale for ω-PRESENTATION, either intuitive or theoretical. The intuitions are:

1: In direct contexts, a term refers to whatever its customary sense presents.

2: In indirect contexts, a term refers to its customary sense.

These are consistent with this additional assumption:

3: In indirect contexts, a term expresses the same sense that it expresses in direct contexts.

The first principle is PRESENTATION and the second generalizes LINK, not by linking the reference of each reembedding with what is expressed by the previous embedding, but by identifying the reference of a sign in any and all levels of embedding with its customary sense. The third principle is the simplification of the theory proposed by Dummett.

Suppose we call this simplified theory SINGLE SENSE. (It is a special implementation of SINGLE SHIFT from section 3.2.) This theory states that each term has a single sense that does not change with embeddings; a term refers to this sense when embedded, and to whatever that sense presents when the term is unembedded. The principles are

CONSTANCY OF SENSE: A sign expresses its customary sense in all contexts.

CONSTANCY OF INDIRECT REFERENCE: An embedded sign refers to its customary sense no matter how many times it is embedded.

PRESENTATION: An unembedded sign refers to whatever is presented by its customary sense.

Formally, they are:

SINGLE SENSE =

C-SENSE:	$expr_n\{X\} = expr_0\{X\}$ for all n.
C-IREF:	$ref_n\{X\} = expr_0\{X\}$ for all $n>0$.
PRES:	$ref_0\{X\} = \Downarrow(expr_0\{X\})$.

SINGLE SENSE is the kind of theory that Dummett 1973 says Frege *should* have adopted. It is apparent that it is not the theory that Frege did adopt. In my opinion, it is one of the two most important versions of the theory of sense and reference. This is because of its simplicity, and also because certain other theories that are apparently quite different from it, and considerably more complex, are outwardly equivalent to it in the sense defined in the previous section; however much they differ from it in terms of the semantical behavior of the parts of sentences, they agree with it for whole sentences occurring autonomously in discourse. This is important from a Fregean perspective because of Frege's insistence (1884) that we should only ask about the meaning of terms in the context of whole sentences. (It is clear that he means autonomously occurring sentences, and that by 'meaning' he means both sense and reference.) I think that many non Fregeans would agree with this perspective; we test a theory by what it tells us about the behavior of whole sentences, together with what we are sure of about the parts. Our surety about the parts is meager, extending at most to the fact that names in unembedded contexts refer to their bearers and that predicates in unembedded contexts are true of certain things and false of certain others. All of the theories under discussion here have no difficulties with these particular facts.

A crucial fact about SINGLE SENSE is this: if two terms agree in their customary senses, then they agree in what they express *and* in what they refer to in all contexts. As a consequence, if two terms agree in customary sense, they will be intersubstitutable everywhere (not counting quotational contexts). This is a crucial piece of data that might be testable. Nonrigid infinite level theories disagree with SINGLE SENSE on this matter, and other Fregean theories do not.

It is interesting to compare SINGLE SENSE with another non Fregean theory, namely, the theory that Carnap (1947) thought Frege should have adopted. Carnap went further than Dummett, holding that the senses *and* references of terms should not shift with context; the only thing that changes in embedded contexts is what is relevant to determining the reference of the whole. Let me use 'refer$_C$' and 'express$_C$' for the semantical relations that Carnap thought Frege should have used. Then Carnap suggests that Frege should have held the following:

> CARNAP:
> A term refers$_C$ to the same thing in every context.
> A term expresses$_C$ the same thing in every context.
> The reference$_C$ of 'F(^[a])' is what you get by applying the reference$_C$ of 'F' to the sense expressed$_C$ by 'a'.
> The sense expressed$_C$ by 'F(^[a])' is what you get by applying the sense expressed$_C$ by 'F' to the sense expressed$_C$ by 'a'.

This eliminates all shifts of reference and expressing, and apparently achieves the same effect as a Fregean theory by letting the references$_C$ of function signs "look at" (apply to) something different than the references$_C$ of the arguments of those functional signs.

It is easy to see that this theory is outwardly equivalent to SINGLE SENSE. We begin by defining the new Carnapian terminology in terms of our old:

$$\text{ref}_C\{X\} =_{df} \text{ref}_0\{X\}.$$
$$\text{expr}_C\{X\} =_{df} \text{expr}_0\{X\}.$$

Then we replace our old principle INDIR by:

$$\text{INDIR}_{\text{CARNAP}}: \quad \text{ref}_C\{\boldsymbol{X}(\wedge[\boldsymbol{Z}])\} = \text{ref}_C\{X\}(\text{expr}_C\{Z\}).$$
$$\text{expr}_C\{\boldsymbol{X}(\wedge[\boldsymbol{Z}])\} = \text{expr}_C\{X\}(\text{expr}_C\{Z\}).$$

It is easy to check that CARNAP is outwardly equivalent to SINGLE SENSE.

5.2 Reduction of Rigid Theories to Single-Sense Theories

If the customary sense of an expression uniquely determines all of its other indirect senses, what difference is there in import between an infinite level rigid theory and one that assigns only one sense to every term? Very little, it turns out. The purpose of this last section is to show

> Given any infinite level rigid theory, there is an outwardly equivalent theory that assigns a single sense to every term.

The argument that follows depends on classifying entities into types in some way so as to distinguish non-senses from senses, functional senses from non-functional senses, customary senses from singly indirect senses, and these from doubly indirect senses, and so on. This can be done in different ways, but the argument can be given in outline without the details, and I think it is more accessible to see it this way.

So suppose that we have a rigid theory formulated in terms of 'ref' and 'expr'. We define new relations, ref*, and expr*, beginning with atomic signs:

$\text{ref}^*\{0,X\} = \text{ref}_0\{X\}$	for any atomic sign X
$\text{ref}^*\{1,X\} = \text{ref}_1\{X\}$	for any atomic non-functional sign X
$\text{ref}^*\{1,F\} = \text{ref}_1\{F\}$	for any atomic functional sign such that $\text{ref}_0\{F\}$ is a function that takes arguments that are not senses

Suppose that F is an atomic functional sign such that $\text{ref}_0\{F\}$ is a function that takes senses of type τ as arguments. Then

$$\text{ref}^*\{1,F\}(x) = \text{ref}_1\{F\}(\Uparrow(x)) \text{ if x is a sense of type } \tau;$$
$$\text{ref}_1\{F\}(x) \text{ otherwise.}$$
$$\text{ref}^*\{n,X\} = \text{ref}^*\{1,X\} \text{ for } n \geq 2, \text{ for any atomic sign X.}$$
$$\text{expr}^*\{n,X\} = \text{ref}^*\{1,X\} \text{ for any atomic sign X.}$$

We define ref* and expr* for complex expressions using analogues of the axioms for Fregean theories (assuming that Y is not of the form ^[Z]):

$$\text{ref}^*\{n,X(Y)\} = \text{ref}^*\{n,X\}(\text{ref}^*\{n,Y\}).$$
$$\text{expr}^*\{n,X(Y)\} = \text{expr}^*\{n,X\}(\text{expr}^*\{n,Y\}).$$
$$\text{ref}^*\{n,X(\wedge[\mathbf{Z}])\} = \text{ref}^*\{n,X\}(\text{ref}^*\{n+1,Z\}).$$
$$\text{expr}^*\{n,X(\wedge[\mathbf{Z}])\} = \text{expr}^*\{n,X\}(\text{ref}^*\{n+1,Z\}).$$

The resulting theory satisfies all of the conditions FUNC, INDIR, AT.0, LINK, ω-DETERMINATION, CONSTANCY OF SENSE, and CONSTANCY OF INDIRECT REFERENCE. Most important, it is outwardly equivalent to the rigid infinite-level Fregean theory.

(This equivalent theory does not satisfy all of the principles of SINGLE SENSE, because we begin with the assumption that both the object language and the metalanguage satisfy the principles of an infinite level rigid theory. We then introduce *new* senses to associate with certain terms of the object language, and the resulting theory associates a single sense with each term. But the metalanguage is unchanged by this reduction, and so we cannot identify the new senses expressed by items in the object language with the old ones already denoted by that-clauses of the metalanguage, for these are the old senses. As a result, certain instances of AT.1, AT.2, ... are not satisfied by the new theory, and thus we do not literally have a version of SINGLE SENSE as previously defined. But the important thing is that the new theory associates exactly one sense with each term of the object language, and the resulting theory is outwardly equivalent to the original.)

This seems to show that if the issue is a choice between the infinite hierarchy of a rigid infinite theory and a theory without the hierarchy, there cannot be any evidence for one over the other, and the choice must be a matter of taste and elegance.

References

Carnap, Rudolf, 1947: *Meaning and Necessity: A Study in Semantics and Modal Logic*. Chicago: University of Chicago Press.

Church, Alonzo, 1951: A Formulation of the Logic of Sense and Denotation. In Paul Henle, H. M. Kallen, and S. K. Langer, eds., *Structure, Method, and Meaning: Essays in Honor of Henry M. Sheffer*. New York: Liberal Arts Press, 3–24.

Church, Alonzo, 1973: Outline of a Revised Formulation of the Logic of Sense and Denotation (Part I). *Noûs* 7, 24–33.

Church, Alonzo, 1974: Outline of a Revised Formulation of the Logic of Sense and Denotation (Part II). *Noûs* 8, 135–56.

Church, Alonzo, 1993: A Revised Formulation of the Logic of Sense and Denotation. Alternative (1). *Noûs* 27, 141–57.

Davidson, Donald, 1965: Theories of Meaning and Learnable Languages. In *Proceedings of the 1964 International Congress for Logic, Methodology and Philosophy of Science*. Amsterdam: North Holland, 383–94.

Dummett, Michael, 1973: *Frege: Philosophy of Language.* London: Duckworth.
Frege, Gottlob, 1884: *Die Grundlagen der Arithmetik, Eine logisch-mathematische Untersuchung über den Begriff der Zahl.* (The Foundations of Arithmetic, A Logico-mathematical Enquiry into the Concept of Number.) Translated as *The Foundations of Arithmetic, A Logico-mathematical Enquiry into the Concept of Number*, translated by J. L. Austin. Evanston, Illinois: Northwestern University Press.
Frege, Gottlob, 1892: "Über Sinn und Bedeutung" ("On Sense and Reference") Translated in Frege 1960.
Frege, Gottlob, 1902: Letter to Russell, 28 December 1902. In Frege 1980, 152–54.
Frege, Gottlob, 1918: "Der Gedanke: Eine logische Untersuchung" ("Thoughts: A Logical Investigation"). Translated in Frege 1984, *Collected Papers on Mathematics, Logic, and Philosophy*, edited by Brian McGuinness. Blackwell.
Frege, Gottlob, 1960: *Translations From the Philosophical Writings of Gottlob Frege*, edited by Peter Geach and Max Black. Blackwell.
Frege, Gottlob, 1980: *Gottlob Frege, Philosophical and Mathematical Correspondance*. Edited by Gottfried Gabriel, Hans Hermes, Friedrich Kambartel, Christian Thiel, and Albert Veraart; translated by Hans Kaal. University of Chicago Press.
Gupta, Anil, 1980: *The Logic of Common Nouns: An Investigation in Quantified Modal Logic.* New Haven: Yale University Press.
Kripke, Saul, 1980: *Naming and Necessity.* Cambridge, Mass.: Harvard University Press.
Linsky, Leonard, 1967: *Referring*. London: Routledge and Kegan Paul.
Linsky, Leonard, 1983: *Oblique Contexts.* Chicago: University of Chicago Press.
Parsons, Terence, 1981: Frege's Hierarchies of Indirect Senses and the Paradox of Analysis. In *Midwest Studies in Philosophy* VI, 37–57.

Frege's Treatment of Indirect Reference

RICHARD L. MENDELSOHN*

Introduction

Frege's story about indirect contexts – that-clauses, like 'Harry believes that' and 'Joan said that' – is widely known and enormously influential. And yet it is only the briefest of sketches. Here is what he says in "On Sense and Reference":

> In reported speech one talks about the sense, e. g., of another person's remarks. It is quite clear that in this way of speaking words do not have their customary reference but designate what is usually their sense. In order to have a short expression, we will say: In reported speech, words are used *indirectly* or have their *indirect* reference. We distinguish accordingly the *customary* from the *indirect* reference of a word; and its *customary* sense from its *indirect* sense. The indirect reference of a word is accordingly its customary sense. ([13], p. 59)

It's really quite elegant, the way he has knitted together the sense/reference distinction with the problem of substitutivity in oblique contexts.[1] But it's not as tight as one might think. For we know what the customary reference of an expression is supposed to be: it's just the thing the word stands for. And we know what the customary sense of an expression is supposed to be: it's, roughly, the meaning of the expression. So we know what the indirect reference is supposed to be, for Frege explicitly identifies the indirect reference with the customary sense. But what is the indirect sense supposed to be?

We can, of course, *speak* of the indirect sense of the sentence 'the cat is on the mat'. We can even say, in the Fregean spirit, that the indirect sense of that sentence contains the mode of presentation of its ordinary

---

* Thanks to Arthur Collins, Jerrold Katz, and the members of my 1994 NEH Summer Seminar on Reference for helpful discussion and criticism of the material in this essay. An earlier version was read as the Dean Kolitch Memorial Lecture at the CUNY Graduate Center.

[1] This is not to endorse his view. On the contrary, the very first sentence of the quoted passage seems to me to be seriously in error.

sense, the thought it expresses, viz. *that the cat is on the mat.* We might even identify the indirect sense of the sentence 'the cat is on the mat' with the customary sense of the name of the thought, '*that the cat is on the mat*'; further, we might contrast it with a different way of introducing that thought, for example, as the thought J. L. Austin etched on the modern philosophical consciousness in his discussion of truth. It is not entirely clear that we can cash this way of speaking in for hard semantics. But even if we can, the problem remains that a sentence can be embedded and reembedded in that-clauses to whatever depth, and, to accommodate these multiple embeddings, we need some *general* method for determining indirect sense.

A good dictionary will provide the customary sense of a word. And, if the entry includes examples of things to which the word applies, it will provide the customary reference as well. But we don't expect to find the indirect sense included in the dictionary entry. Indirect sense should be structurally determined: there should be some way of computing the indirect sense of a word using the customary sense and reference of the word in combination with the properties of the 'that'-operator. But, as Russell observed in "On Denoting," "there is no backward road from denotations to meanings." ([32], p. 50) And, indeed, saying how indirect sense is to be determined has turned out to be a task of considerable difficulty: in fact, no plausible, coherent, systematic, structural way of computing indirect senses has been forthcoming.

Russell, already mentioned, appears to have understood the problem;[2] but it is Carnap who is usually credited with raising the issue of the *infinite hierarchy* for Frege's semantics. In [4], he charged that Frege required an infinite number of distinct names for indirect senses. Davidson made the criticism more pointed: since the indirect sense of a word cannot possibly be a function of its customary sense, he argued, Frege is committed to an infinite number of semantic primitives, an absurd requirement for any natural language.[3] And most philosophers

[2] My current thinking is at odds with the received scholarly view about Russell's criticism of Frege in [32]. I am inclined to believe that it *is* directed at Frege, that it is a cogent criticism of Frege, and that Russell had, in effect, anticipated the later problems others had noted with the infinite hierarchy.

[3] "Neither the languages Frege suggests as models for natural languages nor the languages described by Church are amenable to theory in the sense of a truth definition meeting Tarski's standards. What stands in the way in Frege's case is that every referring expression has an infinite number of entities it may refer to, depending on

do regard this prospect of an infinite hierarchy of indirect senses as a *reductio* of the theory.

Explicitly citing Russell's admonition in "On Denoting," and recoiling from the hierarchy, Michael Dummett has urged that Frege's semantic theory be reshaped so that there is no indirect sense distinct from the customary sense:

> According to Frege, a word does not have a reference *on its own*, 'considered in isolation': it has a reference only in the context of a sentence. It is fully harmonious with this view to hold that, while a word or expression *by itself* has a sense, it does not by itself have a reference at all: only a particular occurrence of a word or expression in a sentence has a reference, and this reference is determined jointly by the sense of the word and the kind of context in which it occurs. The sense of a word may thus be such as to determine it to stand for one thing in one kind of context, and for a different thing in some other kind of context. We may therefore regard an expression occurring in an opaque context as having the same sense as in a transparent context, though a different reference....
>
> With this emendation, there is no such thing as the indirect sense of a word: there is just its sense, which determines it to have in transparent contexts a reference distinct from this sense, and in opaque contexts a referent which coincides with its sense. There is therefore no reason to think that an expression occurring in double *oratio obliqua* has a sense or a reference different from that which it has in single *oratio obliqua*: its referent in double *oratio obliqua* will be the sense which it has in single *oratio obliqua*, which is the same as the sense it has in ordinary contexts, which is the same as its referent in single *oratio obliqua*. This is intuitively reasonable: the replacements of an expression in double *oratio obliqua* which will leave the truth-value of the whole sentence unaltered are – just as in single *oratio obliqua* – those which have the same sense. ([10], pp. 268–9)

Dummett's view, as I understand it, comes to this. The indirect sense of a word just *is* its customary sense. In ordinary contexts, the word stands for its customary reference, but in indirect contexts (at whatever level of indirectness), the word stands for its customary sense.

But Dummett has not quite eliminated the infinite hierarchy. Let us, following Terence Parsons,[4] call an infinite hierarchy *rigid* if expressions that agree in customary sense, agree at every level of indirect

---

the context, and there is no rule that gives the reference in more complex contexts on the basis of the reference in simpler ones. In Church's languages, there is an infinite number of primitive expressions; this directly blocks the possibility of recursively characterizing a truth predicate satisfying Tarski's requirements." ([9], p. 99)

[4] In [29]. I have borrowed much from this very fine analysis of the hierarchy.

sense; otherwise it will be *nonrigid*. Parsons has shown, much to our surprise, that Dummett's two-level reconstruction of Frege is externally equivalent to a *rigid* infinite-level hierarchy.

In a rigid hierarchy, there is a functional relation between customary sense and indirect sense, for, if two expressions have the same customary sense, they have the same indirect sense. This does not mean that each level of indirect sense is identical with customary sense. But it does mean that words having the same customary sense are interchangeable one with the other at any level of indirectness. Dummett's can be viewed as the *smallest* rigid hierarchy, because his functional relation is just identity: customary sense is identical with indirect sense, so all levels collapse, essentially, into the first. But there might be other rigid hierarchies in which, for each i, the sense at level i is distinct from the sense at level $i+1$. Parsons's surprising result is that all rigid hierarchies present equivalent semantic analyses of sentences with multiple embeddings.

The rigid hierarchy, then, constitutes one response to Davidson's charge of absurdity. Some hierarchies are not absurd. In the case of the rigid hierarchy, it might well be that there are infinitely many different indirect senses attached to a given expression. But, because the indirect sense is a function of the customary sense, we don't have to learn them all in order to understand a sentence with multiply-embedded that-clauses. We need only know the customary sense.

There is much sympathy for Dummett's position among those favorable to a Fregean semantics.[5] But it has not won universal agreement. Some Fregeans prefer to take the bull by the horns and deny that the nonrigid hierarchy leads to absurdity in the way Davidson charges. The nonrigid hierarchy has been advocated by Church [6, 7] and Anderson [1, 2], and also by Herbert Heidelberger [19] in his review of [10]. The challenge before them is to explain the structural connection between indirect sense and customary sense, given that the former is clearly not a function of the latter.

The essential idea behind a nonrigid hierarchy is that expressions having the same customary sense, might yet differ in indirect sense. Although 'vixen' and 'female fox' have the same customary sense, a person might not know that they do; the expressions are therefore sub-

[5] Parsons [29], for example, counts himself in the same camp. Forbes [12] contrasts his own position with Dummett's, but counts himself within the broader camp of rigid hierarchists.

stitutable one for the other in any singly-embedded that-clause, but not necessarily in a doubly-embedded one. The nonrigid hierarchist believes he can distinguish the contribution customary sense makes to the semantic interpretation of a deeply embedded sentence from the contribution made my some (at least) of the indirect senses. This is what differentiates him from the rigid hierarchist. The rigid hierarchist believes that it is difficult to pinpoint at which level of indirectness a failure of substitutivity is to be attributed: whatever reason he has for supposing that two expressions differ in indirect sense (at whatever level) is a reason for supposing they differ in customary sense. His solution is to push all differences back to the first level of customary sense. If one can know the customary sense of the expressions 'vixen' and 'female fox', and yet not realize that they have the same sense, then they don't have the same sense.[6]

Both the rigid and the nonrigid hierarchist needs to be able to make judgements of sameness of sense independent of the way in which the terms work in indirect contexts. The rigid hierarchist needs a clear enough conception of the customary sense of a word to avoid the conclusion of Mates's Problem, viz. that no two distinct expressions have the same customary sense. [25] The nonrigid hierarchist needs a clear enough conception of the customary sense of an expression to distinguish it from the indirect sense of the expression. In my own thinking about the notion of sense, I find it very difficult to suppose that we have a clear enough notion of customary sense to avoid Mates's destructive argument that is not also sufficient to distinguish customary sense from indirect sense. On this score, then, I do not find Dummett's position attractive.

This is not to say that I favor the nonrigid hierarchy. I have deep misgivings about such an infinite hierarchy, and I am not convinced that there is a way to make it palatable. But my argument here is against the rigid hierarchy, and, in particular, with Dummett's version. I will argue that while Dummett was busy finessing the issue of infinitely many discrete indirect senses, he became vulnerable to the other horn of Russell's argument, which is the collapse of the sense/reference distinction. My analysis of the problem reveals that the arguments involving substitutivity that Dummett marshals to support his view are inappropriate if his turned out to be the correct view. Moreover, Dummett's reading of Frege appears to require more severe changes to

[6] This is Dummett's argument.

Frege's system than Dummett acknowledges. The upshot of my discussion is that Dummett's proposed reconstruction of Frege is deeply flawed. I am therefore inclined to believe that if Frege's treatment of indirect reference is to be found coherent and fruitful, it must be shown that the Orthodox nonrigid infinite hierarchy poses no absurdity in semantic theory.

The Sense/Reference Story

Let's begin by reminding ourselves of the story about sense and reference.

A proper name, like 'Murphy Brown', has both a sense and a reference. The reference of the name is the woman herself: she is whom you talk about, refer to, mean, if you like, when you ordinarily use the name in conversation. The sense of the name, on the other hand, is, very roughly, whatever it is that enables you to place, pick out, identify, or locate the person you speak about. The sense you attach to the name could be 'the sit-com character played by Candice Bergen' or it could be 'the chief investigative reporter on FYI', or it could be something else.

Frege's syntactical analysis of

(1) Murphy Brown drives a Porsche,

is that the name 'Murphy Brown' combines with the predicate '( ) drives a Porsche' to form a declarative sentence. Frege regards a declarative sentence as a complex name. His semantical analysis of (1) comes in two parts. On the *Bedeutung* side, 'Murphy Brown' refers to the woman and '( ) drives a Porsche' refers to a concept,[7] and the complex name refers to the value of that function for that argument. So, (1) is a name of the True. On the *Sinn* side, 'Murphy Brown' expresses a sense of the woman, '( ) drives a Porsche' expresses a sense-function, and the two combine to form the sense of the whole sentence, the thought or proposition that Murphy Brown drives a Porsche.

I will use 'r(t)' for 'the reference of t', 'e(t)' for 'the sense of t', and the curly braces $\frown$ for the relevant combining of senses or references. Now, abbreviating 'Murphy Brown' by 'b', and '( ) drives a Porsche' by 'P', the semantic analysis of (1) looks like this:[8]

---

[7] I. e., a function whose value is always a truth value.

[8] This way of diagramming Frege's semantic theory is borrowed from Parsons [29].

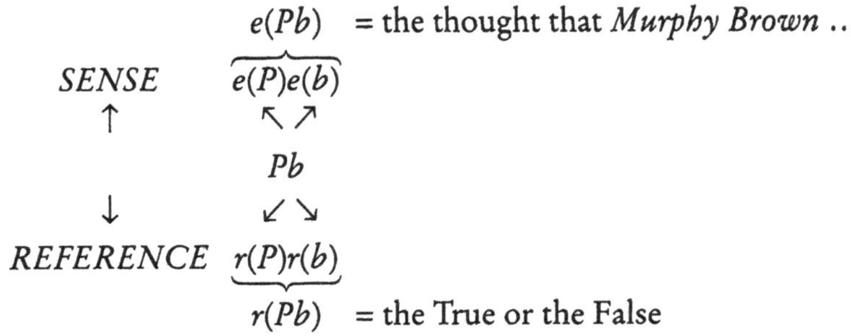

Figure 1

Here is a summary of the semantic principles Frege assumes. Frege characterizes *Sinn* in a number of different ways: as conventional significance, as the common store of knowledge of the referent, as mode of presentation, as an individual's way of picking out an object. Let us not focus on these differences and the unclarities they generate. The central facts are that the sense of a complex is composed out of the senses of its parts, i. e.,

(A) $e(Fa) = e(F)e(a)$;

and the sense of a complex is uniquely determined by the sense of its parts,

(B) If $e(a) = e(b)$, then $e(Fa) = e(Fb)$.

These two principles capture the relation between the sense of a part and the sense of a complex. Now, let's look at the relation between the sense of an expression and its reference. A term refers to what the sense determines; so, although we speak of a term's referring, it is the sense of the term that does the work. We can even say, then, that it is the sense of the term that refers.

(C) $r(t) = r(e(t))$[9].

(C) expresses one part of Frege's view that sense determines reference; the other, the uniqueness of the referent, i. e., the fact that r is a function, is given by

(D) If $e(t) = e(w)$, then $r(t) = r(w)$.

[9] This introduces an ambiguity into the notation, which we take to be harmless. For, we take $r(\,)$ to a be a function which maps *expressions* into their referents as well as a function which maps *senses* into the referents they determine.

Given (C) and (D), we can derive from (A) and (B), Frege's two fundamental principles governing reference:

(E) $r(Fa) = r(F)r(a)$.
(F) If $r(a) = r(b)$, then $r(Fa) = r(Fb)$.

(E) says that all significant parts of the sentence refer. (F) defines the functional relation between the reference of a complex name and the reference of its constituent singular terms. A name is complex for Frege if, and only if, (F) holds for that name; so (F) actually serves as a parsing principle for identifying the significant parts of a sentence. (A) and (E) are frequently identified as *Compositionality* principles; (B) and (F) are frequently identified as *Substitution* principles. (A)–(F) contain the heart of the sense/reference story.

Let's now consider what happens when (1) is embedded in a that-clause, as e. g., in

(2) Ted Koppell believes that Murphy Brown drives a Porsche.

We cannot replace 'Murphy Brown' by just any co-referential singular term and preserve the truth value of the sentence. Nor can we replace the embedded sentence by just any sentence having the same truth value and preserve the truth value of (2). (E) and (F) fail when a declarative sentence is embedded in a propositional attitude context; and since these are derived from (A) and (B), (A) and (B) fail as well.

Frege could not be satisfied with leaving the matter like this. For one thing, he would be abandoning compositionality for a large class of sentences; and compositionality was a compelling idea for Frege (just as it is for many philosophers today). But there is another, more critical, reason. Frege had said that the sense of a declarative sentence is a thought or a proposition: this is what the sentence expresses. Sentence (1) therefore expresses the proposition that Murphy Brown drives a Porsche. Frege had to be able to tell a convincing story that 'expresses that' and 'says that' related a sentence (or person) to a proposition, or else his claim that a sentence expresses a proposition would be incomprehensible. And, of course, the same story would have to be told for the other propositional attitude verbs, because that's the role thoughts or propositions are supposed to play. So, the story Frege told about oblique contexts is not an after-thought or an add-on to the basic account; it's an essential component of the picture.

How did Frege handle these oblique contexts? There are two parts to his solution. First, he says that *that* shifts the reference of the words in its scope, and, second, he relativizes reference to the context in which the term occurs. The *customary* reference of the embedded sentence in (2) is its truth value; but it is not referring to its truth value in that context, so replacing it by an equipollent sentence need not preserve the truth value of (2). Just because two names have the same customary reference, i. e., the same reference in one kind of context, they need not have the same reference in every context in which they occur. But if two names have the same reference appropriate for the context in which they occur, then they are substitutable in that context *salva veritate*. The appropriate reference of a term embedded in a that-clause is its *indirect* reference. So, the truth value of (2) is a function of the indirect reference of its constituent 'Murphy Brown', and substituting another term for 'Murphy Brown' that has the same indirect reference should leave the truth value of (2) unchanged. Compositionality is preserved: (2) is regarded as having parts whose reference contributes to determining the reference of the whole.

Let us use $e_0(t)$ and $r_0(t)$ for the customary sense and reference, respectively, of t, and $e_1(t)$ and $r_1(t)$ for the indirect sense and reference, respectively, of t.[10] And let us abbreviate 'Ted Koppell believes that ( )' by 'K'. Then Frege's semantic analysis of (2) looks like this:

$$
\begin{array}{lcl}
 & e_0(KPb) & = \text{the thought that } \textit{Ted Koppell} \ldots \\
 & \overbrace{e_0(K)e_1(Pb)} & \\
\textit{SENSE} & e_0(K)\overbrace{e_1(P)e_1(b)} & \\
\uparrow & \nwarrow\uparrow\nearrow & \\
 & KPb & \\
\downarrow & \swarrow\downarrow\searrow & \\
\textit{REFERENCE} & r_0(K)\underbrace{r_1(P)r_1(b)} & \\
 & \underbrace{r_0(K)r_1(Pb)} & \\
 & r_0(KPb) & = \text{the True or the False}
\end{array}
$$

Figure 2

Let us introduce some more formalization – just enough to clarify the analysis. First, we will use 'Θ' as a metalinguistic expression standing for '*that*', with parentheses when needed to clarify scope. Frege's

[10] $e_i(t)$ and $r_i(t)$ will, then, be the appropriate sense and reference, respectively, of t when embedded in i that-clauses.

view is that the indirect reference of an expression is what the expression refers to inside the scope of *that*. We will express this principle as follows:

(G) $r_1(t) = r_0(\Theta(t))$.

Furthermore, Frege identifies the indirect reference with the customary sense, i. e.,

(H) $r_1(t) = e_0(t)$.

(G) and (H), in effect, *define* $r_1(t)$, the indirect reference of a term t: the indirect reference of an expression is its customary sense, which is to say that it is the customary reference of that expression inside a that-clause.[11] Next, we assume that indirect reference is compositional, i. e.,

(I) $r_1(Fa) = r_1(F)r_1(a)$.

And this means that

(J) $r_1(\Theta(Fa)) = r_1((\Theta(F))r_1(\Theta(a))$.

So, the compositionality and substitutivity principles necessary for handling that-clauses are

(K) $r_0(F\Theta(t)) = r_0(F)r_1(t)$,

and

(L) If $r_1(a) = r_1(b)$, then $r_0(F\Theta(a)) = r_0(F\Theta(b))$.

(K) tells us that the reference of a complex (which, in this case of a sentence, will be its truth value) is a function of the *appropriate* reference of the part; (L) tells us that if we replace a term by another having the same reference *appropriate for the context in which the term occurs*, then the reference of the complex will remain unchanged. And, since propositions, and senses in general, are detachable, independently existing entities, we have no difficulty speaking about *the customary sense* of a sentence or, what comes to the same thing, about *the indirect reference* of a sentence; so, we have no difficulty identifying (K) and (L) as precisely the principles we want to link the *appropriate* reference of the part with the reference of the complex.

---

[11] This is ambiguous. We are not saying that the indirect reference is the customary reference of the expression that happens to be inside a that clause, for that would collapse the distinction. We are saying that the indirect reference is the customary reference of the expression formed by concatenating that expression with '*that*'.

Some Loose Ends

There are some loose ends in the story that I want to identify, even though I can't tie them up neatly.

The singular term 'Murphy Brown' and the predicate '( ) drives a Porsche' are parts of sentence (1): the sentence is constructed by concatenating these items. The semantic story, however, is a bit more complicated. On the *Bedeutung* side, *driving a Porsche* is supposed to be a function that maps Murphy Brown into a truth value. There is no implication that the reference of the part (viz., Murphy Brown), is a part of the reference of the complex (viz., the True).[12] How do matters stand at the level of *Sinn*? We have seen that the sense of the predicate combines with the sense of Murphy Brown to form a thought. Is this combining function/argument-combining? If so, there would be no more reason to suppose that the sense of 'Murphy Brown' was part of the thought expressed by (1) than there is to suppose that Murphy Brown is part of the True. And there would be no reason to suppose that the proposition expressed by (1) had a structure that mimicked the structure of the sentence. Frege, however, adopted the part/whole reading, so that the thought expressed is a *structured proposition*: the sense of the singular term and the sense of the predicate are both parts of the thought.[13]

A proposition is the *Bedeutung* of a sentence in a that-clause. Which story holds here – function/argument or part/whole? If the former, there is no assurance that the sentence refers inside a that-clause to the very same thing it customarily expresses. But, this is surely wrong: the proposition referred to in (2), i. e., the one Ted Koppell believes, must be the very same as the one expressed by (1). The same story must be told each time. We've already decided that it's part/whole at the level of sense. So, it must be part/whole at the level of reference. But, then, we don't have

---

[12] See the relevant section in [13] and Dummett's comments in [10], pp. 158–9. I discuss these issues in [26].

[13] This passage taken from his "Notes for Ludwig Darmstaedter" is typical:

> We can regard a sentence as a mapping of a thought: corresponding to the whole-part relation of a thought and its parts we have, by and large, the same relation for the sentence and its parts. ([17], p. 255)

Dummett argues for this view in [10]. Parsons [29], who takes the sense of a function-expression to be a *function*, and not just incomplete in some way analogical to the reference of function-expressions, does not appear to hold this view. The structured proposition view has been ably argued by Richard [31]; the idea that propositions are unstructured has been argued by Stalnaker [34].

the function applied to the argument to yield a value, which is the way *referring* is supposed to work. I have no solution to this puzzle.

There is another, closely related, puzzle in Frege's treatment of indirect reference. If we can refer in one context to the same thing expressed in another, what does the distinction between referring to something and expressing it come to?

In the simplest case, (1), there is no mistaking the reference of 'Murphy Brown', i. e., the woman, with the sense of the name – however you choose to take it. They are very different kinds of things. We tend to distinguish referring and expressing by these different kinds of things. But in the cases we are considering, we have a proposition each time. So, we cannot account for the difference between referring and expressing by appealing to a difference in the kind of thing referred to or expressed. But we are also unable to appeal to a difference in the way in which the proposition is engaged, if I might put it that way. In the straightforward case, again, when we talk of the reference of (1), we have a function *applied to* an argument, yielding a value; when we talk of sense, on the other hand, a function is not applied to an argument, but is bracketed or exhibited inert as part of the thought.[14] But that distinction has been eliminated when we consider expressing and referring to a proposition: for either we require a function/argument analysis for each or we require a part/whole analysis for each. So, there's no way to mark the distinction between referring and expressing this way. Frege cannot even appeal to some pre-analytic intuition about referring to help him out here. To be sure, he does: he says, in the passage quoted at the very beginning, that in indirect speech, we mean to talk about the sense of the expressions inside the that-clause. But it certainly does not ring true to my intuitions that when I assert (2), I mean to talk about the meaning of the name 'Murphy Brown'. I would sooner say that I mean to be speaking about Murphy Brown: she's the one I claim Ted Koppell believes to drive a Porsche. *Referring* in this context has been lifted far off its intuitive moorings and is almost entirely an internally defined, theoretical notion for Frege. So, it is far from obvious *how* to make the distinction between referring to and expressing a proposition; indeed it is far from obvious whether there is such a distinction.[15]

---

[14] See Kaplan [20] for this way of characterizing the distinction.

[15] Searle, in defending Frege against Russell's objections in [32], says:

> Russell's arguments suffer from unclarity and minor inconsistencies throughout and I have tried to restate them in a way which avoids these. But even in their restated form, they are faulty. Their faults spring from an initial mis-statement of

One last point. Russell also held a structured proposition view, but unlike Frege, Russell thought that Murphy Brown, not the sense of 'Murphy Brown', was contained in the proposition expressed by (1). Now, it is well known that Frege explicitly rejected Russell's idea that an object could be part of a thought.[16]

Why did he think this? Certainly, he believed that some objects, in his technical sense of 'object', were parts of thoughts. The sense of 'Murphy Brown' is an object in this technical sense, and it can surely be part of a thought. Perhaps Frege thought that Murphy Brown was the wrong type of object to be a part of a thought, a physical object, not a mental object. I cannot rule this out as a factor. But I don't think it was decisive, because Frege also denied that a logical object, like the number 2, was part of the thought expressed by '2+2=4'. The number 2 is what the numeral '2' refers to, not what it expresses. So, it does not appear to be the materiality that is so important. The most plausible explanation I can come up with is this: Frege takes a proposition to be a representation – it is a mode of presentation of the True – and the elements of the proposition are also representations – of the objects and concepts they refer to. An item that is not representing something cannot be a significant part of a proposition. So, since Murphy Brown is not representing anything, it does not belong in the thought expressed by (1); furthermore, including Murphy Brown in a proposition about Murphy Brown confuses the signified with the signifier. So, even if a sense is referred to, that sense cannot be part of the thought expressed, but only a sense of that sense, something that represents it. Frege is, therefore, assuming a NO SELF-REFERENCE PRINCIPLE: no sense can refer to itself, i. e., for $i = 0$ or $i = 1$,

(M) $e_i(t) \neq r_i(t)$.

So, in particular, the indirect reference (i. e., the customary sense) of an expression must be distinct from its indirect sense, since, on his view,

---

> Frege's position, combined with a persistent confusion between the notions of *occurring as a part of a proposition* (being a constituent of a proposition) and *being referred to by a proposition*. The combination of these two leads to what is in fact a denial of the very distinction Frege is trying to draw and it is only from this denial, not from the original thesis, that Russell's conclusions can be drawn. ([33], p. 342)

Searle is quite right in underscoring Frege's desire to distinguish being a constituent of a proposition and being referred to by a (part of a) proposition. But, I have argued, it is not obvious that Frege had successfully made the distinction when propositions themselves were the subject of discourse. And these are just the cases that exercised Russell. (See [5]). It is no misreading on Russell's part to point this out.

[16] See the correspondence between Frege and Russell in [18].

the customary sense is what the indirect sense presents. Further evidence that Frege assumes a principle like (M) is that some such principle is needed to get the hierarchy off the ground.[17]

The Infinite Hierarchy

Frege says nothing in his published writings about how the story of indirect sense and reference is to be extended to doubly-embedded sentences, and to even more deeply embedded sentences. But in a letter to Russell dated December 28, 1902, Frege says that in

(3) The thought that all thoughts belonging to class M are true does not belong to class M,

the second occurrence of 'M' has its customary reference, but the first occurrence of 'M', which is in the underlined part, has its indirect reference. He contrasts (3) with

(4) The thought that the thought that all thoughts belonging to class M are true does not belong to class M.

And he says:

> Since 'M' has different meanings in its two occurrences in [(3)], there must also be a difference in the meaning of 'M' in [(4)]. It can be said that in the twice-underlined part it has an indirect meaning of the second degree, whereas in the once-underlined part it has an indirect meaning of the first degree. ([18], p. 153)[18]

Clearly, Frege believes that the reference of the term when doubly embedded must be different from its reference when singly embedded.

The picture that emerges from Frege's remarks is the one many philosophers thought Frege held; in fact Parsons [29] calls it the *Orthodox* view. Each term t has two sequences, E_t and R_t, associated with it as follows:

---

[17] I am inclined to think that Frege is actually assuming something stronger, which I term a *hereditary* no self-reference principle, namely, for $j \leq i$,
(N) $e_i(t) \neq r_j(t)$.
See [3], [12], and [29] for good discussions of the principles needed for the hierarchy.

[18] We have reproduced the example as Frege gives it in the text. In particular, (3) is a sentence while (4) is not. Parsons [29] and Linsky [23] have drawn our attention to this passage.

$$\begin{array}{lcccccc} E_t & <e_0(t), & e_1(t), & e_2(t), & \ldots, & e_i(t), & \ldots > \\ & \downarrow \; \searrow\!\!\searrow & \downarrow \; \searrow\!\!\searrow & \downarrow & & \downarrow & \\ R_t & <r_0(t), & r_1(t), & r_2(t), & \ldots, & r_i(t), & \ldots > \end{array}$$

Figure 4

$e_i(t)$ refers to $r_i(t)$, and for $i > 0$, $r_i(t) = e_{i-1}(t)$. Each time a term is embedded in another that-clause, it shifts its reference one notch up to an item distinct from any it refers to in less deeply embedded contexts. So, whereas in (2), the singly-embedded 'Murphy Brown' stands for its indirect reference r_1 ('Murphy Brown'), in

(5) Myles Silverberg said that Ted Koppell believes that Murphy Brown drives a Porsche,

being doubly embedded, it must stand for the distinct indirect indirect reference r_2 ('Murphy Brown'). Let us abbreviate 'Myles said that ( )' by 'M'. Then Frege's semantic analysis of (5) looks like this:

$$\begin{array}{lcl} & e_0(MKPb) & = \text{the thought that } \textit{Myles said that} \ldots \\ & \overbrace{e_0(M)e_1(KPb)} & \\ & e_0(M)\overbrace{e_1(K)(e_2(Pb)} & \\ SENSE & e_0(M)e_1(K)\overbrace{(e_2(P)e_2(b))} & \\ \uparrow & \nwarrow\!\nwarrow \;\; \nearrow\!\nearrow & \\ & MKPb & \\ \downarrow & \swarrow\!\swarrow \;\; \searrow\!\searrow & \\ REFERENCE & r_0(M)r_1(K)\underbrace{r_2(P)r_2(b)} & \\ & r_0(M)\underbrace{r_1(K)r_2(Pb)} & \\ & \underbrace{r_0(M)r_1(KPb)} & \\ & r_0(MKPb) & = \text{the True or the False} \end{array}$$

Figure 5

Carnap objected to Frege's story. He complained that

> Frege's method leads ... to an infinite number of entities of new and unfamiliar kinds; and, if we wish to be able to speak about all of them, the language must contain an infinite number of names for these entities. ([4], p. 130)

The proof he offered, as a number of commentators have pointed out, is faulty;[19] As a matter of fact, Frege's argument is also faulty, and for roughly the same reason. The problem with the argument is this: Frege fails to establish a distinct indirect reference of the second degree. Let me rewrite (3) and (4) as

(6) The thought that all thoughts belonging to class M_1 are true does not belong to class M_2,

and

(7) The thought that the thought that all thoughts belonging to class M_3 are true does not belong to class M_4.

respectively. Now, Frege says that the two occurrences of 'M' in (6) cannot refer to the same thing, i. e.,

(8) $M_1 \neq M_2$

and also that the two occurrences of 'M' in (7) cannot refer to the same thing, i. e.,

(9) $M_3 \neq M_4$.

(8) seems to be apparent. For, as Frege says,

(10) $M_2 \neq M_4$,

since one is extensional and the other not, and

(11) $M_1 = M_4$,

because both occurrences appear to be embedded to the same degree. But (9) follows only if Frege has something like a hereditary NO SELF-REFERENCE Principle to fall back on. But he does not explicitly articulate such a principle.[20]

Carnap's objection, unfortunately, set philosophers down a strange, and irrelevant, path. The task was set first, to find out which Fregean principles would generate such an infinite hierarchy – this became a problem because Carnap's argument for the hierarchy failed; [12, 24] and second, to find out which principles could be safely jettisoned so as to avoid the infinite hierarchy. [12] The problem was thus seen as a straightforward cardinality problem: too many senses (in fact, infinitely

[19] See [24] for a criticism of Carnap.

[20] Thanks to John Justice for identifying an error in a previous version of this paper.

many), and too many names of them (in fact, infinitely many). But why is this a problem? And why should this be a problem for Carnap? Carnap, surely, admitted infinitely many intensions into the system described in [4]; and he certainly included enough names to talk about all of them. So, why should *he* be disturbed about them?

I think, however, that the issue of the infinite hierarchy Carnap raised has been misunderstood. The issue is not one of how many senses or how many names we need. The issue is whether there is some regular – perhaps algorithmic – way of determining the sense of an expression when embedded *i*-times in that-clauses. Even if it turned out that there were infinitely many indirect senses required, we would not be particularly troubled if we had some rule for computing the *i*-th element of the sequence E_t. Given $e_0(t)$, the customary sense of t, we want to be able to determine $e_1(t)$, the indirect sense of t, and then to determine $e_2(t)$, the doubly indirect sense of t, and so on.

This is the problem Russell saw: given the customary sense of a term, compute its indirect sense. How can this be done? The customary sense is supposed to be what the indirect sense refers to. Therefore, we are being asked to compute the sense from the reference: we are being asked to forge a backward road from denotation to meaning. Not only must it be a unique sense (for it is a function), but somehow, we should be able to figure out what this sense is simply by examining its referent. Now, if we accept the idea that there is no backward road, then we have no regular way of figuring out the items in the sequence E_t: we have, in effect, infinitely many semantic primitives, as Davidson would say. This is not a comfortable position for a semantic theory to be in. The obvious way of getting around this problem is to hold that the customary sense of a term, so to speak, self-presents itself. And this is the route Dummett takes.

Collapsing the Hierarchy

In my characterization of the sequence E_t, I assumed that e_i was unique. In this, I was simply following Frege's lead when he spoke of *the* indirect sense of a term. Is this assumption justified? Couldn't there be more than one indirect sense for a given sense? There are strong reasons for thinking so: surely, a sense can be presented in different ways, just as any object, say, the Evening Star, can be.

Consider, for example, the following two identities:

(12) The proposition that Giorgione was so-called because of his size = the proposition that Giorgione was so-called because of his size,

and

(13) The proposition that Giorgione was so-called because of his size = the proposition that is expressed by sentence (2) in Quine's "Reference and Modality."

These two identities clearly differ in sense: (13) is much more informative than (12). And the reason (13) is nontrivial is that I have picked out one and the same proposition in two different ways: first, directly, as *the proposition that Giorgione was so-called because of his size*, and second, indirectly, as *the proposition expressed by sentence (2) in Quine's "Reference and Modality."* Propositions, which are the senses of complete declarative sentences, can be presented in different ways just like planets and numbers. There remains, then, this problem: the clear Fregean intuition that senses, like other objects, can be presented in different ways cannot require us to reject talk of *the* indirect sense of a term.

A nominalization like

(14) the proposition that Giorgione was so-called because of his size,

appears to be a rigid designator: it designates the same object, a particular proposition, in every possible world. By contrast,

(15) the proposition that is expressed by sentence (2) in Quine's "Reference and Modality,"

is clearly not a rigid designator: it is only a contingent fact that *that* proposition is expressed by that sentence. But (14) does not embody a rigid *description* of the proposition; to the contrary, it seems to be *directly referential* in Kaplan's sense. [20] This means that the proposition is presented to us without mediation, without any intervening sense: it is self-presenting. To be sure, then, although a proposition can be presented in different ways, when a sentence occurs in a that-clause, there is no further mode of presentation of the proposition expressed that we identify as the indirect sense, but only the proposition itself: the indirect sense just is the customary sense. This is Dummett's idea.

Here is Dummett's argument [10] to this effect.

Let us suppose that 'is similar to' has the same customary sense as 'resembles', so that we can substitute one for the other *salva veritate* when singly embedded in a that-clause. So,

(16) Barry thinks that Harvard is similar to Oxford,

and

(17) Barry thinks that Harvard resembles Oxford,

must have the same truth value. If these two expressions were nonetheless to differ in indirect sense, then

(18) Ayrton knows that Barry thinks that Harvard is similar to Oxford,

and

(19) Ayrton knows that Barry thinks that Harvard resembles Oxford,

need not have the same truth value. "But this," Dummett claims, "is contrary to intuition."

> The only case in which it might seem plausible to say that Ayrton knew that Barry thought that Harvard resembled Oxford, but did not know that he thought they were similar, is that in which Ayrton is ignorant of, or mistaken about, the sense of the word 'similar': but, if we admit this as a legitimate counter-example, then we likewise ought to deny that it follows from Barry's thinking that Harvard resembles Oxford that he thinks they are similar; and, if we deny this, we reject Frege's whole theory of senses as indirect referents. ([11], p. 92)

That is, whatever reason we had for supposing that the two terms differed in indirect sense would equally be a reason for supposing that they differed in customary sense; so if two terms had the same customary sense, they would have to have the same indirect sense.

The very same line of argument, however, that Dummett has been laying down for us, leads to the view that the customary sense of a term is the same as its indirect sense. For, consider the argument to show that if (16) and (17) have the same truth value, then (18) and (19) must have the same truth value. We could only make sense of (18) being true while (19) is false by supposing that Ayrton was mistaken about the sense of the word 'similar'. That is, we supposed that he did not realize that

(20) the indirect sense of 'similar' = the indirect sense of 'resembles';

and we could not see how he could fail to realize this without failing to realize

(21) the sense of 'similar' = the sense of 'resembles'.

There was no way we could attribute the error to (20), without attributing it to (21). So, there was no way we could distinguish the contribution of the higher level sense

(22) the indirect sense of 'similar'

from the lower level sense

(23) the sense of 'similar'.

So, the very same reasons which led Dummett to require that (20) is true when (21) is true, also leads him to require that (22) and (23) are the same thing.

Here is how the story is supposed to work. We consider our original sentence, (1). (1) expresses its customary sense and refers to its customary reference. When embedded in (2), however, it refers to its customary sense. 'Ted Koppell believes that ( )' is not embedded, so it refers to its customary reference, a function that maps the customary sense of (1), i. e., the thought it expresses, into a truth value. When (2) itself gets embedded, as in (5), the doubly-embedded (1) does not change reference, but the singly-embedded 'Ted Koppell believes that ( )' is shifted to refer to its customary sense, and 'Myles Silverberg said that ( )', which is not embedded, refers to its customary reference, a function that maps the thought that Ted Koppell believes that Murphy Brown drives a Porsche into a truth value. The semantic picture for (2) is the same as before (see Figure 2). But the semantic picture for (5) is somewhat different from Figure 5.

Dummett's rejection of indirect sense amounts, as I have presented it, to identifying indirect sense and customary sense. In extensional contexts, an expression refers to its customary reference; but in that-clauses, no matter how deeply embedded, the expression refers to its customary sense. The principle Dummett consciously rejects is that sense determines reference: for, Dummett holds that it is not sense alone, but sense plus context, that determines reference. The other principle he rejects, is the NO SELF-REFERENCE PRINCIPLE (M): for the indirect sense just is the customary sense, and so, in a singly indirect context, the indirect sense and reference are the same.

It is time now for Russell to drop his other shoe.

We get, instead,

$$e_0(MKPb) \qquad = \text{the thought that } \textit{Myles} \ldots$$

$$\overbrace{e_0(M)e_1(KPb)}$$

$$e_0(M)\overbrace{e_1(L)e_1(Pb)}$$

$$e_0(M)e_1(L)\overbrace{(e_1(P)e_1(b))}$$

$$\nwarrow\nwarrow \quad \nearrow\nearrow$$

$$MKPb$$

$$\swarrow\swarrow \quad \searrow\searrow$$

$$r_0(M)r_1(L)\underbrace{r_1(P)r_1(b)}$$

$$r_0(M)\underbrace{r_1(L)r_1(Pb)}$$

$$\underbrace{r_0(M)r_1(KPb)}$$

$$r_0(MKPb) \qquad = \text{the True or the False}$$

Figure 6

Russell's Other Shoe

Russell [32] claimed that if we tried to forge a logical relation between sense and reference, Frege's account of indirect contexts would lead to semantic anomalies. In particular, he claimed that the two sentences,

(24) The center of mass of the solar system is a point,

and

(25) The sense of 'the center of mass of the solar system' is a point,

would turn out to express the same proposition. These two, of course, do not express the same proposition; they do not even have the same truth value. For, while (24) is true, (25) is most certainly false: a sense is not a point. Both sentences are well-formed and meaningful – meaningful even for Frege, since on numerous occasions, he explicitly distinguished senses from other things, like ideas. [13, 16]

Let's try to reconstruct Russell's argument. Each of the following is true and unproblematic.

(26) 'The center of mass of the solar system is a point' expresses the proposition that the center of mass of the solar system is a point.

(27) 'The sense of "the center of mass of the solar system" is a point' expresses the proposition that the sense of "the center of mass of the solar system" is a point.

Now, since, in an indirect context, a term shifts its reference to its sense, in

(28) The proposition that the center of mass of the solar system is a point,

the expression 'the center of mass of the solar system' refers to its sense, namely,

(29) the sense of 'the center of mass of the solar system';

and in

(30) the proposition that the sense of 'the center of mass of the solar system' is a point,

the expression 'the sense of "the center of mass of the solar system"' refers to its sense, namely,

(31) the sense of 'the sense of "the center of mass of the solar system"'.

We argued, in the previous section, that iterated senses collapse, and so, it would seem, (29) and (31) are the same, so the propositions (28) and (30) are the same.

But what does the collapse of the iterated senses come to? In our reconstruction of Russell's argument, we said that the collapse entailed that (29) and (31) were the same. This means that we were assuming

(32) the sense of 'the sense of t' = the sense of t

Now, (32) is not a straightforward iteration of the function 'the sense of'. Moreover, (32) is seriously objectionable. For, what (32) says is that t and 'the sense of t' have the same sense. But how can this be? If they had the same sense, they would have the same reference. But, let t be 'Giorgione was so-called because of his size'. The reference of t is a truth value; the reference of 'the sense of t' is a proposition. So, (32) does not appear to be the right way of understanding the collapse of the hierarchy argued for in the previous section; and therefore, our reconstruction of Russell's argument, which depends on (32) is unacceptable.

It is perhaps worth noting at this point that

(33) the sense of the sense of t = the sense of t

also fails to capture correctly the collapse of the hierarchy argued for in the previous section. To be sure, we have here a straightforward iteration of senses. But, as we remarked earlier, a sense can have more than one sense. Let t be 'Giorgione was so-called because of his size'. Now, the sense of t will be a proposition, *the proposition that Giorgione was so-called because of his size*. But there is no unique object that is the sense of this (i. e., the sense of the sense of t): this is why (12) and (13) differ in cognitive value. (33) is just what Dummett seeks to avoid in his collapse of the hierarchy.

But Russell did not couch his argument using standard philosophical quotation marks. He supposed that when an expression was enclosed in quotation marks, one referred, not to the expression, but to its meaning. So, Russell's use of quotation marks is closer to Kaplan's meaning quotes.[21] Without going into details about this interpretation of Russell, I will simply propose Frege's reference shifter *that* as the device to do Russell's work of referring to meanings or senses. And, the collapsing of indirect senses argued for in the previous section comes to this:

(O) $r_0(\Theta\Theta(Fa)) = r_0(\Theta(Fa))$.

Together with (G) and (J) identified earlier, (O) will enable the argument to go through.

We assume

(34) 'the center of mass of the solar system is a point' expresses *that*(the center of mass of the solar system is a point).

From which it follows, by principle (J), that

(35) 'the center of mass of the solar system is a point' expresses *that*(the center of mass of the solar system)*that*(is a point).

Next, we assume the truism

(36) '*that*(the center of mass of the solar system) is a point' expresses *that*(*that*(the center of mass of the solar system) is a point).

---

[21] See [21], pp. 120–121.

And, once again, we use principle (J) to get

(37) '*that*(the center of mass of the solar system) is a point' expresses *that that*(the center of mass of the solar system) *that*(is a point).

Now, from principle (O), we have

(38) *that*(the center of mass of the solar system) = *that that*(the center of mass of the solar system).

So,

(39) *that* the center of mass of the solar system is a point = *that*(*that*(the center of mass of the solar system) is a point).

It therefore follows, that since these are the very same proposition, the two sentences

(40) the center of mass of the solar system is a point,
(41) *that*(the center of mass of the solar system) is a point,

express the very same proposition.[22,23]

Russell is vindicated!

Conclusion

We have reached a truly disastrous result for Dummett and for Frege. Let me close by trying to see if I can explain what has happened.

Let's simplify Figure 4, where we introduced the relation between E_t, the sequence of senses, and R_t, the sequence of references, for a term t. Let's suppose that we have only two items in each sequence, the customary and the indirect sense or reference:

---

[22] The steps in the argument are more readily accessible when put in symbols:
34 '*Fa*' expresses ΘFa
35 '*Fa*' expresses $\Theta F\Theta a$
36 '$F\Theta a$' expresses $\Theta F\Theta a$
37 '$F\Theta a$' expresses $\Theta F\Theta\Theta a$
38 $\Theta\Theta a = \Theta a$
Now, (34) and (36) are truisms; (35) and (37)are each sanctioned by (I); (38) is sanctioned by (O). By (38), the items to the right of 'expresses' are the same in (35) and (37), so the items on the left, in each case – which becomes (40) and (41) – express the same proposition.

[23] This argument has an ancestor in [3]. See [27].

$$\begin{array}{lcc} E_t & < e_0(t), & e_1(t) > \\ & \downarrow & \downarrow \\ R_t & < r_0(t), & r_1(t) > \end{array}$$

Figure 7

Now, as we look at Figure 7, it appears that reference crops up twice, once in the sequence R_t, and once again in the relation between the items in the sequence E_t and the items in the sequence R_t that we represented by the downward arrow. Furthermore we appear to have a relativized notion in the sequence R_t, and an unrelativized notion represented by the arrows. Figure 7 is obviously misleading, and we must show how the two notions of reference are reduced to one. There are two ways of doing this.

One way is to take the indirect reference of t, $r_1(t)$, to be the customary reference of the indirect sense of t, $r_0(e_1(t))$. This reduces the two referring relations to the arrow:

$$\begin{array}{lcc} E_t & < e_0(t), & e_1(t) > \\ & \downarrow & \downarrow \\ R_t & < r_0(e_0(t)), & r_0(e_1(t)) > \end{array}$$

Figure 8

The other is to reduce the two referring relations to the items in the lower sequence, like so:

$$\begin{array}{lcc} E_t & < e_0(t), & e_1(t) > \\ & \downarrow & \\ R_t & < r_0(e_0(t)), & r_1(e_0(t)) > \end{array}$$

Figure 9

On the first view, that given in Figure 8, we are supposing that the customary reference of the indirect sense of t is the indirect reference. But, then, the customary sense and the indirect sense of t would have to be distinct so that the customary reference and the indirect reference of t are distinct. This seems to be the picture Frege had in mind. It is this picture that seems to lead to infinitely many semantic primitives; it is this picture, as I understand him, that Dummett seeks to jettison.

In Figure 9, the indirect reference of t is not the customary reference of the indirect sense of t; it is the indirect reference of t, i. e., what t

refers to *in a certain context*.[24] $r_1(e_0(t))$ refers to the indirect reference *only in an indirect context*. If we could refer to the indirect reference in a nonembedded context, then

(42) $r_0(e_1(t)), = r_1(e_0(t)),$

would be true and the distinction between sense and reference collapses. *This* is how we collapsed the distinction with (O).

Did we do something unfair? I don't think so. Dummett's argument against the nonrigid hierarchy required that he make judgements about the indirect reference (i. e., customary sense) of an expression in a nonembedded context (as, for example, with (20)–(23)). But sentences of the form,

(43) the indirect reference of t = the indirect reference of w,

are extremely problematic for Dummett since they require that we speak of the indirect reference of an expression in nonembedded contexts. The following also turns out to be problematic,

(44) that Giorgione was so-called because of his size = Quine's favorite proposition,

because one and the same sense is being introduced in an embedded context on the left-hand side of the equals-sign and in a nonembedded context on the right-hand side of the equals-sign. So, there are serious questions about Dummett's position: the picture argued for is that in Figure 9, but the arguments given assume the picture given in Figure 8. Finally, Frege requires that we be able to refer to the indirect reference in a nonembedded context in order to make sense of the relativized substitution principle (L). So, Dummett's way out for Frege would work only if Frege abandoned (I) – which is the heart of Frege's story about oblique contexts.

References

[1] C. Anthony Anderson, "Semantical Antinomies in the Logic of Sense and Denotation," *Notre Dame Journal of Formal Logic* 28 (1987), 99–114.
[2] C. Anthony Anderson, "Some New Axioms for the Logic of Sense and Denotation: Alternative (0)," *Noûs* 14 (1980), 217–234.

[24] Note the irrelevance of $e_1(t)$ in the sequence, and presumably of $e_i(t)$, for $i \geq 1$, if we were to extend the sequence indefinitely. This is what enables Parsons to say that Dummett's semantic analysis is the same as that given by any rigid hierarchy.

[3] Tyler Burge, "Frege and the Hierarchy," *Synthese* 40 (1979), 265-281.
[4] Rudolf Carnap, *Meaning and Necessity: A Study in Semantics and Modal Logic*, 2nd ed., University of Chicago Press (1947: Chicago).
[5] Richard Cartwright, "On the Origins of Russell's Theory of Descriptions," in his *Philosophical Essays*, The MIT Press (1987: Cambridge), pp. 95–134.
[6] Alonzo Church, "Outline of a Revised Formulation of the Logic of Sense and Denotation, II," *Noûs*, 8 (1980), 135–156.
[7] Alonzo Church, "Outline of a Revised Formulation of the Logic of Sense and Denotation, I," *Noûs*, 7 (1980), 24–33.
[8] Alonzo Church, review of [4], *The Philosophical Review* 52 (1943), p. 302.
[9] Donald Davidson, "On Saying That," reprinted in *Inquires into Truth & Interpretation*, Oxford University Press (1984: Oxford), pp. 93–108.
[10] Michael Dummett, *Frege, The Philosophy of Language*, Duckworth (1973: London).
[11] Michael Dummett, *The Interpretation of Frege's Philosophy*, Harvard University Press (1981: Cambridge).
[12] Graeme Forbes, "Indexicals and Intensionality: A Fregean Perspective," *The Philosophical Review* 96 (1987), 3–33.
[13] Gottlob Frege, "On Sense and Reference," in [15], pp. 56–78.
[14] Gottlob Frege, "Concept and Object," in [15], pp. 42–55.
[15] Gottlob Frege, *Translations from the Philosophical Writings of Gottlob Frege*, tr. and ed. Peter Geach and Max Black, Blackwell (1952: Oxford).
[16] Gottlob Frege, "The Thought: A Logical Inquiry," in [22], pp. 507–536.
[17] Gottlob Frege, *Posthumous Writings,* ed. Hermes, Kambartel, Kaulbach, tr. Long, White, The University of Chicago Press (1979: Chicago).
[18] Gottlob Frege, *Philosophical and Mathematical Correspondence*, ed. Gabriel, Hermes, Kambartel, Thiel, Veraart, tr. Hans Kaal, The University of Chicago Press (1980: Chicago).
[19] Herbert Heidelberger, Review of [10], *Metaphilosophy* 6 (1975), 35–43.
[20] David Kaplan, "Demonstratives," in *Themes from Kaplan*, ed. J. Almog, J. Perry, and H. Wettstein, Oxford University Press (1989: New York), pp. 481–564.
[21] David Kaplan, "Quantifying In," reprinted in *Reference and Modality,* ed., Leonard Linsky, Oxford University Press (1971: London), pp. 112–144.
[22] E. J. Klemke, ed., *Essays on Frege,* University of Illinois Press (1968: Urbana).
[23] Leonard Linsky, *Oblique Contexts* (University of Chicago Press: Chicago, 1983).
[24] Leonard Linsky, *Referring* (Humanities Press: New York, 1967).
[25] Benson Mates, "Synonymity," reprinted in *Semantics and the Philosophy of Language,* ed. Leonard Linsky, University of Illinois Press (1952: Urbana), pp. 111–138.
[26] Richard L. Mendelsohn, "Frege on Predication," *Midwest Studies in Philosophy* VI, University Of Minnesota Press (1981: Minneapolis), 69–82.

[27] Terence D. Parsons, "Frege's Hierarchies of Indirect Sense," Handout at the Frege Conference, Munich, 1991.
[28] Terence D. Parsons, "What Do Quotation Marks Name? Frege's Theories of Quotations and That-clauses," *Philosophical Studies* 42 (1982), 315–328.
[29] Terence D. Parsons, "Frege's Hierarchies of Indirect Senses and the Paradox of Analysis," *Midwest Studies in Philosophy* VI, University of Minnesota Press (1981: Minneapolis), pp. 37–57.
[30] W. V. O. Quine, *From a Logical Point of View*, 2nd. ed. rev., Harper & Row (New York: 1961).
[31] Mark Richard, *Propositions and Attitudes*, Cambridge University Press (1990: Cambridge).
[32] Bertrand Russell, "On Denoting," in *Logic and Knowledge: Essays 1901–1950*, ed. Robert C. Marsh, Allen & Unwin (1956: London).
[33] John Searle, "Russell's Objections to Frege's Theory of Sense and Reference," in [22], pp. 337–345.
[34] Robert Stalnaker, *Inquiry*, M.I.T. Press (1984: Cambridge).

Singular Terms (1)

BOB HALE

1. Why criteria are needed

Frege's analysis of language and the ontology that is based upon it – including, centrally, the argument which is at least implicit in *Grundlagen* for the existence of numbers as objects – presupposes the availability of criteria by means of which expressions functioning as proper names [in his broad sense – henceforth singular terms] may be recognised as such and distinguished from expressions of other kinds[1]. *Semantically*, a singular term is any expression the function of which is to convey a reference to a particular object. But it is evident, from even the most superficial consideration of the kind of argument that may be seen as underlying Frege's belief in numbers as objects, that our means of recognising expressions as singular terms should make no *direct* appeal to their having this role. In essence, that argument draws the conclusion that there are objects – numbers – to which certain kinds of numerical expression make reference, from the premiss that expressions of those kinds function as singular terms in various true statements. In effect, the argument reduces the question of the existence of numbers as objects to questions about the truth of certain statements and their logical form. To put it schematically: statements of the kind in question (in this case, purely arithmetic ones) are to be regarded as involving expressions (here, simple numerals and various complex numerical expressions) functioning as singular terms; thus, since the function of a singular term is precisely to effect reference to an object, the truth of those statements requires the existence of objects to which the expressions refer. Clearly, any such argument would be entirely pointless if, in order to determine whether the expressions that interest us do indeed function as singular terms, we had first to resolve the

[1] For a clear and forceful explanation of the basis for the general claim advanced here, see Michael Dummett (1973) pp. 54–8. For its application to number, see Crispin Wright (1983) pp. 10–15 and Bob Hale (1987) pp. 10–14.

question whether there are objects to which they make reference. Criteria are needed which will enable us to recognise an expression as having the characteristic function of a singular term (i. e. as purporting reference to some one object) in virtue of features of its use – its behaviour in complete sentences – that may, in the present context, be taken as undisputed.

I shall here be concerned exclusively with the problem of formulating acceptable criteria for singular termhood, of the general kind required, *inter alia,* to subserve the Fregean argument for numbers as objects. Roughly speaking, by *acceptable* criteria I mean criteria which on the one hand admit as singular terms all (uses of) expressions whose status as such may, in the present context, be taken as (relatively) unproblematic, and which, on the other, exclude (uses of) expressions which clearly do not play that role. I am assuming, in other words, that we are approaching the task, not empty handed, but equipped with a range of clear positive and negative cases which any satisfactory general tests we might propose should fit – the idea being that general tests which meet this adequacy condition may then be applied to give a ruling in less clear, or otherwise problematic, cases. It should be clear that even assuming success in this enterprise, there remain several ways in which the Fregean argument might be challenged. Someone might, for example, grant that what pass as (numerical) singular terms by our proposed tests do indeed function as such in some specified range of statements, but resist Frege's conclusion by rejecting the other, equally essential, premiss – i. e. that the relevant statements are true[2]. Somewhat differently, it might be contended that the best theory we can contrive to explain how singular reference works – for the range of positive cases that can be safely regarded as unproblematic – has the consequence that whilst e. g. simple numerals and other, complex, numerical expressions do indeed qualify as singular terms by the criteria proposed, there is some insuperable obstacle to viewing them as really discharging that function[3]. In particular, any theory which assigned an essential role to causal relations between speaker and object of reference would, it seems, rule out reference to numbers or other ab-

[2] This is, in effect, the line taken by Hartry Field, initially in his (1980) and subsequently in several of the papers now collected together in his (1989).

[3] This covers various more specific sources of scepticism about the possibility of identifying thought about and reference to abstract objects. For details, and attempts to dispose of them, see Crispin Wright (1983) chs. 2 and 3, and Bob Hale (1987) ch. 7.

stracta. A challenge of either kind, if sustained, carries with it, of course, a substantial philosophical obligation – to provide an alternative account of those aspects of our linguistic practice which, taken at face value, involve singular reference to numbers, sets and other abstract objects, and it is a great and very difficult question whether any satisfactory such account can be given. A full discussion of the Fregean argument for numbers as objects would involve confronting these and other doubts and objections; but here I would like to set them aside, with the inevitable acknowledgement that that argument, even when supplemented with acceptable criteria for singular terms, must remain – so far – vulnerable to attack of the kinds mentioned.

2. *The shape of the task*

If, for the moment, we view the situation from the perspective of Frege's hierarchical categorisation of the various logical types of expression, we may note that singular terms form one of the two categories of *complete* expressions, the other being that of *sentences*. In devising criteria for singular terms, we may presuppose a capacity to recognise well-formed sentences as such. There is, accordingly, no objection to our presupposing also a capacity to recognise as such any type of *incomplete* expression for whose characterisation only a capacity to recognise sentences need be assumed. This means, in effect, that we can immediately set aside as lying outside the category of singular terms all those expressions which function as sentential operators. On the other hand, it would obviously be viciously circular to take for granted, in formulating our tests, a capacity to recognise as such expressions belonging to the various other categories of incomplete expression figuring in Frege's hierarchy – centrally, predicates, relational and functional expressions, of the various levels – since the notion of a singular term has to be employed in their characterisation: we have as yet no means of recognising these expressions, at the basic[4] (i. e. first) level, other than as being expressions which result from the omission, from complete expressions of the appropriate type, of (occurrences of) singular terms.

[4] Nor, therefore, do we as yet have any non-question-begging means of recognising higher-level incomplete expressions of these kinds, including – crucially – first- and higher-order quantifiers.

From an intuitive point of view, it is clear that the class of expressions we are seeking to circumscribe is included within the broader class of singular substantival expressions. It is necessary and sufficient for an expression's belonging to this broader class that it be capable, without loss of grammaticality, of supplanting a proper name, in the ordinary sense as opposed to the more inclusive one employed by Frege, in any sentence. It is intuitively clear that the class of singular terms, though included within it, does not coincide with this broader class. In particular, it is a conspicuous feature of English, and many other natural languages, that generality is commonly expressed by words or phrases which can stand where proper names, in the narrow sense, can stand – words such as 'everything', 'something', etc., along with their personal counterparts, and restricted quantifier phrases like 'every philosopher', 'some goldfish', 'a poet', etc.

From Frege's point of view, of course, expressions of this latter sort are to be regarded as incomplete – in these cases, as carrying with them an argument place requiring to be filled by a first level predicate, or by a corresponding bound variable. However, at this stage of our enquiry, we have – as already remarked – no non-circular means of recognising such expressions as incomplete in Frege's sense. Their incompleteness resides, in effect, in their being obtained from complete sentences by omission of a first-level predicate, and we have yet to see how expressions belonging to that category are to be recognised. This complicates our task. It means – if we look at that task in terms of what we are seeking to *exclude* from the category of singular terms – that there are two broad types of expression to be dealt with. There are, first, all those expressions which are grammatically unsuited to positions in sentences occupiable by singular terms: this includes, centrally, predicates and relational expressions in Frege's strict sense. Then there are, second, the various substantival expressions to be encountered in natural language which are able to replace singular terms *salva congruitate* (that is, without destructive effect upon the grammar of a sentence), but which are nevertheless not to be thought of as discharging the semantic function characteristic of *bona fide* singular terms. Here are included the natural language quantifier words and phrases mentioned just now. The division is actually not quite as neat as this suggests, however. For among expressions in this last group are some, such as 'a philosopher' which, whilst they certainly can occupy positions suitable for singular terms, may also stand in positions in sentences not occupiable by singular terms. This is exemplified by the use of such expressions as grammatical complements to the verbs 'to be', 'to become', etc., as in 'Quine is a philosopher'[5]. And further, adjectives standardly occur as grammatical complements, though they of course are unsuited to sentence positions occupiable by singular terms[6].

Our task thus divides into two parts. We need, first, some means of setting aside, as lying outside the category of singular terms, the various kinds of non-substantival expression (including, centrally, predicates in the Fregean sense). And we require, second, tests which will enable us to discriminate, within the residual class of substantival expressions in general, those which function as genuine singular terms from those others which, though resembling the genuine article in surface grammatical form (at least to the extent of being interchangeable with them without destructive effect upon the grammar of sentences), do not.

It might be thought that the first task is very easily discharged. Even if a capacity to stand where a proper name (in the ordinary, narrow, sense) can go is *insufficient* for an expression's being a singular term, can we not simply stipulate that it is at least *necessary* – thereby excluding predicates and other species of incomplete expression right from the start? It is important to see that, and why, this quick way with the question will not do. The objection to proceeding in this way is not that we cannot readily mark off the broader class of substantival expressions: as we have seen, we can do that easily enough, in a purely syntactic way – it comprises exactly those expressions capable of standing, *salva congruitate*, where a proper name (in the ordinary sense) can occur. The point is rather that the demarcation we are seeking to effect (i. e. between singular terms and other kinds of expression) is intended to correspond to a functional difference – a difference in semantic role. We may not simply assume, as we should then be doing, that this (for all we are as yet in position to claim) quite superficial syntactic division coincides with the difference in semantic or functional role in which we are really interested – that we somehow know in advance that non-substantival expressions could not serve to convey reference to particular objects. On the contrary, there is in advance no evident absurdity in the suggestion that verb-phrases, for example, might discharge that function. It is, certainly, arguable that on Frege's view of the matter, it

5 Of course, 'a philosopher' could here be replaced by a singular term, such as 'that man talking to Putnam', but only with an accompanying shift in the sense of 'is' to the 'is' of identity.

6 Unless of course accompanied by an opposite switch in the sense of 'is', from identity to predication.

makes no sense to suppose that one and the same type of non-linguistic entity might be referred to by expressions of such radically different types as singular terms and predicates[7]. But if this is so, it is at best a consequence of Frege's philosophy of language and ontological views, which we have no right to take as a *datum*, at this stage of our enquiry. What is needed is a criterion which relates, in an intelligible way, to the function of singular terms.

3. *Dummett's criteria*

There is, then, a substantial question how a principled distinction is to be effected between singular terms on the one hand and, on the other, predicates and other kinds of incomplete expression, in Frege's sense. This question will occupy us for most of the remainder of the present paper. But before we engage with it, we should take note of what is unquestionably the most promising published proposal – put forward by Michael Dummett – for tackling the second of the two subtasks distinguished above.

Dummett observes that, precisely because expressions other than singular terms may replace the genuine article *salva congruitate*, tests based merely upon considerations of grammaticality will not suffice[8]. The criteria he proposes are broadly inferential in character, the underlying idea being that there are certain simple patterns of inference which are characteristic of singular terms in the sense that when – in an instance of one of these patterns – a certain position is occupied by a singular term, the resulting inference is valid, whereas when that position is filled by an expression other than a singular term, it is not. One obvious thought is that for an expression to be (functioning as) a singular term, the inference commonly known as existential generalisation should be valid with respect to it. Whilst this is unquestionably a neces-

---

[7] Dummett (1973) pp. 174–79.

[8] It is not entirely clear to me whether Dummett means to retain it as a *necessary* condition for an expression to be a singular term that it can replace an ordinary proper name *salva congruitate*. He does introduce his inferential criteria as 'further tests', which suggests that he does mean to retain the purely grammatical test. For reasons indicated in the preceding section, it seems to me that reliance on such purely grammatical considerations should be avoided; in what follows, I shall take Dummett's criteria not to include it.

sary condition[9], it is important to observe that it would not be satisfactory to frame a criterion in these terms, since doing so would presuppose that those who are to employ it have already at their disposal some general means of recognising a sentence B as being the result of existentially generalising with respect to some occurrences of an expression t in a given sentence A. But it is questionable, to say the least, that this could be available to someone who did not already have the capacity to discriminate singular terms. This is not to say that it is illegitimate, in formulating a test, to presuppose an ability to recognise – as valid – inferences which are *in fact* instances of the pattern known as existential generalisation. The point is rather that the test should be framed so as not to presuppose, on the part of those who are to employ it, either possession of the general concept of existential generalisation or (if this is anything different) a general capacity to classify statements as existential generalisations as such. Dummett gets round the apparent difficulty here by framing his tests relative to a particular language, which may be presumed understood by those who are to employ them. In the present case, there is no objection[10] to presupposing (taking the language to be English) a grasp of the use of the indefinite pronoun 'something', and a capacity to recognise as valid such inferences as that from 'Tom loves Mary' to 'Someone loves Mary'. This is Dummett's first test, aimed at excluding 'nothing', 'nobody' and cognate expressions such as 'no elephant':

(I) From any sentence 'A(t)', it shall be possible to infer 'There is something such that A(it)'

Remarking that this will not exclude 'something', and related expressions ('someone', 'some philosopher', 'a crocodile', ...), he adds:

[9] Anyone who wishes to hold that ordinary proper names, and maybe also definite descriptions, function as singular terms when they appear as grammatical subjects in negative existential statements will, of course, regard it as anything but unquestionable. A quick reply is that it is definitive of the enterprise in which I am here engaged that genuine singular terms will satisfy (a suitable formulation of) the existential generalisation condition – it is only if the notion of singular term is so understood that it can subserve the Fregean argument for numbers, for example. It is, of course, perfectly consistent with my view to allow that there may be other, less exacting, notions of singular termhood, for which this condition fails.

[10] This is a bit fast. Actually, it might be felt that framing our criteria relative to a particular language infects the Fregean argument for numbers with an unwanted ontological relativity. For the worry, see Wright (1983) pp. 62–4, and, for an attempt to meet it, Hale (1984).

(II) From two sentences 'A(t)' and 'B(t)', it shall be possible to infer 'There is something such that A(it) and B(it)'

These are insufficient to exclude (all occurrences of) 'everything', and the like, so Dummett adds:

(III) A disjunction 'A(t) or B(t)' of two sentences may be inferred from 'It is true of t that A(it) or B(it)'

Dummett notices that whilst these conditions may suffice to exclude various kinds of expression other than singular terms, *when they stand in places where genuine singular terms can go*, they will not exclude e. g. indefinite noun phrases such as 'a policeman', when they appear in positions not occupiable by singular terms. Thus, in particular, 'a policeman' as it occurs in 'Henry is a policeman' passes (I) – we can infer 'There is something such that Henry is it'; and from the premisses 'Henry is a policeman', 'George is not a policeman', we can infer 'There is something such that Henry is it and George is not it', so (II) appears to fare no better. And whilst there is some awkwardness over the application of (III) to such examples, it is at least doubtful that it will suffice to exclude them. Noting that the rogue cases exploit the possibility of taking 'something' to express higher-level generality, Dummett (eventually) proposes to amend (I) and (II) so as to require that the displayed occurrences of this word be understood as expressing first-level generality, and accordingly adds a supplementary test for discriminating among first- and higher-level uses of 'something'. This supplementary test – the *specification* test, as I shall call it – exploits the fact that when an assertion involving 'something' is made, we can lodge a request for (further) specification. Dummett's thought was that when 'something' expresses second- or higher-level generality, a point may be reached where such a request, though grammatically well-formed, may nevertheless be rejected as not requiring an answer, whereas when it is first-level generality that is expressed, this situation does not arise. To illustrate, suppose I assert:

There is something that Henry is which George is not

Then you may ask: 'What is that?' I must allow that to this (first-round) request, there is an answer, whether I can give it or not. It being unlikely that my claim is based merely on the general conviction that any two human beings differ in some respect or other, I shall very likely be able to supply an answer – say: 'A policeman'. If so, the further request: 'Which policeman?' – though grammatically in good shape – may properly be rejected, as betraying a misunderstanding of my original statement. By contrast, had I asserted, say:

Jones borrowed something from Smith

and to your initial enquiry: 'What?', replied: 'A flame thrower', then you might further ask: 'Which flame thrower?' – and this time the question cannot be rejected as illegitimate – I must agree that there has to be an answer, whether I know it or not.

Against Dummett's criteria as thus formulated, various objections may be brought, and we shall need to consider how far such difficulties as may be raised can be met by modifying the tests. These matters will be taken up in the sequel. But there are some limitations on the efficacy of Dummett's supplementary test for level of generality of which it is sensible to take note at this stage. It is not difficult to see that the specification test cannot be accepted as providing a satisfactory *general* means of determining whether 'something' serves to express first- or rather higher-level generality. For one thing, the test would as it stands oblige us to say that occurrences of 'something' (or 'someone') embedded within opaque contexts are higher-level, when plainly they are not. For example, if someone – Quine, perhaps – asserts: 'Ralph believes that someone is a spy', and is asked: 'Who?', he may reply: 'A member of his department'. But if Quine is further asked: 'Which member?', he can perfectly properly reject the question as not requiring an answer. The play with opacity here is dispensable – essentially the same difficulty can arise over almost any statement of ∀∃- form[11]. The ensuing difficulty for Dummett's overall project is less serious than it may at first appear. The type of case that causes trouble for his first two tests is that where the candidate singular term is an indefinite noun phrase, such as 'a poet', and where we can validly infer 'Something is such that A(it)' (or, in the case of test (II), 'Something is such that A(it) and B(it)'), but where construing the inference so as to be valid depends upon taking 'something' to express second-level generality. A universally applicable test for higher-level generality is not needed to exclude such cases – in them, the relevant occurrences of 'something' do not lie within an opaque construction or within the scope of another quantifier, and it would, it seems, be sufficient for Dummett's purpose to in-

[11] Consider e. g. the sequence 'Every schoolboy idolizes someone' – 'Who?' – 'A famous sportsman' – 'Which famous sportsman?' – where the last request, though grammatically well-formed, can be rejected as not requiring any answer.

corporate the leading idea of the specification test directly into these two tests, by stipulating that the validity of the relevant inferences shall not depend upon their conclusions being construable so that a point can be reached where a grammatically well-formed request for further specification can be rejected as not requiring an answer[12].

It might be thought that the specification test at least works well enough, as a means of discerning expressions of higher-level generality, when the relevant occurrence of 'something' is the principal operator in the test-context. But, setting aside the point that this involves admitting a serious limitation on the test's range of application, even this much is not true. Suppose I assert, as before: 'There is something Henry is that George is not', but in reply to a first-round request, reply: 'Good at squash', or: 'Patient'. Then at this point, you cannot ask: 'Which good at squash?' or 'Which patient?'- these are just gibberish[13]. Upshot – we don't, in this case, reach a point where a *grammatically well-formed* request can be rejected. Yet plainly we have here an expression of second-level generality. The problem does not just affect adjectives, but is quite general. It will arise in every case where the expression in respect of which higher-level existential generalisation is made is non-substantival. The test is thus powerless to pick up e. g. second-level generalisation with respect to a first-level predicate (in Frege's sense)[14]. This does not interfere with the capacity of Dummett's tests (amended as proposed above) to exclude substantival expressions such as 'a poet' occurring in

---

[12] This way of framing the requirement – in terms of a request's being rejectable (whether or not it is actually rejected) as *not requiring an answer* – is to be preferred to other formulations – employed by Dummett, Wright and myself – in terms of rejecting a request as *illegitimate*, or as *betraying a misunderstanding* of the original statement. These other – less precise – formulations invite irrelevant objections to test. In particular, provided the present formulation is taken as canonical, and we are careful to distinguish between rejecting a specification request as not *requiring* an answer and rejecting it because it has *already* been answered, most of the purported counterexamples offered by Linda Wetzel (1990) can be seen to be off target.

[13] Of course, you could ask: 'How good?' or 'How patient?' but these are obviously not requests for further specification in the relevant sense.

[14] There is some slight awkwardness over the expression of higher-level generalisation in these cases, probably deriving from a tendency in English – and as far as I know, in other natural languages – to employ only nominal quantifiers, like 'some*thing*' and 'every*thing*', which require what is quantified over to be represented by substantival expressions. But this is, I think, a fairly superficial limitation, of no logical or philosophical significance. In any case, we can get round it by semantic ascent, say by expressing the second level existential generalisation of 'Jack loves Sally and Jim loves Joan' by 'There is something which is true of <Jack,Sally> and true of <Jim,Joan>'.

complement position; but it does mean that they cannot be relied upon to exclude predicates and other species of incomplete expression[15].

4. *An Aristotelian Criterion?*

As is very well known, Aristotle maintained that whereas any quality has a contrary, a (primary) substance does not. Several philosophers[16] have been attracted by the thought that this might be made

[15] Dummett does not himself claim that they do so – at least, not explicitly. But it seems likely that this will have been his thought: for, had the specification test provided a quite generally applicable means of marking off higher-level generality, it would, when taken in conjunction with the first two inferential tests, have excluded predicates, etc.

[16] Cf. P. F. Strawson (1974) p. 19 ff., P. T. Geach (1975) and Dummett (1973) pp. 63–7. See also Hale (1979) and Wright (1983) sections iii, ix. Dummett considers the possibility of employing an Aristotelian criterion in the course of grappling with the problem for his basic inferential tests posed by higher-level uses of 'something'. He formulates it as follows:

> Suppose that we have a sentence 'S(t,u)' involving two expressions 't' and 'u' both of which pass the tests for proper names which we have already laid down. We now enquire, with respect to 't', whether we may assert 'There is something such that 'S(it, anything) if and only if it is not the case that S(t, that thing)': if we may, then 't' is not a proper name. Likewise, with respect to 'u', we enquire whether we may assert 'There is something such that S(anything, it) if and only if it, is not the case that S(that thing, u)': if we may, then 'u' is not a proper name. (In these schematic sentences, the 'it' relates to 'something', and the 'that thing' to 'anything'.) [op. cit. p. 64].

After some discussion, he concludes that whilst this criterion is "of value in showing [as against Ramsey's scepticism about the existence of a fundamental distinction between subjects and predicates] the rationale of regarding the distinction between proper name and predicate in a singular statement as constituting a distinction between levels of expression ... As an instrument for identifying expressions as proper names it is ... clumsy, involving as it does double generality" [(1973) p. 67] I am somewhat unsure whether Dummett means to retain this criterion alongside the others he discusses, or whether, rather, he thinks we can dispense with it, in the presence of the specification test for level of generality which he immediately goes on to propose. He does claim that despite its cumbrousness, the Aristotelian criterion will "rule out any predicate functioning merely as a predicate, e. g. 'a man' in 'Plato is a man'." But his thought may have been that, once we have a suitable test for the expression of higher-level generality, predicates will in any case be excluded by his tests (I) and (II), when they are understood so as to require that the occurrences of 'something' therein shall be first-level. In fact, neither suggestion is correct. The first founders on a version of the difficulty for the Aristotelian criterion shortly to be aired in the text. The second runs foul of the fact that the specification test either does not apply at all, or, if it does, gives the wrong answers, in cases where the answer to a first-round request for specification is not given in the form of a noun-phrase. Thus if I assert 'Henry is something that George isn't', and in response to your: 'What?', reply: 'Good at chess', there simply is no grammatically well-formed 'which'-ques-

the basis of a criterion of distinction between proper names, or singular terms, on the one hand, and predicates on the other. Transposed from Aristotle's jargon of substance and quality, the thought would be to the general effect that whereas for any given predicate there is always a contradictory predicate, applying to any given object if and only if the original predicate fails to apply, there is not, for singular terms, anything parallel to this – we do not have, for a given singular term, another 'contradictory' singular term, such that a statement incorporating the one is true if and only if the corresponding statement incorporating the other is not true.

An 'Aristotelian' criterion framed along these lines would, it seems clear, possess the general character we should demand of a satisfactory means of marking off singular terms from predicates. The feature of singular terms to which it appeals, in contrast with any purely grammatical distinguishing mark, bears a discernible relation to their semantic function. A given object's possession, or lack, of a certain property has, in general, no tendency to restrict the possibilities for possession, or lack, of that property by other objects. Given that the characteristic function of a singular term is to convey identifying reference to a particular object, the absence of 'contradictory' singular terms may be seen as the reflection, at the level of language, of this fact, i. e. that objects do not, in general, compete with one another for possession of properties. And the fact that properties do, by contrast, compete for possession by objects may be seen as reflected by the possibility of forming, for any given predicate, another which applies to a given object just in case the original fails to apply.

It might, indeed, be thought that such a criterion would accomplish – at one stroke – *all* that we seek, by furnishing a single test by which expressions functioning as singular terms could be separated from all others. But matters are not so straightforward. Although the difficulties here are, I feel sure, independent of the exact detail of our for-

---

tion to be raised. Of course, you can ask: 'How good?', but that it clearly not to the point. When a second-level 'something' generalises on a predicate in the strict sense – as when we generalise from 'Henry loves Mary but George does not' – there is some awkwardness, both about the formulation of the second-level generalisation itself and, accordingly, about that of a specification request. The awkwardness may be surmounted by recourse to some such formulation as 'There is something true of Henry that is not true of George'. And the answer to a first-round request: 'What?' can then be given: 'That he loves Mary'. But now, plainly, we have no space for a well-formed but rejectable further request – 'Which Mary?' is irrelevant, and 'Which that he loves Mary?' is gibberish.

mulation of the proposed criterion, it will help to bring it to light if we think in terms of a specific proposal. Exact formulation is a matter of some delicacy – it is a feature of English (and, so far as my knowledge extends, of other natural languages) that generality is expressed by means of pro*nouns*, rather than pro-*verbs*, or pro-*adjectives*, for example. But this bias in favour of the substantival is not something on which we should trade, in the present context. To achieve a more neutral formulation, we might proceed as follows. Let t be some expression that forms part of a sentence, the remainder of which we can represent by C( ), so that the whole sentence is C(t). I shall use '$\Sigma\alpha$' and '$\Pi\beta$' as substitutional quantifiers, the substitution class for α comprising all and only those expressions which can replace t in C(t) preserving grammaticality, and that of β comprising, similarly, all and only those expressions which can similarly replace C( ) in C(t). Thus a pair consisting of one expression from the α class and one from the β class will always form a well-formed sentence, which we may schematically depict by (α,β). In this notation, we may formulate an Aristotelian criterion purporting to give a necessary and sufficient condition for singular termhood as follows:

(A) t functions as a singular term in C(t) $\leftrightarrow \neg\Sigma\alpha\Pi\beta\ ((\alpha,\beta) \leftrightarrow \neg(t,\beta))$

To see how this is intended to work, consider the sentence 'Plato is wise', and suppose first that we are to determine whether or not 'Plato' is a singular term. Then we have to ask whether we can find, or could introduce, an expression α, grammatically congruent with 'Plato', so that no matter how β is chosen, the biconditional

(α,β) iff not-(Plato, β)

holds true. Bearing in mind that the substitution class for β here comprises *all* expressions that can replace 'is wise' *salva congruitate* in 'Plato is wise', our question is whether we can so choose α that

(α, is wise) iff not-(Plato, is wise)
(α, smokes) iff not-(Plato, smokes) etc.

all hold true. And it seems clear enough that we cannot – that there is no logical 'complement' – 'non-Plato', as it were – of the name 'Plato'. In the material mode, there is no object that is wise iff Plato is not, smokes iff Plato does not, etc. Thus, as we should hope and expect, 'Plato' qualifies as a singular term by the proposed test.

So far, so good. The snag comes when we consider whether, by the A-test, 'is wise' in the same sentential context is to be reckoned a singular term. It might seem straightforward that it does not – for can we not simply let α be the contradictory predicate 'is not wise', and do we not then have the desired result, i. e. that, no matter how β is chosen, the biconditional

(is not wise, β) iff not-(is wise, β)

holds true, so that 'is wise' fails to qualify as a singular term?

But a moment's reflection reveals that this is too quick. We are assuming, in effect, that β will always be chosen so as to be a singular term. But the β substitution class does not constrain us to that extent. We can just as well take β to be, say, 'everyone', or 'some philosopher'. And if we do, then the relevant biconditionals fail. That is, neither of

(is not wise, everyone) iff not-(is wise, everyone)
(is not wise, some philosopher) iff not-(is wise, some philosopher)

holds true. It seems clear that no other choice of α will fare any better. The A-test as formulated thus misclassifies predicates as singular terms. Notice that little is to be gained by retreating to a weaker version of the test, purporting to give only a *necessary* condition for being a singular term, i. e.

(A') t functions as a singular term in C(t) $\rightarrow \neg\Sigma\alpha\Pi\beta\ ((\alpha,\beta) \leftrightarrow \neg(t,\beta))$

Whilst this will, of course, no longer rule predicates *in*, it will still fail to rule them *out*. The source of the difficulty is only too plain – when the test is applied to a predicate, the context sentence being formed, say, by attaching it to a proper name (or other intuitive singular term), the β substitution class comprises *all* those expressions which may replace the proper name, *salva congruitate,* with the result that we can discredit any choice of complementary expression by choosing a substantival quantifier phrase as β. The requisite biconditional will then fail, a difference in the truth-conditions of its components being consequent upon scope differences of β relative to negation. It is evident – but worth remarking – that our condition fails (for the same reason) to exclude predicates occurring in non-atomic contexts, such as 'is wise' in 'someone is wise'.

It is not yet clear that the difficulty just exposed is fatal to any attempt to construct a criterion for singular terms on the basis of

Aristotle's contrast alone. One plausible suggestion[17] would be that, whilst that difficulty blocks the construction of a simple, direct test for singular terms, it may prove possible to build a test for *quantifiers* from the materials of the Aristotelian criterion, and then build on that to obtain, successively, further tests for (first-level) predicates and finally singular terms. Specifically, the suggestion runs thus: the general form of our test is:

t functions as a Φ in C(t) iff $\Sigma\alpha\Pi\beta\ (\alpha,\beta) \leftrightarrow \neg(t,\beta)$

where Φ can be 'quantifier', '(first-level) predicate', or 'singular term'. Then:

(1) t functions as a *quantifier* if it passes the test as formulated

(2) t functions as a (first-level) *predicate* if it passes the test when β is not a quantifier

(3) t functions as a *singular term* if it fails the test when β is a (first-level) predicate

The proposal is appealing, not least because it promises to cut through the complications that seem inevitable if we are to protect Dummett's combination of inferential and level-of-generality tests against various objections that have been made to them; and I have argued that something like the Aristotelian test seems to be needed in any case, as a means of excluding various types of non-substantival expression which those tests fail to exclude from the category of singular terms.

There is, however, what appears to be a lethal objection. Suppose we are concerned with a sentence composed of an expression for first-level generality, say 'something', together with an expression for second-level generality – this might be 'somehow'. We can then write the instance of C(t) in which we are interested as

Somehow (something)

if we wish to test whether 'something' functions as a quantifier, and as

Something (somehow)

if we are, rather, concerned with whether 'somehow' functions as a quantifier. Of course, neither of these looks much like an English sentence, but this doesn't matter. Read them both, if you like, as 'something is somehow' [the point of the brackets being merely to indicate

[17] I am indebted to Crispin Wright for this suggestion, and for extensive discussion of it.

which expression is being treated as t and which as C( )] – you could also read the former as 'something has some property' and the latter as 'some property is instantiated by something'. I choose 'somehow' as a convenient non-nominal English expression for (existential) second-level generality, and will use 'anyhow', 'nohow' similarly.

If we want to know whether 'something' functions as a quantifier by part 1 of our test, we have to ask whether there is, or could be, an expression α such that

(1) (α,β) iff not (something,β)

holds, for every choice of β. In particular, our choice of α should be such that

(2) (somehow(α)) iff not (somehow(something))

holds. The obvious choice of α is 'nothing'; and it does at first appear to hold that

(3) (somehow(nothing)) iff not (somehow(something))

But consider: we may, it seems, formalise 'somehow(something)" by either of

(a)$(\exists F)(\exists x)$ Fx or (b) $(\exists x)(\exists F)Fx$

since these are surely equivalent. Remember that the brackets in 'somehow (something)' merely serve to indicate which expression is t and which C( ) – they do not mean that 'something' has wide – or, for that matter, narrow – scope relative to 'somehow'. We may write our new (first-level) quantifier as '$(\emptyset x)$' [read 'for no x...'], and define it by: $(\emptyset x)A$ iff $\neg(\exists x)A$. But plainly, the formulae that result from replacing '$(\exists x)$...x...' by '$(\emptyset x)$...x...' in (a) and (b) respectively – viz. '$(\exists F)(\emptyset x)Fx$' and '$(\emptyset x)(\exists F)Fx$' – are *not* equivalent. The former says that there is somehow that nothing is (i. e. there is some property that nothing has), whereas the latter asserts, more strongly, that nothing is such that there is somehow that it is (i. e. nothing is such that there is some property that it has). Further, whilst the biconditional:

(4) $(\emptyset x)(\exists F)Fx$ iff $\neg(\exists x)(\exists F)Fx$

holds, as required, the biconditional:

(5) $(\exists P)(\emptyset x)Fx$ iff $\neg(\exists F)(\exists x)Fx$

fails: there being somehow that nothing is plainly compatible with the very weak assertion that there is somehow that something is [i. e. with

$(\exists F)(\exists x)Fx]$, and so cannot entail its negation. Reverting to our near-English original, we can, I think, express this difficulty by saying that 'nothing' is a suitable complementary quantifier for 'something' *only* if 'somehow(something)' and 'somehow(nothing)' are understood with 'something' and 'nothing' as having *wide* scope relative to 'somehow'. When, contrariwise, 'somehow' is taken as having wide scope, the required equivalence fails. And this means that 'nothing' is not, after all, a suitable complement for 'something', since there is a way to choose β – namely, as 'somehow( )' understood so that the displayed expression enjoys wide scope relative to what fills the gap – so that the required equivalence fails.

It may be observed that 'somehow(nothing)' is ambiguous: on one reading – tantamount to taking 'nothing' as dominant – it says that nothing is anyhow, while on another – taking 'somehow' as dominant – it makes the weaker statement that there is somehow that nothing is. Perhaps the difference comes out more clearly if the first reading is glossed 'nothing is such that there is some property that it has' [which is naturally understood as equivalent in turn to 'nothing has any property'], and the second as 'there is some property that nothing has'. The difficulty then is that the biconditional (3) holds when 'somehow (nothing)' is read in the first way, but fails when it is read in the second. Since the stage 1 test requires that α be chosen so that *however* β is chosen – *salva congruitate* – the corresponding biconditional of the form (1) holds true, it appears that 'nothing' does not constitute a suitable α, and that – unless some other expression for which this difficulty does not arise can be found to serve as α – 'something' fails to qualify as a quantifier.

It should be clear, without going through the entirely parallel details, that 'somehow' will likewise fail the test for functioning as a quantifier in 'something(somehow)'. Here, the only obvious choice for α is 'nohow'. But again, the relevant biconditional:

(6) something(nohow) iff not (something(somehow))

appears to hold only when the left hand side is read with 'nohow' as dominant [as asserting, in effect, that no property is instantiated], and to fail when it is read with 'something' as dominant [as asserting, in effect, that something has no properties].

It may at first appear that the objection turns essentially on the involvement of higher-level generality. The stage 1 test appears to work smoothly when applied to examples like:

(Something) moves
(Everything) moves

and the like. Accordingly, it might be suggested that we can rescue the test by first excluding higher-level generality by means of the specification test. There are several problems with this suggestion. Two – neither of which is clearly insuperable – are that (i) we do not as yet have a quite generally effective test for telling whether 'something' functions as expressing higher-level generality – the specification test is reliable only when the relevant occurrence of 'something' is as principal operator, and (ii) that we have no comparable test, even of this limited application, for 'everything', 'nothing' and other expressions of generality. A third – and perhaps more damaging – point is that the effectiveness of the specification test is still further limited, to those cases where the answer to a first-round request for specification takes the form of a substantival expression, so that a further request of the form 'which so and so?' is grammatically in order. In consequence, we lack an effective means of discerning second-level uses of 'something', when the position generalised is apt for filling by a predicate or relational expression. But even if these difficulties could somehow be met, the proposed rescue is doomed – for, contrary to its assumption, the play with higher-level generality is quite inessential. The original problem is – in essence – a problem about *scope*, not about *level* of generality at all. Thus an exactly parallel difficulty arises over an example like:

Somebody loves (nobody)

where, although the natural choice of α is 'somebody', the required equivalence:

Somebody loves (somebody) iff not (somebody loves (nobody))

holds only when the original is read with wide scope 'nobody' – i. e. as meaning 'nobody is such that somebody loves him/her', and fails when it is read with narrow scope 'nobody' – i. e. as meaning 'there is somebody who loves nobody'.

What conclusion should we draw from these considerations? It would be premature to conclude that the idea underlying our efforts to construct an Aristotelian criterion is simply misguided – that it does not, after all, correspond to a fundamental difference between expressions serving to effect identifying reference to particular objects on the one hand, and those standing for first- or higher-level properties on the other. The situation is rather that our attempt to transform that idea into a definite criterion, applicable to natural languages, runs afoul of the fact that such languages – in contrast with formal languages constructed according to Frege's principles – may very well contain expressions which serve for the expression of generality (and so stand, by Fregean lights, for second- or higher-level properties) but are – in surface form, at least – grammatically congruent with singular terms. It thus appears that we could dispose of the difficulty, could we but impose a suitable restriction on the β substitution class. Intutively, what is wanted is a general restriction which, when our candidate t is a (first-level) predicate, ensures that the corresponding β class does not include substantival expressions functioning as quantifiers. But if that is all that stands in the way of achieving a workable formulation of the Aristotelian test, then an obvious remedy is to hand. For whilst the Dummettian tests do not, by themselves, provide the means of discriminating singular terms from expressions of *all* other kinds, they may accomplish the more modest goal of marking off singular terms within their grammatical congruence class, thus distinguishing them from substantival expressions functioning as (first-order) quantifiers. In view of this, the following composite proposal suggests itself:

(1) A *substantival* expression t functions as a singular term in a sentential context 'A(t)' iff t passes Dummett's inferential tests (I)–(III), where (I) and (II) are modified so as to incorporate the substance of the specification test

(2) Where t is any expression, and A( ) any suitable context:
t functions as a singular term in A(t) $\rightarrow \neg\Sigma\alpha\Pi\beta\ ((\alpha,\beta) \leftrightarrow (t,\beta))$ where the β substitution class comprises all expressions grammatically congruent with 'A( )', except any that fail our stage (1) tests.

I hope that I have made a good case for the second part of this proposal. Full defence of the proposal requires, of course, the provision of grounds for thinking that Dummett's criteria, or some well-motivated modifications of them, are in good order. This task is addressed in a sequel to the present paper[18,19].

[18] Hale (1994).

[19] I thank others taking part in the conference for several helpful suggestions. I have been helped also by discussion of an earlier draft of some of this material at a conference on the philosophy of mathematics held in St. Andrews, March 1991. Special thanks are due, as usual, to Crispin Wright for much extremely helpful discussion.

Works cited

Michael Dummett (1973), *Frege: Philosophy of Language*, Duckworth, London.
Hartry Field (1980), *Science without Numbers*, Blackwell, Oxford.
Hartry Field (1989), *Realism, Mathematics and Modality*, Blackwell, Oxford.
P. T. Geach (1975), 'Names and Identity' in S. Guttenplan (ed.) *Mind & Language: Wolfson College Lectures 1974*, Oxford University Press.
Bob Hale (1979), 'Strawson, Geach and Dummett on singular terms and predicates' *Synthese* 42.
Bob Hale (1984), 'Frege's Platonism' *The Philosophical Quarterly* 34, reprinted in *Frege: Tradition and Influence*, edited by Crispin Wright, Blackwell, Oxford.
Bob Hale (1987), *Abstract Objects*, Blackwell, Oxford.
Bob Hale (1994), 'Singular Terms' in Brian McGuinness and Gianluigi Oliveri eds. *The Philosophy of Michael Dummett*, Kluwer.
P. F. Strawson (1974), *Subject and Predicate in Logic and Grammar*, Methuen, London.
Linda Wetzel (1990), 'Dummett's criteria for singular terms' *Mind* Vol. 99, No. 394.
Crispin Wright (1983), *Frege's Conception of Numbers as Objects*, Aberdeen University Press.

Notes on the Contributors

George Boolos was Professor of Philosophy at the Massachusetts Institute of Technology. He was Vice-President of the Association for Symbolic Logic and an editor of *The Journal of Symbolic Logic*. His publications include *Computability and Logic* (with R. C. Jeffrey) (1974), *The Unprovability of Consistency: An Essay in Modal Logic* (1979) and *The Logic of Provability* (1993). He is also the author of numerous articles on logic and the philosophy of mathematics and on Frege. He died on 27 May, 1996.

Tyler Burge is Professor of Philosophy at the University of California at Los Angeles. He has published extensively on the philosophies of language, logic and mind. He is also the author of several articles on Frege's philosophy.

Michael Dummett is Fellow of New College, Oxford and Emeritus Wykeham Professor of Logic at the University of Oxford. In 1976 he delivered the William James Lectures at Harvard University. Among his many publications are *Frege: Philosophy of Language* (1973, 1981), *Elements of Intuitionism* (1976), *Truth and Other Enigmas* (1978), *The Interpretation of Frege's Philosophy* (1981), *Frege: Philosophy of Mathematics* (1991), *Frege and Other Philosophers* (1991), *The Logical Basis of Metaphysics* (1991), *The Seas of Language* (1993).

Gottfried Gabriel is Professor of Philosophy at the University of Jena. He has published *Definitionen und Interessen. Über die praktischen Grundlagen der Definitionslehre* (1972), *Fiktion und Wahrheit. Eine semantische Theorie der Literatur* (1975), *Zwischen Logik und Literatur. Erkenntnisformen der Dichtung, Philosophie und Wissenschaft* (1991) and *Grundprobleme der Erkenntnistheorie. Von Descartes zu Wittgenstein* (1993).

Bob Hale is Professor of Metaphysical Philosophy at the University of Glasgow. His principal research interests lie in the epistemology and ontology of mathematics and related areas of philosophy of language and metaphysics. His book *Abstract Objects* was published in 1987 by Blackwell. He is currently editing, jointly with Crispin Wright, the Blackwell companion to the Philosophy of Language, to which he is also contributing various chapters.

Richard G. Heck graduated from the Massachusetts Institute of Technology in 1991 and is currently Assistant Professor of Philosophy at Harvard University. His chief interests lie in the philosophies of language, logic and mathematics. He has published articles on Frege in *The Journal of Symbolic Logic, Noûs* and *Mind*.

Franz von Kutschera is Professor of Philosophy at the University of Regensburg. Among his many publications are *Elementare Logik* (1967), *Sprachphilosophie* (1975), *Wissenschaftstheorie* (1972), *Grundfragen der Erkenntnistheorie* (1981), *Grundlagen der Ethik* (1982), *Der Satz vom ausgeschlossenen Dritten* (1985), *Gottlob Frege* (1989), *Ästhetik* (1989), *Vernunft und Glaube* (1990).

Richard Mendelsohn is Professor of Philosophy at the Graduate Center of the City University of New York. He has published several articles on the philosophy of language, and especially on the philosophy of Frege.

Terence Parsons is Professor of Philosophy at the University of California at Irvine. He was co-editor of *The Journal of Philosophical Logic* from 1989 to 1993. He has published two books – *Nonexistent Objects* (1980) and *Events in the Semantics of English* (1990) – and many articles, some of which deal with Frege's logic and semantics.

Eva Picardi is Associate Professor of Philosophy at the University of Bologna. She has published in the history of logic and in contemporary philosophy of language. Her publications include *Assertibility and Truth: A Study of Fregean Themes* (1981), *Linguaggio e analisi filosofica* (1992) and *La chimica dei concetti* (1994).

Michael Resnik is Distinguished Professor of Philosophy at the University of North Carolina at Chapel Hill. His principal research interests lie in the philosophy of mathematics, the philosophy of logic and the theory of rationality. Among his most important publications are: *Frege and the Philosophy of Mathematics* (1980), *Choices* (1987), "Mathematics as a Science of Patterns" (1981, 1982), "Second-Order Logic Still Wild!" (1988).

Matthias Schirn is Professor of Philosophy at the University of Munich. He has held visiting positions in Oxford, Cambridge, Santiago de Compostela, Minneapolis, Campinas, Buenos Aires, Mexico City and other universities in Europe and North and South America. He has published on topics in the philosophies of language, logic and mathematics, and in epistemology. His book *Gottlob Frege: Grundlagen der Logik und Mathematik* is forthcoming from Walter de Gruyter.

Peter Simons is Professor of Philosophy at the University of Leeds. He has also taught in Austria, Switzerland and the United States. He is currently the editor of *History and Philosophy of Logic*. He has published *Parts. A Study in Ontology* (1987), *Philosophy and Logic in Central Europe from Bolzano to Tarski. Selected Essays* (1992) and numerous articles.

Christian Thiel is Professor of Philosophy at the University of Erlangen-Nürnberg. He is chiefly interested in the foundations of mathematics and its history and in 16th and 19th century philosophy. He has published *Sinn und Bedeutung in der Logik Gottlob Freges* (1965), *Grundlagenkrise und Grundlagenstreit* (1972), *Philosophie und Mathematik* (1995) and several articles on Frege's logic. His centenary edition of Frege's *Die Grundlagen der Arithmetik* appeared 1986.

William W. Tait is Professor of Philosophy at the University of Chicago. He has been a member of the Institute for Advanced Study in Princeton and a Visiting Professor of Mathematics at the University of Aarhus, Denmark. His most important articles on logic and the philosophy of mathematics include "Intensional Interpretations of Functionals of Finite Type I" (1967), "Normal Derivability in Classical Logic" (1968), "Finitism" (1981), "Truth and Proof: The Platonism of Mathematics" (1986) and "The Law of Excluded Middle and the Axiom of Choice" (1994).

Crispin Wright is Professor of Logic and Metaphysics at the University of St. Andrews, and formally Nelson Professor of Philosophy at the University of Michigan. He has held visiting positions at the University of Pennsylvania, Princeton University and other institutions. He is the author of *Wittgenstein on the Foundations of Mathematics* (1980), *Frege's Conception of Numbers as Objects* (1983), *Realism, Meaning and Truth* (1986) and *Truth and Objectivity* (1992). He is also the author of numerous articles on the philosophies of language, mathematics and mind, and on the theory of knowledge.

Index of Subjects

a priori 10–12
abstraction 74–75, 117
analyticity (analytic) 2, 114, 162–163, 315–316
ancestral (of a relation) 207–211, 220, 222, 229
- definition of the 209
- strong 202, 209, 217
- weak 202, 209
arithmetic
- Fregean 201, 206, 211, 216, 223–226, 228
- second-order 201
ascription of number 115, 121, 125–127, 129–132, 135, 146
assertion 310, 317, 324
Axiom V 5, 7–11, 13, 34–36, 134, 150, 161, 166–168, 200–201, 236, 242–247, 249–250, 253–255

Basic Law V *see* Axiom V
Begriffsschrift, logical system of 2, 281–285, 298

category
- basic ontological 190
- of explanation 181
class of Relations 16
cognitive value 311, 324
communication 308, 323–324
compositionality 311–312, 316
compositionality principle 417
concept 268–269
conditional 282, 289–290, 298
conservativeness 33, 175
consistency 54–55, 62, 162
consistency proof 254–255
content stroke 282, 284–285, 298
context
- indirect 375, 410
- opaque 446
context principle 135–142
contingency 176
- austere 181
- barren 197
- brute 180
- strong 183
course-of-values (of a function) 5–8, 10–13, 169, 262–267, 269–270, 272–273, 275, 281, 286, 288, 291, 297–298
course-of-values term 7–10, 13
criteria for singular terms
- Dummett's 446
- Aristotelian 448

Dedekind-Peano-Axioms 34, 200, 224–226
default value 295, 298
definite description operator 288–289, 298
definition by cases 294–298
dispensability programme 177
distinguishing characteristic 87
domain
- all-encompassing 9, 22
- of first-order quantification 9, 12
- of quantification 36, 256, 258–259
- quantitative 14, 17

elucidation (Erläuterung) 5, 346
epistemology 29, 330–336
equality 103
equinumerosity 3, 117, 135, 143, 151, 157, 168–169
Equivalence (T) 3, 150, 158–169
existence 348, 353–357, 361–362
expression
- complete 440
- incomplete 440
- Trojan 275
extension
- of a concept 2–3, 5–6, 10, 77, 143, 153–158, 160–161, 168
- of a relation 16

fiction 61, 64–65
fictionalism 177
formalism 316–317
free creation 76
Frege's Theorem 201
Fregean argument for numbers as objects 439
function
- first-level 12, 240–241
- primitive 5, 282, 286–287, 290, 298
- primordial 263, 270–272, 274

- second-level 5, 9, 240–241
functionality (of a relation) 202–204, 212–214, 216, 219–222

gap formation (principle) 133, 139–141, 266, 270, 277
generality
- first-level 446
- higher-level 446
- second-level 446
geometry 338–339
- Euclidean 20, 26–28
- non-Euclidean 20, 22, 27–28
Grundgesetze der Arithmetic, logical system of 168, 280–282, 285, 287, 298

hierarchy 372
- infinite 411
- nonrigid 413
- rigid 413
horizontal (function) 262, 264–265, 270, 274, 282, 285–287, 290, 292, 298
Hume's principle 201–202, 204–205, 217, 224, 228, 243–245, 247, 249–250, 268

identity 103, 148, 285–287
impredicativity 275
inconsistency 8, 28, 34, 36, 161, 234, 239, 244, 251
indefinitely extensible concept 36, 234, 236, 258–260
induction 341–342
- definition of a function by 218–219
infinite
- Dedekind 84
- simply 75–76, 84
intersubstitutivity 385
intuition 20, 25, 347, 353
- spatial 20, 25–26

judgeable content 284, 298
judgement
- analytical 331
- apodictic 336–337, 340–341
Julius-Caesar problem 3, 8–9, 148, 153, 161, 166, 249
justification (Rechtfertigung, Begründung) 341–346, 349, 359, 362–364, 366

knowledge
- geometrical 19
- psychology of 345

level of generality 446
literalism 186
logic
- and epistemology 332, 335–346
- first-order 2, 35, 161, 255
- many-valued 291–294, 298
- modal 335–337
- second-order 2–3, 6, 161–162, 168, 202, 205, 216–217, 228, 244

meaning 372
- lexical 333–334

natural deduction 301
naturalism 309, 327
necessity 315, 319
negation 282, 287, 290, 294–296, 298
number
- cardinal 71, 116, 153, 167
- natural 72, 145–147, 158
- ordinal 71, 143, 259–260
- real 13–16, 18, 71–72
numerical equation 130–131, 158–160, 165–167
numerical operator 149–152, 158, 161
numerical predication, principle of 115, 121, 127, 130–132

objectivity (objective) 25–26, 119, 308, 321–322, 328, 349, 351–354, 357, 360–362
ordered pair 205–208, 210–211, 216, 220
ordered pair axiom 205, 207–208, 211

pedigree 266, 276–277
permutation argument 7, 28, 37, 168–169, 265
philosophy of language 29, 330–334
platonism 39, 118–119, 349–353, 356, 359, 361–362, 364, 367
posit 54, 63, 65, 67
positival class 17–18
positive class 17–18
postulation 48
postulationalism 45, 56, 60
priority thesis 139–142
proof (Beweis) 338–339, 341–346
- indirect 342
- of referentiality 13, 28
proper name 310–312, 321–323, 333–334
provability 334–336, 344–345
psychologism 80

quantification 254–256, 258–260
- first-order 235, 240, 246, 253, 255

- second-order 36, 234–236, 238–247, 251–252, 254–256
- objectual interpretation of 35–36, 239–241, 254–255
- substitutional interpretation of 35–36, 239–241, 254–255
quantification range *see* domain of quantification

ratios of quantities 14–15, 18
realism 45, 56–57, 59–60, 309, 314, 317–320, 323, 325
reason 349–352, 362, 366–367
recursion equation 214–215
recursion theorem for ω 34, 200, 214
reference 65–67, 373
- customary 410
- indirect 374, 410
referential indeterminacy 3, 7, 9–10, 158–169
referentiality 265–266, 273–275, 277
Russell's paradox 36, 249–250, 263–265, 267, 270, 274

Schlußweise 302
self-evidence 163, 347–348, 363–364
sense 333–334, 371
- customary 410
- indirect 410
set 268–269
skepticism 326–328
specification test (for singular terms) 445
structural property 93
structuralism 52, 58–59
structure 51
structured proposition 420
substitution principle 417

third realm 320–325, 348–350, 360, 362, 364, 367
transcendentalism 313–316, 318, 326–328
truth 309, 312, 315, 317–320, 323, 325, 328, 372
- disquotational 58
- logical 162, 165–166
truth-value 4–5, 7–10, 12, 286–287, 291, 297–298

unit class 7–8, 10–11
use of number words
- adjectival 122, 129, 131, 133
- substantival 129, 131
- syncategorematic 125, 127–128, 146

value-range (of a function) 236, 244, 249, 254–256
value-range term 243, 245–246, 251
variable
- first-order 8, 150, 239
- second-order 239–240, 246

Wertverlauf *see* course-of-values and value-range

Zermelo-Russell antinomy *see* Russell's paradox

Index of Names

Aczel, P. 36, 264–265, 280n
Adeleke, S. A. 17n, 18
Addison, J. 170
Angelelli, I. 2n, 222n, 261, 319n
Aristoteles 78, 82–83, 90, 100n, 448, 452
Avenarius, R. 308n
Awodey, S. 70n
Austin, J. L. 103, 315n, 330n, 411

Baker, G. P. 261
Balaguer, M. 68n
Bartlett, J. M. 261, 275
Baumann, J. J. 30
Baumgarten, A. G. 331
Bealer, G. 61n, 68n
Bell, D. A. 284n
Beltrami, E. 23
Beman, W. W. 200n
Benacerraf, P. 85n, 87–88, 115n, 119, 143–144, 159n, 176n, 185–186, 199n, 349n
Black, M. 4n
Blakeley, T. J. 274n
Blanchette, P. 170
Boolos, G. 3n, 4, 34–36, 160–163, 165, 166n, 167, 170, 201, 205, 228n, 234, 253–256, 258–260, 268–269, 275, 280n
Bolzano, B. 71, 90, 101, 103, 105, 308n
Bostock, D. 136n, 166n
Brouwer, L. E. J. 72
Bublak, R. 170
Burge, T. 39–40, 281
Burgess, J. P. 3n
Buzzetti, D. 308n
Bynum, T. W. 319n, 330n

Cantor, G. 13–14, 29–32, 45, 47–49, 70–71, 73–80, 82, 84–85, 89, 95–96, 100–101, 102n, 104–106, 109–112, 160, 166n, 167, 307n, 308n, 313, 319
Carnap, R. 124n, 152, 402, 406, 424–426
Carroll, L. 61
Carruthers, P. 191
Cartwright, R. 201n, 236n, 242n
Carus, A. 70
Chihara, C. 115n, 170
Church, A. 280, 372, 411n, 413
Clark, P. 253n
Cocchiarella, N. B. 265
Cohen, B. 170
Cohen, H. 336
Collins, A. 410n
Consuegra, F. R. 170
Corcoran, J. 38n
Cremer, M. 170
Cresswell, M. 170
Currie, G. 119, 139n, 281, 326, 331n
Curry, H. B. 243

Darmstaedter, L. 320
Davidson, D. 396, 411, 413, 426
Dedekind, R. 14, 29–31, 34, 45, 47–49, 51–53, 56, 58, 70–72, 75–76, 78–80, 82, 84–92, 94–96, 100, 105n, 119, 200, 228
Demopoulos, W. 29n
Descartes, R. 97–98
Detlefsen, M. 170
Dufour, C. 264
Dummett, M. 17n, 18, 28n, 31, 34–36, 41–42, 47n, 70–73, 79, 80, 84n, 85–88, 89n, 92–94, 104n, 105, 108, 123n, 124n, 131–132, 138n, 141n, 142, 146, 150n, 152n, 155n, 161n, 164n, 191, 200n, 228n, 234–237, 239–242, 244–246, 251, 256, 280n, 281, 312n, 316, 320, 321n, 331n, 333n, 345n, 403, 405, 412–415, 420n, 426–430, 432–433, 435, 438n, 443–447, 448n, 452, 456

Erdmann, B. 308, 309n, 314
Euclid 102, 193

Ferriani, M. 308n
Field, H. 29, 33, 60, 170, 174–181, 184–187, 189–191, 198, 239, 439n
Forbes, G. 413n
van Fraassen, B. 174
Fred, I. 170
Fricke, R. 24n
Furth, M. 240, 314n, 330n

Gabriel, G. 2n, 39, 275n, 310n, 326, 330n, 345n, 346n
Gauss, C. F. 72
Geach, P. T. 280n, 321n, 448n
Gentzen, G. 38, 301, 304

Giaretta, P. 170
Gödel, K. 56, 119, 176n, 222n
Goldfarb, W. 170
Goodman, N. 174n, 237
Greimann, D. 42n, 170
Gupta, A. 390

Haaparanta, L. 37n, 341n
Hacker, P. M. S. 261
Hale, B. 32–33, 42, 68n, 136n, 170, 176n, 178, 189n, 196n, 438n, 439n, 444n, 448n, 456n
Hallett, M. 70n
Hankel, H. 46–47, 91, 119
Heck, R. G. 6, 13n, 33–34, 42n, 70n, 88n, 170, 200, 239n, 242n, 247–248
Hegel, G. W. F. 353
Heidegger, M. 309n
Heidelberger, H. 413
van Heijenoort, J. 86, 317n
Heine, E. 13, 71, 73, 107n, 317
Hellman, G. 29, 55n, 58, 59n, 60
von Helmholtz, H. 22–23, 27, 307
Hermes, H. 2n, 310n, 316n, 330n
Hilbert, D. 29–30, 45, 47, 49, 52–56, 58, 70, 72, 86, 316–317, 335n
Hinst, P. 261, 275
Hintikka, J. 37n, 170, 261
Hodes, H. 32, 120, 135–136, 138–139, 158n, 161
Hrushovski, E. 236n
Hume, D. 78, 103, 105
Hürter, T. 42n
Husserl, E. 75, 79, 96, 106, 309n

Irvine, A. D. 189n
Isaacson, D. 170

Jamieson, D. 61n, 68n
Jané, I. 170
Jevons, W. S. 30, 99
Jourdain, P. E. B. 310–311
Justice, J. 425n

Kaal, H. 310n
Kambartel, F. 2n, 310n, 316n, 330n
Kanamori, A. 170
Kant, I. 23n, 24–25, 27, 81, 90, 313–317, 327, 331, 335, 337, 343, 353
Katz, J. 170, 410n
Kaplan, D. 421n, 427
Kaulbach, F. 2n, 316n, 330n
Keferstein, H. 86
Kelley, J. L. 238, 241, 257
Kerry, B. 308
Kitcher, P. 39, 309–310, 314, 318, 331n
Klein, F. 23n, 24n, 26
Kossak, E. 47–48, 78, 104
von Kutschera, F. 18n, 38, 264
Kripke, S. A. 334, 371
Kronecker, L. 89–90, 307, 308n, 319

Lange, F. A. 308n
Leeds, S. 68n
Leibniz, G. W. 30, 97–98, 99n, 332, 337
Liebmann, H. 47
Leśniewski, S. 46, 291
Levison, A. B. 341n
Linsky, L. 70n, 372, 423n
Lipschitz, R. 73, 82, 107
Loar, B. 127n, 170
Locke, J. 30
Long, P. 316n
Lotze, R. H. 331n, 335, 336n, 341
Lück, U. 170
Łukasiewicz, J. 283, 297

Mach, E. 325–326
Maddy, P. 59n, 62, 176n
Manders, K. 70n
Marion, M. 228n
Mates, B. 237, 414
McGee, V. 248n
McGuinness, B. 310n, 330n
McLarty, C. 65, 68n
McLaughlin, P. 330n
Mehrtens, H. 261
Mendelsohn, R. 41–42, 170
Menn, S. 70n
Menzler-Trott, E. 261
Mill, J. S. 30
Moore, A. W. 265, 280n
de Morgan, A. 293

von Neumann, J. 250
Neumann, P. 17n, 18
Newton, I. 72, 89, 91

Orayen, R. 170
Ostrowski, A. 24n

Parsons, C. D. 3, 84n, 170, 236, 244
Parsons, T. 7, 35, 40–41, 61, 68n, 161n, 170, 200n, 235n, 236n, 245–246, 250n, 265, 275, 281n, 291n, 297, 394, 397, 412, 413n, 415n, 420n, 423, 435n
Pasch, M. 24n
Pears, D. 323n

Picardi, E. 38–39, 308n, 345n
de Pierris, G. 345n
Plato 71, 100
Prauss, G. 334n
Price, M. 70n
Putnam, H. 120, 176n, 185–186

Quine, W. v. O. 174n, 183, 280n, 427

Ramsey, F. P. 448n
Reichenbach, H. 23n
Rein, A. 265, 280n
Reinhardt, W. 68n
Rescher, N. 292n
Resnik, M. D. 29–30, 37, 119, 127n, 152n, 170, 261, 265–266, 268n, 273, 275, 280n, 281n
Reyes, G. E. 70n
Richard, M. 420n
Ricketts, T. G. 170, 347n, 361n, 362n
Richter, M. M. 264
Riemann, B. 26–27, 71
Ritter, U. 42n
Rosado Haddock, G. E. 170
Rosen, G. 191n
Rucker, R. 61
Ruffino, M. 170
Russell, B. 35, 42, 45, 78–79, 94, 142, 247, 275n, 280, 310–311, 320–321, 323, 328, 411, 413, 421n, 422–423, 426, 430–433

Schiffer, S. 170
Schirn, M. 124n, 136n, 139n, 141n, 145n, 152n, 154n, 156n, 157n, 158n, 164n, 174n, 330n, 334n
Scholz, B. 68n
Scholz, H. 308n
Schröder, E. 30, 73, 78, 92, 102, 104, 106
Schroeder-Heister, P. 265, 280n, 281n
Schweitzer, H. 261
Schulte, J. 340n
Schütte, K. 304
Searle, J. 421, 422n
Seidenfeld, T. 70n
Shapiro, S. 59n
Sieg, W. 70n
Sigwart, C. 308, 309n
Simons, P. 18n, 37–38, 170, 288n
Sluga, H. D. 39, 119, 139n, 170, 313–314, 318, 331n, 355n, 356n
Smiley, T. 170
Sobociński, B. 280n
Socrates 100
Stalnaker, R. 420n
Stanley, J. 228n, 242n
von Staudt, G. 26
Stein, G. 367
Stein, H. 70n, 85n, 88n, 170
Stolz, O. 46
Strawson, P. F. 448n
Summerell, F. 330n

Tait, B. 170
Tait, W. W. 30–31, 84n
Tarski, A. 38, 242, 248, 264, 291, 387–388, 411n, 412n
Thiel, C. 2n, 34, 36–37, 269, 274n, 275, 280n, 281n, 310n, 315n, 330n
Thomae, J. K. 13, 30, 73, 82, 107, 317
Torretti, R. 170

Ueberweg, F. 110

Veraart, A. 2n, 308n, 310n, 330n
Villegas-Forero, L. 170

Weierstraß, K. T. 13–14, 71
Weiner, J. 346n, 356n, 357n
Wetzel, L. 447n
White, R. 316n
Wilson, M. 28n
Windelband, W. 313
Wittgenstein, L. 70n, 81n, 310, 332, 343n
Woodruff, P. 170
Wright, C. 3n, 4, 32–33, 68n, 122n, 123, 161, 166n, 167n, 170, 176n, 178, 180n, 181n, 196, 222n, 243, 253, 438n, 439n, 444n, 447n, 448n, 452n, 456
Wrigley, M. 170, 330n
Wundt, W. 308, 319n, 325

Zermelo, E. 32, 74n, 109, 235, 250–251, 280, 307n

Bei Fragen zur Produktsicherheit wenden Sie sich bitte an:
If you have any questions regarding product safety,
please contact:

Walter de Gruyter GmbH
Genthiner Straße 13
10785 Berlin
productsafety@degruyterbrill.com